Aircraft Systems

Aircraft Systems
Mechanical, electrical, and avionics subsystems integration

Third Edition

Ian Moir
Allan Seabridge

John Wiley & Sons, Ltd

Other Wiley Editorial Offices

John Wiley & Sons Inc., 111 River Street, Hoboken, NJ 07030, USA

Jossey-Bass, 989 Market Street, San Francisco, CA 94103-1741, USA

Wiley-VCH Verlag GmbH, Boschstr. 12, D-69469 Weinheim, Germany

John Wiley & Sons Australia Ltd, 42 McDougall Street, Milton, Queensland 4064, Australia

John Wiley & Sons (Asia) Pte Ltd, 2 Clementi Loop #02-01, Jin Xing Distripark, Singapore 129809

John Wiley & Sons Canada Ltd, 6045 Freemont Blvd, Mississauga, ONT, L5R 4J3

Wiley also publishes its books in a variety of electronic formats. Some content that appears in print
may not be available in electronic books.

Library of Congress Cataloging in Publication Data

Moir, I. (Ian)
   Aircraft systems : mechanical, electrical, and avionics subsystems integration / Ian Moir,
   Allan Seabridge.
      p.   cm.
   Includes bibliographical references and index.
   ISBN 978-0-470-05996-8 (cloth : alk. paper)
   1. Aeronautics—Systems engineering.   2. Airplanes, Military—Design and construction.
   3. Airplanes—Equipment and supplies.   I. Seabridge, A. G. (Allan G.)   II. Title.
   TL671.M59 2008
   629.135—dc22

                                                           2008001330

British Library Cataloguing in Publication Data

A catalogue record for this book is available from the British Library

ISBN 978-0-470-05996-8

Typeset in 10.5/12.5pt Palatino by Integra Software Services Pvt. Ltd, Pondicherry, India

To Mike Woodhead
1944 to 2007

Professor of Systems Engineering
at Loughborough University

An inspiration to all systems engineers
and sadly missed

Contents

Foreword

The Aerospace and Defence industry has been at the forefront of systems engineering for many decades. The imperatives of commercial success and / or military need have compelled those in the Industry to seize the opportunities offered by taking a systems engineering approach to solve a variety of complex problems.

The insights offered by use of computer based modelling techniques, which have the capacity to represent multiple complex systems, their interdependencies, interactions and their inputs and outputs have propelled the exploitation of systems engineering by those in Aerospace and Defence. The approach is not confined to those mechanical and electrical systems for which stand alone systems models can be constructed. Rather, it is put to its best use when considering a major product or service as a system made up of many subsystems. For example, the optimisation of aircraft layout involving trade-offs between structural aspects, aerodynamic design, electronic and mechanical system performance as well as integrity can be achieved. Carried out in a balanced way, this can be the most powerful tool used by the Engineering teams in the process of defining a light, cheap to manufacture, reliable and high performance aircraft.

In stark terms, success or failure in the Aerospace and Defence sector is determined by the approach taken in the development of systems and how well or otherwise the systems or their interactions are modelled, understood and optimised. The most obvious output from such a process is the resulting system performance, for example how fast your aircraft can fly and what it can see using its radar. In addition however, the dimensions of cost and elapsed time to develop and build a system, together with its inherent reliability throughout its life, are also all critically dependent on effective systems engineering from the outset. Projects, and sometimes entire businesses, will succeed or flounder on the basis of how well the systems engineering approach has informed decision making relating to the definition of responsibilities between, for example, customers and suppliers, industrial partners or members of an alliance or team. Effective systems engineering will help to expose where the natural boundaries are between areas of activity which in turn informs the definition of suitable contractual boundaries and terms and conditions of a contract. The ultimate benefit of this approach is more effective assignment of responsibilities, enduring contracts and, most importantly, safer systems.

The ultimate consequence of having a culture within an organisation that centres on Systems Engineering is that the inherent approach spills over into other aspects of the activity across the enterprise involved. Obvious benefits in manufacturing process optimisation sit alongside the creation of business information management systems and other tools each playing a part in the quest for an organisation to make the best use of its resources, skills and funding. All of this contributes to the drive for predictable business performance and business success.

This book exemplifies the need to apply a systems engineering approach to the aircraft systems as well as the avionics systems deployed by the aircraft and weapons systems in the performance of its military role. The performance and inter-relationship of all systems are paramount in meeting the air vehicle specification requirements, which in many future offensive air vehicles will be unmanned. The authors have described the Aircraft Systems that emerge from the application of Systems Engineering to show the benefits to individual systems performance and whole aircraft design and integration. Examples of solutions in commercial and military aircraft are given, which complement the systems described in companion volumes.

The forthcoming More-Electric Aircraft and More-Electric Engine technologies as described in various places within this text herald the approach of innovative and highly integrated technologies for many of the aircraft systems that will serve both civil and military applications in the future. The book has much to recommend it as a place mark in time in relation to the ultimate maturity and application of these technologies.

Nigel Whitehead, Group Managing Director – Military Air Solutions, BAE SYSTEMS

Series Preface

The field of aerospace is wide ranging and covers a variety of products, disciplines and domains, not merely in engineering but in many related supporting activities. These combine to enable the aerospace industry to produce exciting and technologically challenging products. A wealth of knowledge is contained by practitioners and professionals in the aerospace fields that is of benefit to other practitioners in the industry, and to those entering the industry from University.

The Aerospace Series aims to be a practical and topical series of books aimed at engineering professionals, operators, users and allied professions such as commercial and legal executives with in the aerospace industry. The range of topics spans design and development, manufacture, operation and support of aircraft as well as infrastructure operations, and developments in research and technology. The intention is to provide a source of relevant information that will be of interest and benefit to all those people working in aerospace.

About the Authors

Ian Moir After 20 years in the Royal Air Force as an engineering officer, Ian went on to Smiths Industries in the UK where he was involved in a number of advanced projects. Since retiring from Smiths he is now in demand as a highly respected consultant. Ian has a broad and detailed experience working in aircraft avionics systems in both military and civil aircraft. From the RAF Tornado and Apache helicopter to the Boeing 777, Ian's work has kept him at the forefront of new system developments and integrated systems in the areas of more-electric technology and system implementations. He has a special interest in fostering training and education in aerospace engineering.

Allan Seabridge was until recently the Chief Flight Systems Engineer at BAE SYSTEMS at Warton in Lancashire in the UK. In over 30 years in the aerospace industry his work has latterly included the avionics systems on the Nimrod MRA 4 and Lockheed Martin Lightning II (Joint Strike Fighter) as well as a the development of a range of flight and avionics systems on a wide range of fast jets, training aircraft and ground and maritime surveillance projects. Spending much of his time between Europe and the US, Allan is fully aware of systems developments worldwide. He is also keen to encourage a further understanding of integrated engineering systems. An interest in engineering education continues with the design and delivery of systems and engineering courses at a number of UK universities at undergraduate and postgraduate level.

Acknowledgements

This book has taken a long time to prepare and we would not have completed it without the help and support of colleagues and organisations who willingly gave their time and provided information with enthusiasm.

We would especially like to thank Gordon Leishman who reviewed the rotorcraft chapter and Geoff Poole, Dr Craig Lawson and Roy Langton who reviewed the entire manuscript. We are indebted to their valuable comments.

The following organisations kindly provided information and images:

AGUSTA WESTLAND
Airbus (UK)
BAE SYSTEMS
Bell Helicopter
BF Goodrich
Boeing
Boeing/Bell
Claverham/Hamilton Sundstrand
DoD photograph by TSGT Edward
    Boyce
Dunlop Aerospace International
Engineering Arresting Systems Corp
Flight Refuelling/Cobham

High Temp Engineering/Cobham
Honeywell
Honeywell Aerospace Yeovil
NASA
Northrop Grumman
Parker Aerospace
Raytheon
Rolls Royce
Rolls Royce/Turbomeca

Smiths Group/GE Aviation
US Air Force
US Air Force photograph by
    Senior Airman Darnall Cannady

We would like to thank the staff at John Wiley who took on this project from a previous publisher and guided us to a satisfactory conclusion.

List of Abbreviations

A 429	ARINC 429 Data Bus
A 629	ARINC 629 Data Bus
A 664	ARINC 664 100Mbits/sec Fast Switched Ethernet
AC	Advisory Circular (FAA)
AAWWS	Airborne Adverse Weather Weapons System (Apache)
AC	Advisory Circular (FAA)
AC	Alternating Current
ACE	Actuator Control Electronics
ACM	Air Cycle Machine
ACMP	AC Motor Pump
ACT	Active Control Technology
A/D	Analogue to Digital
ADM	Air Data Module
ADP	Air Driven Pump
ADU	Actuator Drive Unit
ADV	Air Defence Variant (Tornado)
AFCS	Automatic Flight Control System
AFDC	Autopilot Flight Director Computer (B777)
AFTI	Advanced Fighter Technology Integration (F-16)
AIAA	American Institute of Aeronautics & Astronautics
Aj	Jet Pipe Area
AMAD	Airframe-Mounted Accessory Drive
AMB	Active Magnetic Bearing
Amp or A	Ampere
AoA	Angle of Attack
APB	Auxiliary Power Breaker
APU	Auxiliary Power Unit
ARINC	Air Radio Inc
ART	Actuator Remote Terminal (B-2 flight control system)
ASCB	Avionics Standard Communications Bus
ASI	Airspeed Indicator
ASIC	Application Specific Integrated-Circuit

ASM	Air Separation Module
AS/PCU	Air Supply/Pressurisation Control Unit (B777)
ATA	Air Transport Association
ATC	Air Traffic Control
ATF	Advanced Tactical Fighter
ATM	Air Transport Management
ATP	Advanced Turbo-Prop
ATR	Air Transport Radio
AUW	All-Up Weight
AVM	Airplane Vibration Monitoring
B	Blue (as in blue hydraulic system)
BAES	BAE SYSTEMS
Batt	Battery
BC	Bus Controller (MIL-STD-1553B)
BCF	Bromo-Chloro-diFluoro-Methane
BCRU	Battery Charger Regulator Units (regulated TRUs used on the A380)
BIT	Built-In Test
BOV	Blow Off Valve
BPCU	Bus Power Control Unit
BSCU	Brake System Control Unit
BTB	Bus Tie Breaker
BTMU	Brake Temperature Monitoring Unit
C	Centigrade
C	Centre
C	Collective
CAA	Civil Aviation Authority
CANbus	Commercial-Off-The-Shelf data bus (originally designed by Bosch for automobile applications)
CASA	Construcciones Aeronauticas Socieda Anonym
CBLTM	Control-By-LightTM (Raytheon proprietary fibre optic bus)
CCA	Common Cause Analysis
CCB	Converter Control Breaker (B777)
CCR	Common Computing Resource (B787)
CDA	Concept Demonstration Aircraft
CDR	Critical Design Review
CDU	Cockpit Display Units
CG	Centre of Gravity
CHRG	Charger
CMA	Common Mode Analysis
CNS	Communications, Navigation, Surveillance
COTS	Commercial Off-The-Shelf
CPIOM	Common Processor Input/Output Module (A380 avionics IMA)

CSAS	Control Stability Augmentation System
CSD	Constant Speed Drive
CT	Current Transformer
CTC	Cabin Temperature Control
CTOL	Conventional Take-Off & Landing
CV	Carrier Variant
DATAC	Digital Autonomous Terminal Access Communication (forerunner to ARINC 629)
D/A	Digital to Analogue
DC	Direct Current
DECU	Digital Engine Control Unit
Def Stan	Defence Standard
Dem/Val	Demonstration/Validation
DFCC	Digital Flight Control Computer (AFTI F-16)
DTD	Directorate of Technical Development
DTI	Department of Trade & Industry
DVO	Direct Vision Optics
E1	E1 Electrical Channel (A380)
E2	E2 Electrical Channel (A380)
E3	E3 Electrical Channel (A380)
EAI	Engine Anti-Ice
EAP	Experimental Aircraft Programme
EASA	European Aviation Safety Authority
EBHA	Electrical Backup Hydraulic Actuator (A380)
EC	European Community
ECAM	Electronic Crew Alerting & Monitoring
ECS	Environmental Control System
EDP	Engine Driven Pump
EE	Electrical Equipment (as in EE Bay)
EEC	Electronic Engine Controller
E^2PROM	Electrically Erasable Programmable Read Only Memory
EFA	European Fighter Aircraft
EFAB	Extended Forward Avionics Bay
EFIS	Electronic Flight Instrument System
EFPMS	Engine Fuel Pump and Metering System
EGT	Exhaust Gas Temperature
EHA	Electro-Hydrostatic Actuator
EICAS	Engine Indication & Crew Alerting System
ELCU	Electronic Load Control Unit
ELMS	Electrical Load Management System (B777)
EMA	Electro-Mechanical Actuator
EMP	Electrical Motor Pump
EMI	Electro-Magnetic Interference
EPC	External Power Contactor

EPMS	Electrical Power Management System (AH-64C/D Apache)
EPROM	Electrically Programmable Read Only Memory
EPU	Emergency Power Unit
ERA	Electrical Research Agency
ESS	Essential
ESS	Environmental Stress Screening
ETOPS	Extended Twin OperationS
EU	Electronics Unit
EU	European Union
EUROCAE	European Organisation for Civil Aviation Equipment
EXT or Ext	External
FAA	Federal Aviation Authority
FAC	Flight Augmentation Computer
FADEC	Full Authority Digital Engine Control
FAR	Federal Aviation Regulations
FBW	Fly-By-Wire
FC	Flight Control
FCC	Flight Control Computer
FCDC	Flight Control Data Concentrator
FCMC	Fuel Control and Monitoring Computer (A340-500/600)
FCP	Fuel Control Panel
FCPC	Flight Control Primary Computer
FCS	Flight Control System
FCSC	Flight Control Secondary Computer
FDC	Fuel Data Concentrator (A340-500/600)
FCU	Fuel Control Unit
FHA	Functional Hazard Analysis
FITEC	Farnborough International Technology Exploitation Conference (1998)
FLIR	Forward Looking Infra Red
FMC	Flight Management Computer
FMEA	Failure Modes & Effects Analysis
FMES	Failure Modes & Effects Summary
FMGEC	Flight Management Guidance & Envelope Computer (A330/A340)
FMQGS	Fuel Management & Quantity Gauging System (Global Express)
FMS	Flight Management System
FOB	Fuel On Board
FQIS	Fuel Quantity Indication System
FQPU	Fuel Quantity Processor Unit (B777)
FSCC	Flap/Slat Control Computers (A380)
FSD	Full Scale Development
FSDG	Fan Shaft Driven Generator

FSEU	Flap Slats Electronics Unit (B777)
ft	Feet
FTA	Fault Tree Analysis
G	Green (as in green hydraulic system)
G or Gen	Generator
GA	General Aviation
G&C	Guidance & Control
GCB	Generator Control Breaker
GCU	Generator Control Unit
GE	General Electric (US)
GEC	General Electric Company
GLY	Galley
GND	Ground
gpm	Gallons per minute
GPS	Global Positioning System
GPU	Ground Power Unit
GR	Ground Reconnaissance
HISL	High Intensity Strobe Lights
HP	High Pressure
HPSG	High Pressure Starter Generator
hp	Horse Power
HUMS	Health & Usage Management System
Hyd	Hydraulic
Hz	Hertz
IAP	Integrated Actuator Package
IC	Integrated Circuit
IDEA	Integrated Digital Electric Airplane
IDG	Integrated Drive Generator
IDS	InterDictor Strike (Tornado)
IEE	Institution of Electrical Engineers
IEEE	Institute of Electrical & Electronic Engineers
IFE	In-Flight Entertainment
IET	Institute of Engineering & Technology (formerly IEE)
IFPC	Integrated Flight & Propulsion Control
IFSD	In-Flight ShutDown
IMA	Integrated Modular Avionics
IMechE	Institution of Mechanical Engineers
INS	Inertial Navigation System
INV	Inverter
I/O	Input/Output
IPN	Iso-Propyl Nitrate
IPT	Integrated Product Team
IPU	Integrated Power Unit

IR	Infra Red
IRS	Inertial Reference System
ISA	International Standard Atmosphere
ISA	Instruction Set Architecture
JAA	Joint Airworthiness Authority
JAR	Joint Aviation Regulation
JET A	JET A Aviation Fuel (also known as JET A-1)
JET B	JET B Aviation Fuel
J/IST	Joint Strike Fighter/Integrated Subsystems Technology
JP-4	Aviation fuel used by the US Air Force
JP-5	Aviation fuel used by the US Navy
JSF	Joint Strike Fighter (F-35 Lightning II)
K	Kelvin
kg	Kilogram
kN	Kilo Newton
kPa	Kilo Pascal
KT or kt	Knot
kVA	Kilo Volt-Ampere
L	Lift
L	Left
LAF	Load Alleviation Function
LAN	Local Area Network
LB or lb	Pound
LH	Left Hand
LHX or LH	Light Helicopter
LOX	Liquid Oxygen
LP	Low Pressure
LRM	Line Replaceable Module
LRU	Line Replaceable Unit
LVDT	Linear Variable Differential Transformer
M	Mach Number
m	Metre
mA	Milli Ampere
MA	Markov Analysis
MAC	Mean Aerodynamic Chord
MAU	Modular Avionics Unit (Honeywell EPIC system)
MBB	Messerschmit Bolkow Blohm
MCDU	Multipurpose Control & Display Unit
MDC	Miniature Detonation Cord
MCU	Modular Concept Unit
MDHC	McDonnell Douglas Helicopter Company (now Boeing)
MEA	More-Electric Aircraft

MEE	More-Electric Engine
MECU	Main Engine Control Unit
MEL	Minimum Equipment List
MFD	Multi-Function Display
MFOP	Maintenance Free Operating Period
MHz	Mega Hertz
MIL-H	Military Handbook
MLI	Magnetic Level Indicator
MIL-STD	Military Standard
ml	Millilitre
MLC	Main Line Contactor
MLI	Magnetic Level Indicator
mm	Millimetre
MR	Maritime Reconnaissance
m/s	Metres/second
MN	Mega Newton
MSOC	Molecular Sieve Oxygen Concentrator
MSOV	Modulating Shut-Off Valve
N	North Pole
NADC	Naval Air Development Center
NASA	National Space & Aerospace Agency
Nav	Navigation
NH or N2	Speed of rotation of engine HP shaft
Ni-Cd	Nickel-Cadmium
NGS	Nitrogen Generation System
HL or N1	Speed of rotation of engine LP shaft
NOTAR	NO TAil Rotor
NRV	Non-Return Valve
Nx	Lateral Acceleration
Ny	Longitudinal Acceleration
Nx	Normal Acceleration
OBOGS	On-Board Oxygen Generating System
OBIGGS	On-Board Inert Gas Generating System
OEM	Original Equipment Manufacturer
Ox	Pitch Axis
Oy	Roll Axis
Oz	Yaw Axis
P	Pressure
P	Pitch
POA	Power Optimised Aircraft (EC More-Electric Technology Programme)
Ps or Po	Ambient Static Pressure
Pc	Pressure Capsule

PCU	Power Control Unit
PDE	Power Drive Electronics (AFTI F-16)
PDU	Power Drive Unit
PDC	Power Distribution Center
PDR	Preliminary Design Review
PEM	Power Electronics Module
PEPDC	Primary Electrical Power Distribution Centre (A380)
PFC	Primary Flight Computer (B777)
PFCS	Primary Flight Control System (B777)
PMA	Permanent Magnet Alternator
PMG	Permanent Magnet Generator
PNVS	Pilot Night Vision System (Apache)
PRV	Pressure Reducing Valve
PRSOV	Pressure Reducing Shut-Off Valve
PSEU	Proximity Switch Electronics Unit (B777)
PSSA	Preliminary System Safety Analysis
psi	Pounds/Square Inch
Pt	Dynamic Pressure
PTFE	Poly-Tetra-Fluoro-Ethylene
PTU	Power Transfer Unit
PSU	Power Supply Unit
PWR	Power

'Q' feel	A pitch feel schedule used in aircraft flight control systems based upon $\frac{1}{2}\rho V^2$

R	Right
R	Roll
R & D	Research & Development
RAeS	Royal Aeronautical Society
RAF	Royal Air Force
RAT	Ram Air Turbine
RDCP	Refuel/Defuel Control Panel (Global Express)
RFI	Request For Information
RFP	Request For Proposal
RIU	Remote Interface Unit
RJ	Regional Jet
ROM	Read Only Memory
RPDU	Remote Power Distribution Units
RT	Remote Terminal (MIL-STD-1553B)
RTCA	Radio Technical Committee Association
RTZ	Return-To-Zero
RVDT	Rotary Variable Differential Transformer

S	South Pole
SAARU	Secondary Attitude Air data Reference Unit
SAE	Society of Automobile Engineers

SCR	Silicon Controlled Rectifier (Thyristor)
SDR	System Design Review
SEC	Spoiler Elevator Computer (A320)
SEPDB	Secondary Electrical Power Distribution Box (A380)
SEPDC	Secondary Electrical Power Distribution Centre (A380)
SFENA	Société Française d'Equipments pour la Navigation Aerienne
SFCC	Slat/Flap Control Computers (A330/A340)
SG	Specific Gravity (Density of water = 1)
shp	Shaft horse Power
SIM	Serial Interface Module (A629)
SMP	Systems Management Processor (EAP)
SMTD	STOL Manoeuvre Technology Demonstrator (F-15)
SOL	Solenoid
SOV	Shut-Off Valve
sq m	Square Metre
SSA	System Safety Analysis
SSPC	Solid State Power Controller
SRR	System Requirements Review
SSR	Software Specification Review
STBY	Standby
STC	Supplementary Type Certificate
STOL	Short Take-Off and Landing
STOVL	Short Take-Off Vertical Landing
SV	Servo-Valve
SVCE	Service
T	Temperature
Tamb	Ambient Air Temperature (°K)
Tram	Ram Air Temperature
Trec	Recovery Air Temperature
TADS	Target Acquisition & Designator System (Apache)
TBT	Turbine Blade Temperature
TCD	Total Contents Display
TCL	Thrust Control Lever
TEOS	Technologies for Energy Optimised Aircraft Equipment Systems
T/EMM	Thermal/Energy Management Module
T/F	Transformer
TGT	Turbine Gas Temperature
THS	Tailplane Horizontal Stabiliser
TPMU	Tyre Pressure Monitoring Unit
TRU or TR	Transformer Rectifier Unit
UAV	Unmanned Air Vehicle
UCS	Utilities Control System
UK	United Kingdom
UMS	Utilities Management System

US	United States
USG	US Gallon (1 USG = 0.8 Imperial Gallon)
UTIL	Utility
UV	Ultra-Violet
U/V	Under Voltage
V	Velocity
V	Volts
VAC	Volts Alternating Current
VDC	Volts Direct Current
VDU	Visual Display Unit
VIB	Vibration
VIGV	Variable Inlet Guide Vane
VF	Variable Frequency
VOR	VHF Omni-Range
VLSI	Very Large Scale Integrated-Circuit
VMS	Vehicle Management System
VSCF	Variable Speed Constant Frequency
V/STOL	Vertical/Short Take-Off & Landing
VSV	Variable Stator Vane
VTOL	Vertical Take-Off & Landing
W	Weight
W	Watts
WWII	World War II
Y	Yaw
Y	Yellow (as in yellow hydraulic system)
ZFW	Zero Fuel Weight
ZSA	Zonal Safety Analysis

Introduction

Since the Second Edition of *Aircraft Systems* was published six years ago a few but not many major new aircraft projects have emerged. At the time of writing, the Airbus A380 is approaching certification and entry into service (the first aircraft being delivered to Singapore Airlines in October 2007), the Lockheed Martin F-35 Lightning II (previously known as JSF) is well established on its flight test programme and the Boeing 787 is months away from first flight and the Airbus A350XWB final design is emerging. However, with the development of these new aircraft the introduction of new technologies abounds and the use of avionics technology to integrate systems at the aircraft and subsystems level has gained considerable pace.

The use of Commercial-of-the-Shelf (COTS) digital data buses is increasingly being adopted; 100 Mbits/sec AFDX/ARINC 664 is used as the aircraft level data bus on both A380 and B787; JSF uses 800 Mbits/sec IEEE 1394b as the integration data bus for the Vehicle Management System (VMS). Ethernet buses of 10 Mbits/sec and CANbus commercial derivatives up to 1 Mbits/sec can be found in many aircraft systems. Often these buses are employed in a deterministic form, that is, their performance is constrained so that it always responds in a repeatable fashion and no optimisation is permitted.

Many More-Electric Aircraft (MEA) features and technologies are found in these three major programmes and significant research, development and demonstration programmes are underway to drive the technology further forward for both aircraft and engine applications. In a real sense some of these developments are challenging the way that aircraft systems are engineered for the first time since World War II. A key enabler in many of these developments is the advent of high power, reliable power electronics. Indeed so influential are these developments, and their effect upon systems integration so widespread, that a new section has been provided in Chapter 10 – Advanced Systems – to provide the reader with an overview of the new technology involved.

The emergence of Unmanned Air Vehicles or Unmanned Air Systems has resulted in at least 600 different platforms from 250 companies in 42 nations at the time of writing [1,2]. The continuing development of unmanned air systems will pose challenges in both military and commercial markets for effective solutions for sensors and general systems. The question of autonomous operation and certification for use in controlled air space will continue to tax people for some time.

It will be noted that this edition contains descriptions of legacy aircraft systems. This is intentional because many of these aircraft types, and much of this technology, remains in service today and will for some time to come. It also serves as a useful historical source of the development of systems. It should also be noted that the lifetime of aircraft is increasing, while the lifecycle of technology is reducing. This means that obsolescence is an issue that will need to be considered in modern developments, especially those using commercial off the shelf systems driven predominantly by commercial and domestic technology.

There has been an increased awareness of environmental issues, both in the use of materials and in the emission of contaminants and pollution. These issues are being addressed by international agreements and protocols, and by measures by industry to reduce the use of banned and restricted materials. This poses some interesting issues when the platforms in service today are due for removal from service or when accidents occur – there will inevitably be contamination from 'heritage' materials. The issue has been addressed to some extent in this edition, although not fully. It has been left to emerging legislation to provide authoritative guidelines. It is the duty of the competent systems engineer to become familiar with Safety, Health & Environmental legislation and the impact on system design.

Systems Integration

It is the integration of major aircraft systems and the increased interrelationship and interdependence between them that is driving the increasing adoption of high-speed digital data buses. Figures 0.1, 0.2 and 0.3 illustrate at a top level the power generation (hydraulic and electrical), environmental control and fuel systems of a modern combat aircraft. These are complex systems within themselves; however, it is the interrelationship between them that gives the vehicle its fighting edge, as well as causing many of the development headaches. Digital data buses greatly facilitate the interchange of data and control that characterises the functional integration of these systems; on more recent aircraft these data buses also carry a significant amount of health monitoring and maintenance data. The ease with which component and subsystem performance information can be gathered and transmitted to a central or distributed computing centre has led to the emergence of prognostics and health monitoring systems that do much more than simply record failures. They now examine trends in system performance to look for degradation and incipient failures in order to schedule cost-effective maintenance operations. This is an important aspect of improvement in the maintenance of aircraft systems, reducing the incidence of No Fault Found component replacement actions.

The engines of the typical military fast jet accessory drive shafts that power Aircraft Mounted Accessory Drives (AMADs) are mounted within the airframe as shown in Figure 0.1. In the simplest implementation these accessory drives power Engine Driven Pumps (EDPs) to pressurise the aircraft centralised

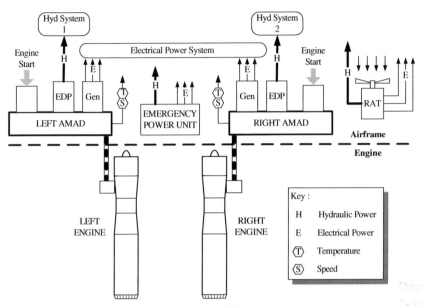

Figure 0.1 Typical military aircraft top-level power generation system

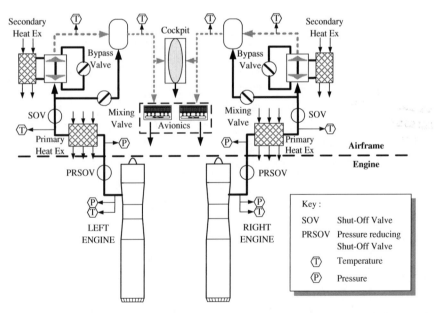

Figure 0.2 Typical military aircraft top-level environmental control system

hydraulic systems. They also drive the electrical power generators that provide electrical power to the electrical distribution system. Most accessory drives will also have an air turbine motor powered by high-pressure air which allows the AMAD and engine to be cranked during engine start, the start process being powered by high-pressure air. Most aircraft also possess an emergency power unit or Ram Air Turbine (RAT) to provide emergency supplies of electrical and hydraulic power.

Once started, the engine provides bleed air for the aircraft systems as well as primary thrust to maintain the aircraft in flight (see Figure 0.2). The generation of electric and hydraulic power has already been described. One of the primary functions of the bleed air extracted from the engine is to provide the means by which the aircraft Environmental Control System (ECS) is driven. Bleed air taken from the engine compressor is reduced in pressure and cooled though a series of heat exchangers and an air cycle machine to provide cool air for the cockpit and the avionics cooling system. Suitably conditioned bleed air is used to pressurise the cockpit to keep the combat crew in a comfortable environment and may also be used to pressurise hydraulic reservoirs and aircraft fuel tanks, among other aircraft systems.

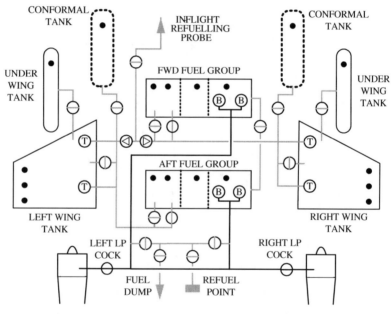

Figure 0.3   Typical military aircraft top-level fuel system

The aircraft fuel system as shown in Figure 0.3 is fundamental to supply fuel to the engines to maintain thrust and powered flight. Fuel feed to the engines is pressurised by using electrically powered booster pumps to prevent fuel cavitation – this is usually an engine HP pump-related problem associated with inadequate feed pressure which is manifest particularly at high altitude. Electrical power is used to operate the transfer pumps and fuel valves that enable the fuel management system to transfer fuel around the aircraft during various phases of flight. In some cases, bleed air, again suitably conditioned, is used to pressurise the external fuel tanks, facilitating fuel transfer inboard to the fuselage tank groups.

Since the Second Edition was published one of the major developments in fuel systems has been the establishment of fuel tank inerting systems as a common requirement for ensuring fuel tank safety of civil aircraft. Boeing is installing Nitrogen Generation Systems (NGS) on all of its current production aircraft and the issues associated with these new requirements are fully described in Chapter 3 – Fuel Systems.

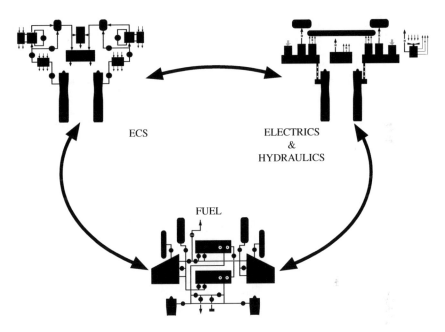

Figure 0.4   Typical military aircraft – systems interaction

From the very short, almost superficial, description of how these major systems interact, it is not difficult to understand how complex modern aircraft systems have become to satisfy the aircraft overall performance requirements. If one system fails to perform to specification then the aircraft as a whole will not perform correctly. Figure 0.4 illustrates in a very simple fashion how these systems functionally interrelate.

Systems Interaction

On more advanced civil and military aircraft a strong functional interaction exists between aircraft CG (controlled by the fuel system) and the performance of the flight control system – aircraft flight performance being critically determined by the CG and flight control stability margins.

A less obvious example of significant interaction between systems is how various systems operate together to reject waste heat from the aircraft. Heat is generated when fluids are compressed and also by energy conversion processes that are not totally efficient. Figure 0.5 depicts the interaction of several major systems – this time within the context of a civil aircraft. The diagram illustrates how a total of eight heat exchangers across a range of systems use the aircraft fuel and ambient ram air as heat sinks into which waste heat may be dumped.

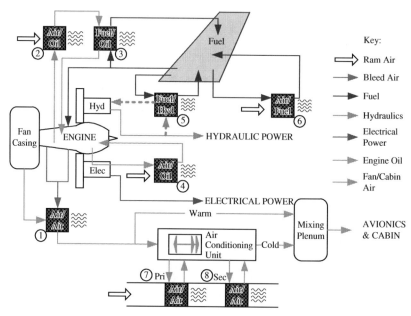

Figure 0.5 Typical civil systems interaction – heat exchange between system (See Colour Plate 1)

Starting with the engine:

1. Air extracted from the engine fan casing is used to cool bleed air tapped off the intermediate or high pressure compressor (depending upon engine type) – Chapter 7, Environmental Control Systems.
2. Air is used to cool engine oil in a primary oil cooler heat exchanger – Chapter 2, Engine Systems.
3. Fuel is used to cool engine oil in a secondary oil cooler heat exchanger – Chapter 7 Engine Systems.
4. The electrical Integrated Drive Generator (IDG) oil is cooled by air – Chapter 5, Electrical Systems.
5. The hydraulic return line fluid is cooled by fuel before being returned to the reservoir – Chapter 4, Hydraulic Systems.
6. Aircraft fuel is cooled by an air/fuel heat exchanger – Chapter 3, Fuel Systems.

7. Ram air is used in primary heat exchangers in the air conditioning pack to cool entry bleed air prior to entering the secondary heat exchangers – Chapter 7, Environmental Control Systems.
8. Secondary heat exchangers further cool the air down to temperatures suitable for mixing with warm air prior to delivery to the cabin – Chapter 7, Environmental Control Systems.

A new chapter has also been introduced to examine the environmental conditions that the aircraft and its systems will be subject to in service. This chapter provides some guidance on how to specify systems to operate in different climatic and environmental contamination conditions and how to ensure that testing is conducted to gather evidence to qualify the systems. This is increasingly important since military aircraft are being deployed to theatres of operation with very different conditions to their home base, and commercial aircraft are flying long routes that may have widely differing conditions at the destination to those prevailing at departure.

The authors hope that this edition has brought together technologies that have emerged since the previous editions and our sincere aim is that readers will practise systems engineering principles in pursing their system analysis and design. The interconnectedness of systems in the modern aircraft means that systems do not stand alone: their performance must be considered in the light of interaction with other systems, and as making a contribution to the performance of the aircraft as a whole.

References

[1] Hughes, David, *Aviation Week & Space Technology*, 12 Feb. 2007.
[2] Unmanned Air Vehicles and Drones Survey, *Aviation Week & Space Technology*, 15 Jan. 2007.

1

Flight Control Systems

1.1 Introduction

Flight controls have advanced considerably throughout the years. In the earliest biplanes flown by the pioneers flight control was achieved by warping wings and control surfaces by means of wires attached to the flying controls in the cockpit. Figure 1.1 clearly shows the multiplicity of rigging and control wires on an early monoplane. Such a means of exercising control was clearly rudimentary and usually barely adequate for the task in hand. The use of articulated flight control surfaces followed soon after but the use of wires and pulleys to connect the flight control surfaces to the pilot's controls persisted for many years until advances in aircraft performance rendered the technique inadequate for all but the simplest aircraft.

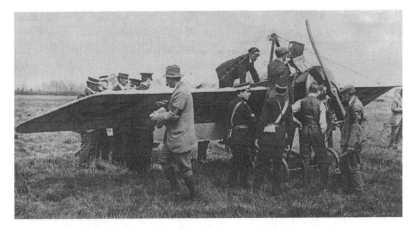

Figure 1.1   Morane Saulnier Monoplane refuelling before the 1913 Aerial Derby (Courtesy of the Royal Aero Club)

Aircraft Systems: Mechanical, electrical, and avionics subsystems integration. 3rd Edition   Ian Moir and Allan Seabridge
© 2008 John Wiley & Sons, Ltd

When top speeds advanced into the transonic region the need for more complex and more sophisticated methods became obvious. They were needed first for high-speed fighter aircraft and then with larger aircraft when jet propulsion became more widespread. The higher speeds resulted in higher loads on the flight control surfaces which made the aircraft very difficult to fly physically. The Spitfire experienced high control forces and a control reversal which was not initially understood. To overcome the higher loadings, powered surfaces began to be used with hydraulically powered actuators boosting the efforts of the pilot to reduce the physical effort required. This brought another problem: that of 'feel'. By divorcing the pilot from the true effort required to fly the aircraft it became possible to undertake manoeuvres which could over-stress the aircraft. Thereafter it was necessary to provide artificial feel so that the pilot was given feedback representative of the demands he was imposing on the aircraft. The need to provide artificial means of trimming the aircraft was required as Mach trim devices were developed.

A further complication of increasing top speeds was aerodynamically related effects. The tendency of many high performance aircraft to experience roll/yaw coupled oscillations – commonly called Dutch roll – led to the introduction of yaw dampers and other auto-stabilisation systems. For a transport aircraft these were required for passenger comfort whereas on military aircraft it became necessary for target tracking and weapon aiming reasons.

The implementation of yaw dampers and auto-stabilisation systems intro-duced electronics into flight control. Autopilots had used both electrical and air driven means to provide an automatic capability of flying the aircraft, thereby reducing crew workload. The electronics used to perform the control func-tions comprised analogue sensor and actuator devices which became capable of executing complex control laws and undertaking high integrity control tasks with multiple lanes to guard against equipment failures. The crowning glory of this technology was the Category III autoland system manufactured by Smiths Industries and fitted to the Trident and Belfast aircraft.

The technology advanced to the point where it was possible to remove the mechanical linkage between the pilot and flight control actuators and rely totally on electrical and electronic means to control the aircraft. Early systems were hybrid, using analogue computing with discrete control logic. The Control and Stability Augmentation System (CSAS) fitted to Tornado was an example of this type of system though the Tornado retained some mechanical reversion capability in the event of total system failure. However the rapid development and maturity of digital electronics soon led to digital 'fly-by-wire' systems. These developments placed a considerable demand on the primary flight control actuators which have to be able to accommodate multiple channel inputs and also possess the necessary failure logic to detect and isolate failures (see Figure 1.2).

Most modern fighter aircraft of any sophistication now possess a fly-by-wire system due to the weight savings and considerable improvements in handling characteristics which may be achieved. Indeed many such aircraft are totally unstable and would not be able to fly otherwise. In recent years this technology

Figure 1.2   Tornado ADV (F 3) Prototype (Courtesy of BAE Systems)

has been applied to civil transports: initially with the relaxed stability system fitted to the Airbus A320 family and A330/A340. The Boeing 777 airliner also has a digital fly-by-wire system, the first Boeing aircraft to do so.

1.2 Principles of Flight Control

All aircraft are governed by the same basic principles of flight control, whether the vehicle is the most sophisticated high-performance fighter or the simplest model aircraft.

The motion of an aircraft is defined in relation to translational motion and rotational motion around a fixed set of defined axes. Translational motion is that by which a vehicle travels from one point to another in space. For an orthodox aircraft the direction in which translational motion occurs is in the direction in which the aircraft is flying, which is also the direction in which it is pointing. The rotational motion relates to the motion of the aircraft around three defined axes: pitch, roll and yaw. See Figure 1.3.

This figure shows the direction of the aircraft velocity in relation to the pitch, roll and yaw axes. For most of the flight an aircraft will be flying straight

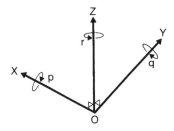

Figure 1.3   Definition of flight control axes

and level and the velocity vector will be parallel with the surface of the earth and proceeding upon a heading that the pilot has chosen. If the pilot wishes to climb, the flight control system is required to rotate the aircraft around the pitch axis (Ox) in a nose-up sense to achieve a climb angle. Upon reaching the new desired altitude the aircraft will be rotated in a nose-down sense until the aircraft is once again straight and level.

In most fixed wing aircraft, if the pilot wishes to alter the aircraft heading then he will need to execute a turn to align the aircraft with the new heading. During a turn the aircraft wings are rotated around the roll axis (Oy) until a certain bank is attained. In a properly balanced turn the angle of roll when maintained will result in an accompanying change of heading while the roll angle (often called the bank angle) is maintained. This change in heading is actually a rotation around the yaw axis (Oz). The difference between the climb (or descent) and the turn is that the climb only involves rotation around one axis whereas the turn involves simultaneous coordination of two axes. In a properly coordinated turn, a component of aircraft lift acts in the direction of the turn, thereby reducing the vertical component of lift. If nothing were done to correct this situation, the aircraft would begin to descend; therefore in a prolonged turning manoeuvre the pilot has to raise the nose to compensate for this loss of lift. At certain times during flight the pilot may in fact be rotating the aircraft around all three axes, for example during a climbing or descending turning manoeuvre.

The aircraft flight control system enables the pilot to exercise control over the aircraft during all portions of flight. The system provides control surfaces that allow the aircraft to manoeuvre in pitch, roll and yaw. The system has also to be designed so that it provides stable control for all parts of the aircraft flight envelope; this requires a thorough understanding of the aerodynamics and dynamic motion of the aircraft. As will be seen, additional control surfaces are required for the specific purposes of controlling the high lift devices required during approach and landing phases of flight. The flight control system has to give the pilot considerable physical assistance to overcome the enormous aerodynamic forces on the flight control surfaces. This in turn leads to the need to provide the aircraft controls with 'artificial feel' so that he does not inadvertently overstress the aircraft. These 'feel' systems need to provide the pilot with progressive and well-harmonised controls that make the aircraft safe and pleasant to handle. A typical term that is commonly used today to describe this requirement is 'carefree handling'. Many aircraft embody automatic flight control systems to ease the burden of flying the aircraft and to reduce pilot workload.

1.3 Flight Control Surfaces

The requirements for flight control surfaces vary greatly between one aircraft and another, depending upon the role, range and agility needs of the vehicle. These varying requirements may best be summarised by giving examples of two differing types of aircraft: an agile fighter aircraft and a typical modern airliner.

The (Experimental Aircraft Programme) EAP aircraft is shown in Figure 1.4 and represented the state of the art fighter aircraft as defined by European Manufacturers at the beginning of the 1990s. The EAP was the forerunner to the European Fighter Aircraft (EFA) or Eurofighter Typhoon developed by the four nation consortium comprising Alenia (Italy), British Aerospace (UK), CASA (Spain) and DASA (Germany).

1.4 Primary Flight Control

Primary flight control in pitch, roll and yaw is provided by the control surfaces described below.

Pitch control is provided by the moving canard surfaces, or foreplanes, as they are sometimes called, located either side of the cockpit. These surfaces provide the very powerful pitch control authority required by an agile high performance aircraft. The position of the canards in relation to the wings renders the aircraft unstable. Without the benefit of an active computer-driven control system the aircraft would be uncontrollable and would crash in a matter of seconds. While this may appear to be a fairly drastic implementation, the benefits in terms of improved manoeuvrability enjoyed by the pilot outweigh the engineering required to provide the computer-controlled or 'active' flight control system.

Roll control is provided by the differential motion of the foreplanes, augmented to a degree by the flaperons. In order to roll to the right, the left foreplane leading edge is raised relative to the airflow generating greater lift than before. Conversely, the right foreplane moves downwards by a corresponding amount relative to the airflow thereby reducing the lift generated. The resulting differential forces cause the aircraft to roll rapidly to the right. To some extent roll control is also provided by differential action of the wing trailing edge flaperons (sometimes called elevons). However, most of the roll control is provided by the foreplanes.

Yaw control is provided by the single rudder section. For high performance aircraft yaw control is generally less important than for conventional aircraft due to the high levels of excess power. There are nevertheless certain parts of the flight envelope where control of yaw (or sideslip) is vital to prevent roll–yaw divergence.

1.5 Secondary Flight Control

High lift control is provided by a combination of flaperons and leading edge slats. The flaperons may be lowered during the landing approach to increase the wing camber and improve the aerodynamic characteristics of the wing. The leading edge slats are typically extended during combat to further increase wing camber and lift. The control of these high lift devices during combat may occur automatically under the control of an active flight control system. The

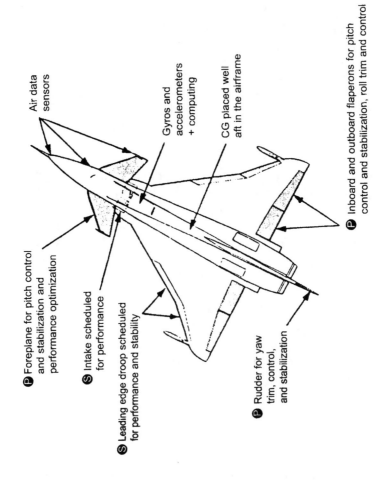

Air data sensors

Gyros and accelerometers + computing

CG placed well aft in the airframe

Ⓟ Foreplane for pitch control and stabilization and performance optimization

Ⓢ Intake scheduled for performance

Ⓢ Leading edge droop scheduled for performance and stability

Ⓟ Rudder for yaw trim, control, and stabilization

Ⓟ Inboard and outboard flaperons for pitch control and stabilization, roll trim and control

Ⓟ Primary controls
Ⓢ Secondary controls

Figure 1.4 Example of flight control surfaces – EAP (Courtesy of BAE Systems)

penalty for using these high lift devices is increased drag, but the high levels of thrust generated by a fighter aircraft usually minimises this drawback.

The Eurofighter Typhoon has airbrakes located on the upper rear fuselage. They extend to an angle of around 50 degrees, thereby quickly increasing the aircraft drag. The airbrakes are deployed when the pilot needs to reduce speed quickly in the air; they are also often extended during the landing run to enhance the aerodynamic brake effect and reduce wheel brake wear.

1.6 Commercial Aircraft

1.6.1 Primary Flight Control

An example of flight control surfaces of a typical commercial airliner is shown in Figure 1.5. Although the example is for the Airbus Industrie A320 it holds good for similar airliners produced by Boeing. The controls used by this type of aircraft are described below.

Pitch control is exercised by four elevators located on the trailing edge of the tailplane (or horizontal stabiliser in US parlance). Each elevator section is independently powered by a dedicated flight control actuator, powered in turn by one of several aircraft hydraulic power systems. This arrangement is dictated by the high integrity requirements placed upon flight control systems. The entire tailplane section itself is powered by two or more actuators in order to trim the aircraft in pitch. In a dire emergency this facility could be used to control the aircraft, but the rates of movement and associated authority are insufficient for normal control purposes.

Roll control is provided by two aileron sections located on the outboard third of the trailing edge of each wing. Each aileron section is powered by a dedicated actuator powered in turn from one of the aircraft hydraulic systems. At low airspeeds the roll control provided by the ailerons is augmented by differential use of the wing spoilers mounted on the upper surface of the wing. During a right turn the spoilers on the inside wing of the turn, that is the right wing, will be extended. This reduces the lift of the right wing causing it to drop, hence enhancing the desired roll demand.

Yaw control is provided by three independent rudder sections located on the trailing edge of the fin (or vertical stabiliser). These sections are powered in a similar fashion to the elevator and ailerons. On a civil airliner these controls are associated with the aircraft yaw dampers. These damp out unpleasant 'Dutch roll' oscillations which can occur during flight and which can be extremely uncomfortable for the passengers, particularly those seated at the rear of the aircraft.

1.6.2 Secondary Flight Control

Flap control is effected by several flap sections located on the inboard two-thirds of the wing trailing edges. Deployment of the flaps during take-off or landing extends the flap sections rearwards and downwards to increase wing

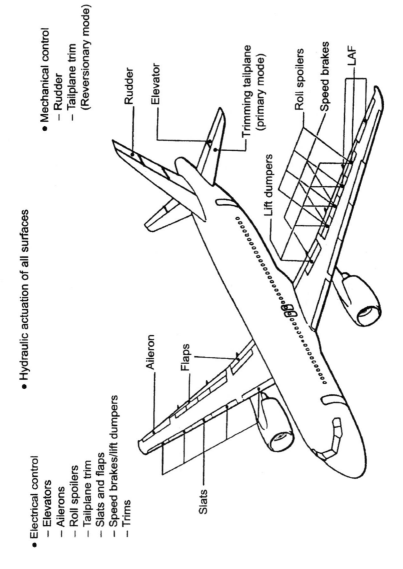

• Electrical control
  – Elevators
  – Ailerons
  – Roll spoilers
  – Tailplane trim
  – Slats and flaps
  – Speed brakes/lift dumpers
  – Trims

• Hydraulic actuation of all surfaces

• Mechanical control
  – Rudder
  – Tailplane trim
    (Reversionary mode)

Rudder

Elevator

Trimming tailplane
(primary mode)

Roll spoilers

Speed brakes

LAF

Lift dumpers

Aileron

Flaps

Slats

Figure 1.5 Example of flight control surfaces – commercial airliner (A320)
(Courtesy of Airbus (UK))

area and camber, thereby greatly increasing lift for a given speed. The number of flap sections may vary from type to type; typically for this size of aircraft there would be about five per wing, giving a total of ten in all.

Slat control is provided by several leading edge slats, which extend forwards and outwards from the wing leading edge. In a similar fashion to the flaps described above, this has the effect of increasing wing area and camber and therefore overall lift. A typical aircraft may have five slat sections per wing, giving a total of ten in all.

Speed-brakes are deployed when all of the over-wing spoilers are extended together which has the effect of reducing lift as well as increasing drag. The effect is similar to the use of air-brakes in the fighter, increasing drag so that the pilot may adjust his airspeed rapidly; most airbrakes are located on rear fuselage upper or lower sections and may have a pitch moment associated with their deployment. In most cases compensation for this pitch moment would be automatically applied within the flight control system.

While there are many identical features between the fighter and commercial airliner examples given above, there are also many key differences. The greatest difference relates to the size of the control surfaces in relation to the overall size of the vehicle. The fighter control surfaces are much greater than the corresponding control surfaces on an airliner. This reflects its prime requirements of manoeuvrability and high performance at virtually any cost. The commercial airliner has much more modest control requirements; it spends a far greater proportion of flying time in the cruise mode so fuel economy rather than ultimate performance is prime target. Passenger comfort and safety are strong drivers that do not apply to the same degree for a military aircraft.

1.7 Flight Control Linkage Systems

The pilot's manual inputs to the flight controls are made by moving the cockpit control column or rudder pedals in accordance with the universal convention:

- Pitch control is exercised by moving the control column fore and aft; pushing the column forward causes the aircraft to pitch down, and pulling the column aft results in a pitch up
- Roll control is achieved by moving the control column from side to side or rotating the control yoke; pushing the stick to the right drops the right wing and vice versa
- Yaw is controlled by the rudder pedals; pushing the left pedal will yaw the aircraft to the left while pushing the right pedal will have the reverse effect

There are presently two main methods of connecting the pilot's controls to the rest of the flight control system. These are:

- Push-pull control rod systems
- Cable and pulley systems

An example of each of these types will be described and used as a means of introducing some of the major components which are essential for the flight control function. A typical high lift control system for the actuation of slats and flaps will also be explained as this introduces differing control and actuation requirements.

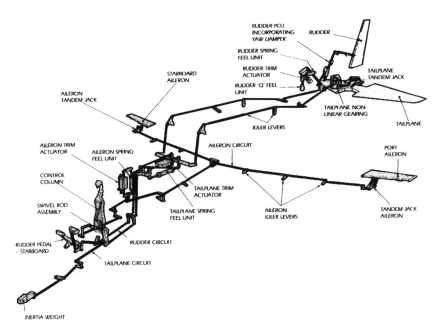

Figure 1.6 Hawk 200 push-pull control rod system (Courtesy of BAE Systems)

1.7.1 Push-Pull Control Rod System

The example chosen for the push-pull control rod system is the relatively simple yet high performance BAE Hawk 200 aircraft. Figure 1.6 shows a simplified three-dimensional schematic of the Hawk 200 flight control which is typical of the technique widely used for combat aircraft. This example is taken from British Aerospace publicity information relating to the Hawk 200 see reference [1]. The system splits logically into pitch–yaw (tailplane and rudder) and roll (aileron) control runs respectively.

The pitch control input is fed from the left hand or starboard side (looking forward) of the control column to a bell-crank lever behind the cockpit. This connects in turn via a near vertical control rod to another bell-crank lever which returns the control input to the horizontal. Bell-crank levers are used to alter the direction of the control runs as they are routed through a densely packed aircraft. The horizontal control rod runs parallel to a tailplane trim actuator/tailplane spring feel unit parallel combination. The output from these units is fed upwards into the aircraft spine before once again being translated by another bell-crank lever. The control run passes down the left side of the fuselage to the rear of the aircraft via several idler levers before entering a nonlinear gearing mechanism leading to the tandem jack tailplane power control unit (PCU). The idler levers are simple lever mechanisms which help to support the control run at convenient points in the airframe. The hydraulically powered PCU drives the tailplane in response to the pilot inputs and the aircraft manoeuvres accordingly.

The yaw input from the rudder pedals is fed to a bell-crank lever using the same pivot points as the pitch control run and runs vertically to another bell-crank which translates the yaw control rod to run alongside the tailplane trim/feel units. A further two bell-cranks place the control linkage running down the right-hand side of the rear fuselage via a set of idler levers to the aircraft empennage. At this point the control linkage accommodates inputs from the rudder trim actuator, spring feel unit and 'Q' feel unit. The resulting control demand is fed to the rudder hydraulically powered PCU which in turn drives the rudder to the desired position. In this case the PCU has a yawdamper incorporated which damps out undesirable 'Dutch roll' oscillations.

The roll demand is fed via a swivel rod assembly from the right hand of port side (looking forward) of the control column and runs via a pair of bell-crank levers to a location behind the cockpit. At this point a linkage connects the aileron trim actuator and the aileron spring feel unit. The control rod runs aft via a further bell-crank lever and an idler lever to the centre fuselage. A further bell-crank lever splits the aileron demand to the left and right wings. The wing control runs are fed outboard by means of a series of idler levers to points in the outboard section of the wings adjacent to the ailerons. Further bell-cranks feed the left and right aileron demands into the tandem jacks and therefore provide the necessary aileron control surface actuation.

Although a simple example, this illustrates some of the considerations which need to be borne in mind when designing a flight control system. The interconnecting linkage needs to be strong, rigid and well supported; otherwise fuselage flexing could introduce 'nuisance' or unwanted control demands into the system. A further point is that there is no easy way or route through the airframe; therefore an extensive system of bell-cranks and idler levers is required to support the control rods. This example has also introduced some of the major components which are required to enable a flight control system to work while providing safe and pleasant handling characteristics to the pilot. These are:

- Trim actuators in tailplane (pitch), rudder (yaw) and aileron (roll) control systems
- Spring feel units in tailplane (pitch), rudder (yaw) and aileron (roll) control systems
- 'Q' feel unit in the rudder (yaw) control system
- Power control units (PCUs) for tailplane, rudder and aileron actuation

1.7.2 Cable and Pulley System

The cable and pulley system is widely used for commercial aircraft; sometimes used in conjunction with push-pull control rods. It is not the intention to attempt to describe a complete aircraft system routing in this chapter. Specific examples will be outlined which make specific points in relation to the larger aircraft (see Figure 1.7).

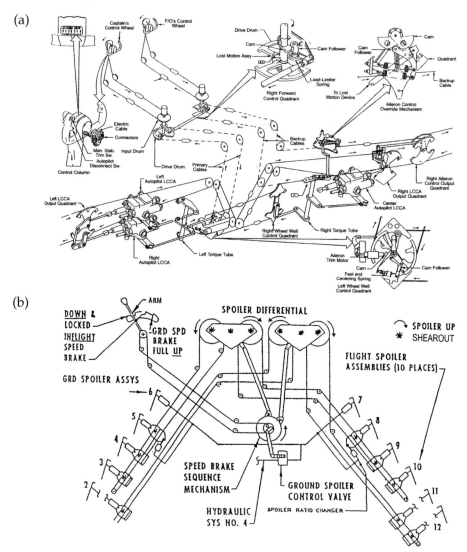

Figure 1.7 Examples of wire and pulley aileron control system (Courtesy of Boeing)

Figure 1.7a shows a typical aileron control system. Manual control inputs are routed via cables and a set of pulleys from both captain's and first officer's control wheels to a consolidation area in the centre section of the aircraft. At this point aileron and spoiler runs are split both left/right and into separate aileron/spoiler control runs. Both control column/control wheels are synchronised. A breakout device is included which operates at a predetermined force in the event that one of the cable runs fails or becomes jammed.

Control cable runs are fed through the aircraft by a series of pulleys, idler pulleys, quadrants and control linkages in a similar fashion to the push-pull rod system already described. Tensiometer/lost motion devices situated throughout the control system ensure that cable tensions are correctly maintained and lost motion eliminated. Differing sized pulleys and pivot/lever arrangements allow for the necessary gearing changes throughout the control runs. Figure 1.7a also shows a typical arrangement for control signalling in the wing. Figure 1.7b shows a typical arrangement for interconnecting wing spoiler and speedbrake controls. Trim units, feel units and PCUs are connected at strategic points throughout the control runs as for the push-pull rod system.

1.8 High Lift Control Systems

The example chosen to illustrate flap control is the system used on the BAE 146 aircraft. This aircraft does not utilise leading edge slats. Instead the aircraft relies upon single section Fowler flaps which extend across 78% of the inner wing trailing edge. Each flap is supported in tracks and driven by recirculating ballscrews at two locations on each wing. The ballscrews are driven by transmission shafts which run along the rear wing spar. The shafting is driven by two hydraulic motors which drive into a differential gearbox such that the failure of one motor does not inhibit the drive capability of the other. See Figure 1.8 for a diagram of the BAE 146 flap operating system.

As well as the flap drive motors and flap actuation, the system includes a flap position selector switch and an electronic control unit. The electronic control unit comprises: dual identical microprocessor based position control channels; two position control analogue safety channels; a single microprocessor based safety channel for monitoring mechanical failures. For an excellent system description refer to the technical paper on the subject prepared by Dowty Rotol/TI Group reference [2].

The slat system or leading edge flap example chosen is that used for the Boeing 747-400. Figure 1.9 depicts the left wing leading edge slat systems. There is a total of 28 flaps, 14 on each wing. These flaps are further divided into groups A and B. Group A flaps are those six sections outside the outboard engines; group B flaps include the five sections between inboard and outboard engines and the three sections inside the inboard engines. The inboard ones are Kreuger flaps which are flat in the extended position, the remainder are of variable camber which provide an aerodynamically shaped surface when extended. The flaps are powered by power drive units (PDUs); six of these drive the group A flaps and two the group B flaps. The motive power is pneumatic with electrical backup. Gearboxes reduce and transfer motion from the PDUs to rotary actuators which operate the drive linkages for each leading edge flap section. Angular position is extensively monitored throughout the system by rotary variable differential transformers (RVDTs).

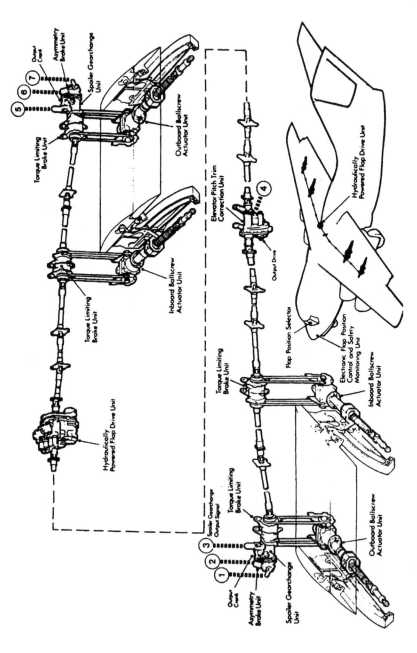

Figure 1.8 BAE 146 flap operating system (Courtesy of Smiths Group – now GE Aviation)

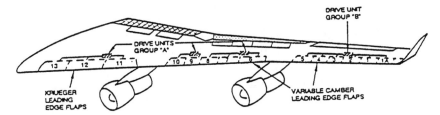

Figure 1.9 Boeing 747-400 leading edge flap system (Courtesy of Boeing)

1.9 Trim and Feel

The rod and pulley example for the BAE Hawk 200 aircraft showed the interconnection between the pilot's control columns and rudder bars and the hydraulically powered actuators which one would expect. However the diagram also revealed a surprising number of units associated with aircraft trim and feel. These additional units are essential in providing consistent handling characteristics for the aircraft in all configurations throughout the flight envelope.

1.9.1 Trim

The need for trim actuation may be explained by recourse to a simple explanation of the aerodynamic forces which act upon the aircraft in flight. Figure 1.10 shows a simplified diagram of the pitch forces which act upon a stable aircraft trimmed for level flight.

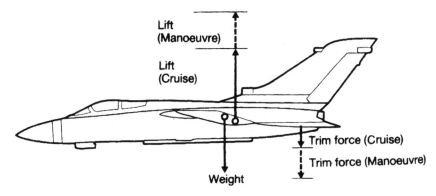

Figure 1.10 Pitch forces acting in level flight

The aircraft weight usually represented by the symbol W, acts downwards at the aircraft centre-of-gravity or CG. As the aircraft is stable the CG is ahead of the centre of pressure where the lift force acts (often denoted by the symbol L) and all aerodynamic perturbations should be naturally damped. The distance between the CG and the centre of pressure is a measure of how stable and also how manoeuvrable the aircraft is in pitch. The closer the CG and centre of pressure, the less stable and more manoeuvrable the aircraft. The converse is true when the CG and centre of pressure are further apart.

Examining the forces acting about the aircraft CG it can be seen that there is a counter-clockwise moment exerted by a large lift force acting quite close to the pivot point. If the aircraft is not to pitch nose-down this should be counterbalanced by a clockwise force provided by the tailplane. This will be a relatively small force acting with a large moment. If the relative positions of the aircraft CG and centre of pressure were to remain constant throughout all conditions of flight then the pilot could set up the trim and no further control inputs would be required.

In practice the CG positions may vary due to changes in the aircraft fuel load and the stores or cargo and passengers the aircraft may be carrying. Variations in the position of the aircraft CG position are allowed within carefully prescribed limits. These limits are called the forward and aft CG limits and they determine how nose heavy or tail heavy the aircraft may become and still be capable of safe and controllable flight. The aerodynamic centre of pressure similarly does not remain in a constant position as the aircraft flight conditions vary. If the centre of pressure moves aft then the downward force required of the tailplane will increase and the tailplane angle of incidence will need to be increased. This requires a movement of the pitch control run equivalent to a small nose-up pitch demand. It is inconvenient for the pilot constantly to apply the necessary backward pressure on the control column, so a pitch actuator is provided to alter the pitch control run position and effectively apply this nose-up bias. Forward movement of the centre of pressure relative to the CG would require a corresponding nose-down bias to be applied. These nose-up and nose-down biases are in fact called nose-up and nose-down trim respectively.

Pitch trim changes may occur for a variety of reasons: increase in engine power, change in airspeed, alteration of the fuel disposition, deployment of flaps or airbrakes and so on. The desired trim demands may be easily input to the flight control system by the pilot. In the case of the Hawk the pilot has a four-way trim button located on the stick top; this allows fore and aft (pitch) and lateral (roll) trim demands to be applied without moving his hand from the control column.

The example described above outlines the operation of the pitch trim system as part of overall pitch control. Roll or aileron trim is accomplished in a very similar way to pitch trim by applying trim biases to the aileron control run by means of an aileron trim actuator. Yaw or rudder trim is introduced by the separate trim actuator provided; in the Hawk this is located in the rear of the aircraft. The three trim systems therefore allow the pilot to offload variations in load forces on the aircraft controls as the conditions of flight vary.

1.9.2 Feel

The provision of artificial 'feel' became necessary when aircraft performance increased to the point where it was no longer physically possible for the pilot to apply the high forces needed to move the flight control surfaces. Initially with servo-boosting systems, and later with powered flying controls, it became necessary to provide powered assistance to attain the high control forces required. This was accentuated as the aircraft wing thickness to chord ratio became much smaller for performance reasons and the hinge moment available was correspondingly reduced. However, a drawback with a pure power assisted system is that the pilot may not be aware of the stresses being imposed on the aircraft. Furthermore, a uniform feel from the control system is not a pleasant characteristic; pilots are not alone in this regard; we are all used to handling machinery where the response and feel are sensibly related. The two types of feel commonly used in aircraft flight control systems are spring feel and 'Q' feel.

Typically the goal is to provide a fairly constant 'Stick force per g' over the full flight envelope. In this regard, the feel system is further complicated with variable geometry aircraft such as the Tornado since aircraft response in pitch and roll varies dramatically with wing sweep. The feel system must therefore take into account both Q and wing sweep.

Spring feel, as the name suggests, is achieved by loading the movement of the flight control run against a spring of a predetermined stiffness. Therefore when the aircraft controls are moved, the pilot encounters an increasing force proportional to the spring stiffness. According to the physical laws spring stiffness is a constant and therefore spring feel is linear unless the physical geometry of the control runs impose any nonlinearities. In the Hawk 200, spring feel units are provided in the tailplane, aileron and rudder control runs. The disadvantage of spring feel units is that they only impose feel proportional to control demand and take no account of the pertaining flight conditions.

'Q' feel is a little more complicated and is more directly related to the aerodynamics and precise flight conditions that apply at the time of the control demand. As the aircraft speed increases the aerodynamic load increases in a mathematical relationship proportional to the air density and the square of velocity. The air density is relatively unimportant; the squared velocity term has a much greater effect, particularly at high speed. Therefore it is necessary to take account of this aerodynamic equation; that is the purpose of 'Q' feel. A 'Q' feel unit receives air data information from the aircraft pitot-static system. In fact the signal applied is the difference between pitot and static pressure, (known as Pt-Ps) and this signal is used to modulate the control mechanism within the 'Q' feel unit and operate a hydraulic load jack which is connected into the flight control run.

In this way the pilot is given feel which is directly related to the aircraft speed and which will greatly increase with increasing airspeed. It is usual to use 'Q' feel in the tailplane or rudder control runs; where this method of feel is used depends upon the aircraft aerodynamics and the desired handling or safety features. The disadvantage of 'Q' feel is that it is more complex and only becomes of real use at high speed. Figure 1.11 is a photograph of a 'Q' feel unit supplied by Dowty for the BAE Harrier GR5 and McDonnell Douglas AV-8B aircraft. This unit is fitted with an electrical solenoid so that the active part of the system may be disconnected if required. This unit is designed to operate with an aircraft 20.7 MN/sq m (3000 psi) hydraulic system pressure.

Figure 1.11 'Q' feel unit for GR5/AV8B (Courtesy of Smiths Group – now GE Aviation)

The rudder control run on Hawk 200 shown in Figure 1.6 uses both spring and 'Q' feel. It is likely that these two methods have been designed to complement each other. The spring feel will dominate at low speed and for high deflection control demands. The 'Q' feel will dominate at high speeds and low control deflections.

1.10 Flight Control Actuation

The key element in the flight control system, increasingly so with the advent of fly-by-wire and active control units, is the power actuation. Actuation has always been important to the ability of the flight control system to attain its specified performance. The development of analogue and digital multiple control lane technology has put the actuation central to performance and integrity issues. Addressing actuation in ascending order of complexity leads to the following categories:

- Simple mechanical actuation, hydraulically powered
- Mechanical actuation with simple electromechanical features
- Multiple redundant electromechanical actuation with analogue control inputs and feedback

The examination of these crudely defined categories leads more deeply into systems integration areas where boundaries between mechanical, electronic, systems and software engineering become progressively blurred.

1.10.1 Simple Mechanical/Hydraulic Actuation

Conventional Linear Actuator

The conventional linear actuator used in powered flight controls would be of the type show in Figure 1.12. This type of actuator would usually be powered by one of the aircraft hydraulic systems – in this case the blue channel is shown. In functionally critical applications a dual hydraulic supply from another aircraft hydraulic system may be used. A mechanically operated Servo Valve (SV) directs the hydraulic supply to the appropriate side of the piston ram.

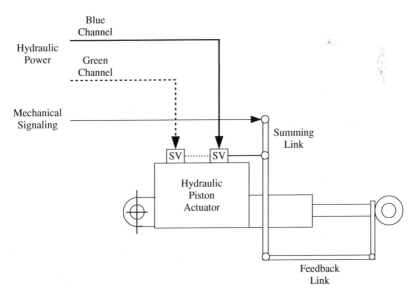

Figure 1.12 Conventional linear actuator

As the pilot feeds a mechanical input to the flight control actuator, the summing link will rotate about the bottom pivot, thus applying an input to the servo valve. Hydraulic fluid will then flow into one side of the ram while exiting the opposite side resulting in movement of the ram in a direction dependent upon the direction of the pilot's command. As the ram moves, the feedback link will rotate the summing link about the upper pivot returning the servo valve input to the null position as the commanded position is achieved.

The attributes of mechanical actuation are straightforward; the system demands a control movement and the actuator satisfies that demand with a power assisted mechanical response. The BAE Hawk 200 is a good example of a system where straightforward mechanical actuation is used for most of the flight control surfaces. For most applications the mechanical actuator is able to accept hydraulic power from two identical/redundant hydraulic systems. The obvious benefit of this arrangement is that full control is retained following loss of fluid or a failure in either hydraulic system. This is important even in a simple system as the loss of one or more actuators and associated control surfaces can severely affect aircraft handling. The actuators themselves have a simple reversion mode following failure, that is to centre automatically under the influence of aerodynamic forces. This reversion mode is called aerodynamic centring and is generally preferred for obvious reasons over a control surface freezing or locking at some intermediate point in its travel. In some systems 'freezing' the flight control system may be an acceptable solution depending upon control authority and reversionary modes that the flight control system possesses.The decision to implement either of these philosophies will be a design decision based upon the system safety analysis.

Mechanical actuation may also be used for spoilers where these are mechanically rather than electrically controlled. In this case the failure mode is aerodynamic closure, that is the airflow forces the control surface to the closed position where it can subsequently have no adverse effect upon aircraft handling. Figure 1.13 illustrates the mechanical spoiler actuator supplied by

Figure 1.13 BAE 146 spoiler actuator (Courtesy of Claverham/Hamilton Sundstrand)

Claverham for the BAE 146. This unit is simplex in operation. It produces thrust of 59.9 kN (13 460 lb) over a working stroke of 15 mm (0.6 inch). It has a length of 22.4 mm (8.8 inch) and weighs 8.3 kg (18.2 lb). The unit accepts hydraulic pressure at 20.7 MN/sqm (3000 psi).

1.10.2 Mechanical Actuation with Electrical Signalling

The use of mechanical actuation has already been described and is appropriate for a wide range of applications. However the majority of modern aircraft use electrical signalling and hydraulically powered (electro-hydraulic) actuators for a wide range of applications with varying degrees of redundancy. The demands for electro-hydraulic actuators fall into two categories: simple demand signals or autostabilisation inputs.

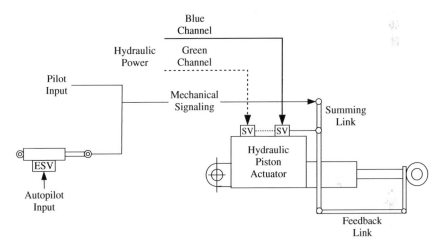

Figure 1.14 Conventional linear actuator with autopilot interface

As aircraft acquired autopilots to reduce pilot work load then it became necessary to couple electrical as well as mechanical inputs to the actuator as shown in Figure 1.14. The manual (pilot) input to the actuator acts as before when the pilot is exercising manual control. When the autopilot is engaged electrical demands from the autopilot computer drive an electrical input which takes precedence over the pilot's demand. The actuator itself operates in an identical fashion as before with the mechanical inputs to the summing link causing the Servo-Valve (SV) to move. When the pilot retrieves control by disengaging the autopilot the normal mechanical link to the pilot through the aircraft control run is restored.

Simple electrical demand signals are inputs from the pilots that are signalled by electrical means. For certain noncritical flight control surfaces it may be easier, cheaper and lighter to utilise an electrical link. An example of this is the airbrake actuator used on the BAE 146; simplex electrical signalling is used and in the case of failure the reversion mode is aerodynamic closure.

In most cases where electrical signalling is used this will at least be duplex in implementation and for fly-by-wire systems signalling is likely to be quadruplex; these more complex actuators will be addressed later. An example of duplex electrical signalling with a simplex hydraulic supply is the spoiler actuators on Tornado. There are four actuators fitted on the aircraft, two per wing, which are used for roll augmentation.

In general, those systems which extensively use simplex electrical signalling do so for autostabilisation. In these systems the electrical demand is a stabilisation signal derived within a computer unit. The simplest form of autostabilisation is the yaw damper which damps out the cyclic cross-coupled oscillations which occur in roll and yaw known as 'Dutch roll'. The Hawk 200 illustrated this implementation. Aircraft which require a stable platform for weapon aiming may have simplex autostabilisation in pitch, roll and yaw; an example of this type of system is the Harrier/AV-8A. A similar system on the Jaguar uses simplex autostabilisation in pitch and roll.

1.10.3 Multiple Redundancy Actuation

Modern flight control systems are increasingly adopting fly-by-wire solutions as the benefits to be realised by using such a system are considerable. These benefits include a reduction in weight, improvement in handling performance and crew/passenger comfort. Concorde was the first aircraft to pioneer these techniques in the civil field using a flight control system jointly developed by GEC (now Finmeccanica) and SFENA.[3] The Tornado, fly-by-wire Jaguar and EAP have extended the use of these techniques; the latter two were development programmes into the regime of the totally unstable aircraft. In the civil field the Airbus A320 and the Boeing 777 introduced modern state-of-the-art systems into service. For obvious reasons, a great deal of care is taken during the definition, specification, design, development and certification of these systems. Multiple redundant architectures for the aircraft hydraulic and electrical systems must be considered as well as multiple redundant lanes or channels of computing and actuation for control purposes. The implications of the redundancy and integrity of the other aircraft systems will be addressed. For the present, attention will be confined to the issues affecting multiple redundant electro-hydraulic actuation.

A simplified block schematic diagram of a multiple redundant electro-hydraulic actuator is shown in Figure 1.15. For reasons of simplicity only one lane or channel is shown; in practice the implementation is likely to be quadruplex, i.e. four identical lanes. The solenoid valve is energised to supply hydraulic power to the actuator, often from two of the aircraft hydraulic systems. Control demands from the flight control computers are fed to the servo valves. The servo valves control the position of the first-stage valves that are mechanically summed before applying demands to the control valves. The control valves modulate the position of the control ram. Linear variable

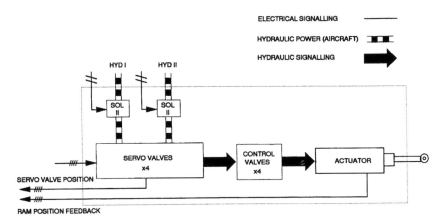

Figure 1.15 Simplified block schematic diagram of a multiple redundant electrically signalled hydraulic actuator

differential transformers (LVDTs) measure the position of the first-stage actuator and output ram positions of each lane and these signals are fed back to the flight control computers, thereby closing the loop. Two examples of this quadruplex actuation system are given below: the Tornado quadruplex taileron and rudder actuators associated with the Control Stability Augmentation System (CSAS) and the EAP flight control system. Both of these systems are outlined at system level in reference [1]. The description given here will be confined to that part of the flight control system directly relevant to the actuator drives.

The Tornado CSAS flight control computation is provided by pitch and lateral computers supplied by GEC (now part of Finmeccanica) and Bodenseewerk (now Thales). The pitch computer predominantly handles pitch control computations and the lateral computer roll and yaw computations though there are interconnections between the two (see Figure 1.16a). There are three computing lanes; computing is analogue in nature and there are a number of voter-monitors within the system to vote out lanes operating outside specification. The combined pitch/roll output to the taileron actuators is consolidated from three lanes to four within the pitch computer so the feed to the taileron actuators is quadruplex. The quadruplex taileron actuator is provided by Fairey Hydraulics (now Hamilton Sundstrand) and is shown in Figure 1.16b. This actuator provides a thrust of 339.3 kN (76 291 lb) over a working stroke of 178 mm. The actuator is 940 mm (37.0 in) long and weighs 51.0 kg and operates with the two aircraft 4000 psi hydraulic systems. The rudder actuator similarly receives a quadruplex rudder demand from the lateral computer, also shown in Figure 1.14b. The rudder actuator is somewhat smaller than the taileron actuator delivering a thrust of 80.1 kN. The CSAS is designed so that following a second critical failure it is possible to revert to a mechanical link for pitch and roll. In these circumstances the rudder is locked in the central position.

The Tornado example given relates to the analogue system that comprises the CSAS. The EAP flight control system (FCS) is a quadruplex digital computing

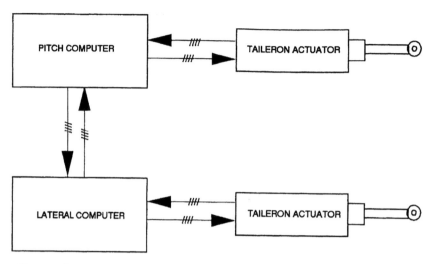

Figure 1.16a Tornado Taileron/Rudder CSAS drive interface

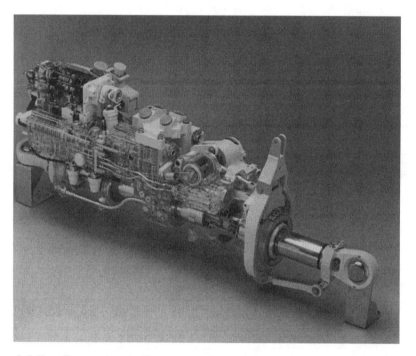

Figure 1.16b Tornado taileron and rudder actuators (Courtesy of Claverham/Hamilton Sundstrand)

system in which control computations are undertaken in all four computing lanes. The system is quadruplex rather than triplex as a much higher level of integrity is required. As has been mentioned earlier the EAP was an unstable aircraft and the FCS has to be able to survive two critical failures. Figure 1.17a shows the relationship between the flight control computers

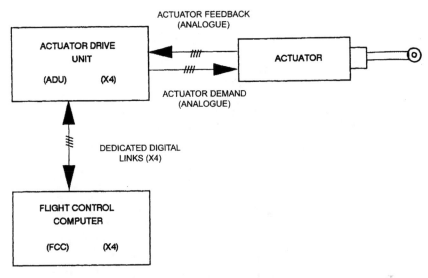

Figure 1.17a   EAP actuator drive configuration

(FCCs), Actuator Drive Units (ADUs) and the actuators. The foreplane actuators are fed quadruplex analogue demands from the quadruplex digital FCCs. Demands for the left and right, inboard and outboard flaperons and the rudder are fed in quadruplex analogue form from the four ADUs. The ADUs receive the pitch, roll and yaw demands from the FCCs via dedicated serial digital links and the digital to analogue conversion is carried out within the ADUs. The total complement of actuators supplied by Dowty (now GE Aviation) for the EAP is as follows:

- Quadruplex electrohydraulic foreplane actuators: 2
- Quadruplex electrohydraulic flaperon actuators:

   - outboard flaperons – 100 mm working stroke: 2
   - inboard flaperons – 165 mm working stroke: 2

- Quadruplex electrohydraulic rudder actuators – 100 mm working stroke: 1
  (Figure 1.17b.)

All seven actuators are fed from two independent hydraulic systems.

The EAP flight control system represented the forefront of such technology of its time and the aircraft continued to exceed expectations following the first flight in August 1986 until the completion of the programme. Further detail regarding the EAP system and the preceding Jaguar fly-by-wire programme may be found in a number of technical papers which have been given in recent years references [3–8]. Most of these papers are presented from an engineering perspective. The paper by Chris Yeo, Deputy Chief Test Pilot at British Aerospace at the time of the fly-by-wire programme, includes an overview of the aircraft control laws reference [5].

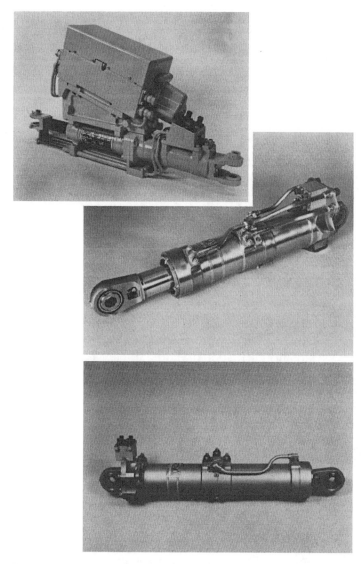

Figure 1.17b   EAP foreplane, flaperon and rudder actuators (Courtesy of Smiths Group – now GE Aviation)

1.10.4 Mechanical Screwjack Actuator

The linear actuators described so far are commonly used to power aileron, elevator and rudder control surfaces where a rapid response is required but the aerodynamic loads are reasonably light.There are other applications where a relatively low speed of response may be tolerated but the ability to apply or withstand large loads is paramount. In these situations a mechanical screwjack is used to provide a slow response with a large mechanical advantage. This is employed to drive the Tailplane Horizontal Stabilator or Stabiliser (THS), otherwise known years ago as a 'moving tailplane'. The THS is used to trim an aircraft in pitch as airspeed varies; being a large surface it moves slowly over small angular movements but has to withstand huge loads. The mechanical screwjack shown in Figure 1.18 often has one or two aircraft hydraulic system supplies and a summing link that causes SVs to move in response to the mechanical inputs. In this case the SVs moderate the pressure to hydraulic motor(s) which in turn drive the screwjack through a mechanical gearbox. As before the left-hand portion of the jack is fixed to aircraft structure and movement of the screwjack ram satisfies the pilot's demands, causing the tailplane to move, altering tailplane lift and trimming the aircraft in pitch. As in previous descriptions, movement of the ram causes the feedback link to null the original demand, whereupon the actuator reaches the demanded position.

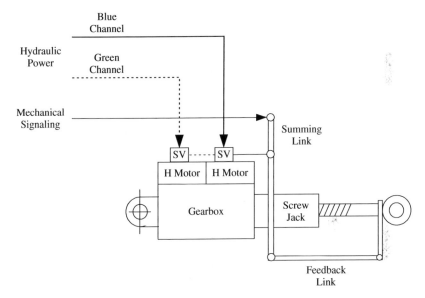

Figure 1.18   Mechanical screwjack actuator

1.10.5 Integrated Actuator Package (IAP)

In the UK, the introduction of powerful new AC electrical systems paved the way for the introduction of electrically powered power flying controls. Four channel AC electrical systems utilised on the Avro Vulcan B2 and Handley Page Victor V-Bombers and the Vickers VC10 transport aircraft utilised flight control actuators powered by the aircraft AC electrical system rather than centralised aircraft hydraulic systems.

Figure 1.19 shows the concept of operation of this form of actuator known as an Integrated Actuator Package (IAP). The operation of demand, summing and feedback linkage is similar to the conventional linear actuator already described. The actuator power or 'muscle' is provided by a three-phase constant speed electrical motor driving a variable displacement hydraulic

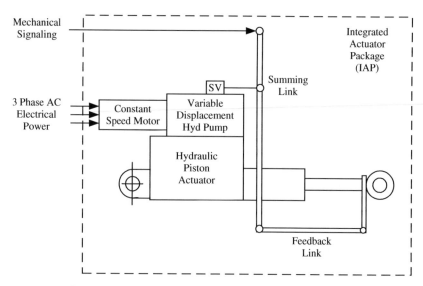

Figure 1.19 Integrated actuator package

pump. The hydraulic pump and associated system provides hydraulic pressure to power the actuator ram.

The variable displacement hydraulic pump is the hydraulic pressure source for the actuator. A bi-directional displacement mechanism which is controlled via a servo valve determines the pumps flow and hence actuator velocity. As with the linear actuator, a feedback mechanism nulls off the input to the servo valve as the desired output position is achieved.

Therefore when the actuator is in steady state, the pump displacement is set to the null position but the pump continues to rotate at a constant speed imposing a significant 'windage' power loss which is a significant disadvantage with this design. The more modern integrated actuator designs, specifically the Electro-Hydrostatic Actuator (described later) eliminates this problem.

Figure 1.20 depicts an overview of a typical IAP used on the Vickers VC-10 flight control system. A total of 11 such units were used in the VC-10 system to powr each of the following flight control surfaces:

- Ailerons: 4 sections
- Elevators: 4 sections
- Rudder: 3 sections

The power consumption of each of the IAPs is in the region of 2.75 kVA and are still flying today in the Royal Air Force's VC-10 Tanker fleet. The units are powered by a constant frequency, split-parallel, 115 VAC three-phase electrical system.

The Avro Vulcan B-2 also used IAP to power the primary flight control surfaces. Being a large delta aircraft this system had an unusual configuration comprising eight elevons powered by IAPs located on the trailing edge of the delta wing plus two on the aircraft rudder. The elevons provided a combined elevator and aileron function to control the aircraft in pitch and roll. Figure 1.21

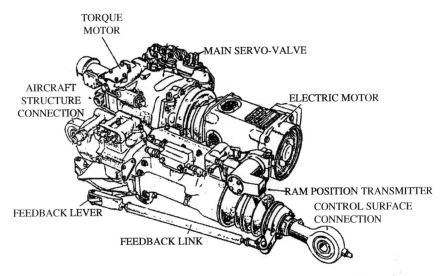

Figure 1.20 Integrated actuator package (VC-10)

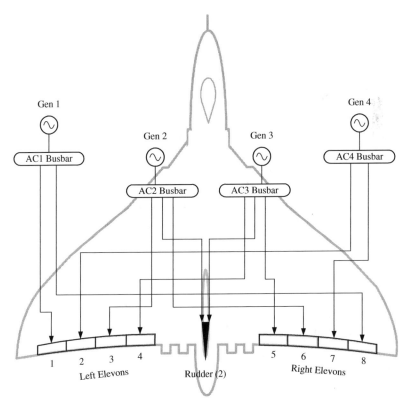

Figure 1.21 Avro Vulcan B-2 FCS architecture using IAPs

illustrates how the total complement of ten power flight control units were powered by the four aircraft AC buses.

1.10.6 Advanced Actuation Implementations

The actuation implementations described so far have all been mechanical or electro-hydraulic in function using servo valves. There are a number of recent developments that may supplant the existing electro-hydraulic actuator. These newer types of actuation are listed below and have found application in aircraft over the past 10–15 years:

• Direct drive actuation
• Fly-by-Wire (FBW) actuation
• Electro-Hydrostatic Actuator (EHA)
• Electro-Mechanical Actuator (EMA)

Direct Drive Actuation

In the electro-hydraulic actuator a servo valve requires a relatively small electrical drive signal, typically in the order of 10–15 mA. The reason such low drive currents are possible is that the control signal is effectively amplified within the hydraulic section of the actuator. In the direct drive actuator the aim is to use an electrical drive with sufficient power to obviate the need for the servo valve/1st stage valve. The main power spool is directly driven by torque motors requiring a higher signal current, hence the term 'direct drive'. Development work relating to the direct drive concept including comparison with Tornado requirements and operation with 8 000psi hydraulic systems has been investigated by Fairey Hydraulics see reference [9]. This paper also addresses the direct digital control of aircraft flight control actuators.

Fly-By-Wire Actuator

The advent of Fly-By-Wire (FBW) flight control systems in civil aircraft commencing with the Airbus A320 introduced the need for a more sophisticated interface between the FCS and actuation. Most first generation FBW aircraft may operate in three distinct modes that may be summarised in general terms as follows:

• *Full FBW Mode.* This mode encompasses the full FBW algorithms and protection and is the normal mode of operation
• *Direct Electrical Link Mode.* This mode will usually provide rudimentary algorithms or possibly only a direct electrical signalling capability in the event that the primary FBW mode is not available
• *Mechanical Reversion Mode.* This provides a crude means of flying the aircraft – probably using a limited number of flight control surface following the failure of FBW and direct electrical link modes

In later implementations such as the Airbus A380 and Boeing 787 no mechanical reversion is provided. The interface with the actuator is frequently achieved by means of an Actuator Control Electronics (ACE) unit that closes the control loop electrically around the actuator rather than mechanical loop closure as hitherto described (see Figure 1.22). The digital FBW or direct link demands from the flight control system are processed by the ACE which supplies an analogue command to the actuator SV. This allows aircraft systems hydraulic power to be supplied to the appropriate side of the ram piston moving the ram to the desired position. In this implementation the ram position is detected by means of a Linear Variable Differential Transducer (LVDT) which feeds the signal back to the ACE where the loop around the actuator is closed. Therefore ACE performs two functions: conversion of digital flight control demands into analogue signals and analogue loop closure around the actuator.

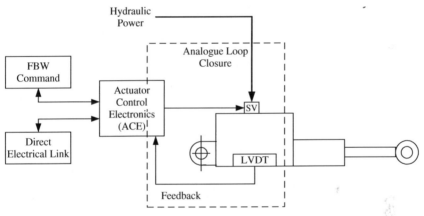

Figure 1.22 Fly-by-wire actuator

Electro-Hydrostatic Actuator (EHA)

The move towards more-electric aircraft has coincided with another form of electrical actuation – the Electro-Hydrostatic Actuator (EHA) which uses state-of-the-art power electronics and control techniques to provide more efficient flight control actuation. The conventional actuation techniques described so far continually pressurise the actuator whether or not there is any demand. In reality for much of the flight, actuator demands are minimal and this represents a wasteful approach as lost energy ultimately results in higher energy offtake from the engine and hence higher fuel consumption.

The EHA seeks to provide a more efficient form of actuation where the actuator only draws significant power when a control demand is sought; for the remainder of the flight the actuator is quiescent (see Figure 1.23). The EHA accomplishes this by using the three-phase AC power to feed power drive electronics which in turn drive a variable speed pump together with a constant displacement hydraulic pump. This constitutes a local hydraulic system for the

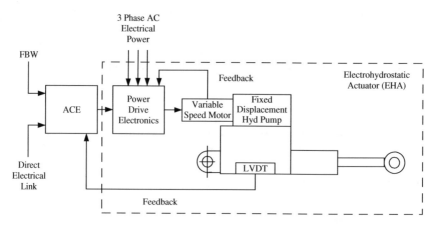

Figure 1.23 Electro-Hydrostatic Actuator (EHA)

actuator in a similar fashion to the IAP; the difference being that when there is no demand the only power drawn is that to maintain the control electronics. When a demand is received from the ACE the power drive electronics is able to react sufficiently rapidly to drive the variable speed motor and hence pressurise the actuator such that the associated control surface may be moved to satisfy the demand. Once the demand has been satisfied then the power electronics resumes its normal dormant state. Consequently power is only drawn from the aircraft buses bars while the actuator is moving, representing a great saving in energy. The ACE closes the control loop around the actuator electrically as previously described.

EHAs are being applied across a range of aircraft and Unmanned Air Vehicle (UAV) developments. The Airbus A380 and Lockheed Martin F-35 Lightning II both use EHAs in the flight control system. For aircraft such as the A380 with a conventional three-phase, 115 VAC electrical system, the actuator uses an in-built matrix converter to convert the aircraft three-phase AC power to 270 VDC to drive a brushless DC motor which in turn drives the fixed displacement pump. The Royal Aeronautical Society Conference, More-Electric Aircraft, 27–28 April 2004, London is an excellent reference for more-electric aircraft and more-electric engine developments where some of these solutions are described.

Aircraft such as the F-35 have an aircraft level 270 VDC electrical system and so the matrix converter may be omitted with further savings in efficiency. Furthermore, electric aircraft/more-electric engine development programmes with civil applications envisage the use of 540 VDC or ±270 VDC systems on the aircraft or engine platform and therefore making similar savings in energy. These developments, including a European Community (EC) funded programme called Power Optimised Aircraft (POA), were described and discussed at the Technologies for Energy Optimised Aircraft Equipment Systems (TEOS) forum in Paris, 28–30 June 2006.

A common feature of all three new actuator concepts outlined above is the use of microprocessors to improve control and performance. The introduction of digital control in the actuator also permits the consideration of direct digital interfacing to digital flight control computers by means of data buses (ARINC 429/ARINC 629/1553B). The direct drive developments described emphasise concentration upon the continued use of aircraft hydraulics as the power source, including the accommodation of system pressures up to 8000 psi. The EMA and EHA developments, on the other hand, lend themselves to a greater use of electrical power deriving from the all-electric aircraft concept, particularly if 270 VDC power is available.

Electro-Mechanical Actuator (EMA)

The electromechanical actuator or EMA replaces the electrical signalling and power actuation of the electro-hydraulic actuator with an electric motor and gearbox assembly applying the motive force to move the ram. EMAs have been used on aircraft for many years for such uses as trim and door actuation; however the power, motive force and response times have been less than that required for flight control actuation. The three main technology advancements that have improved the EMA to the point where it may be viable for flight control applications are: the use of rare earth magnetic materials in 270 VDC motors; high power solid-state switching devices; and microprocessors for lightweight control of the actuator motor [10].

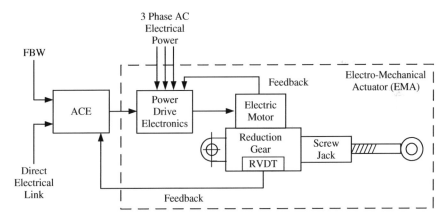

Figure 1.24 Electro-mechanical actuator

As the EHA is the more-electric replacement for linear actuators so the Electro-Mechanical Actuator (EMA) is the more-electric version of the screw-jack actuator as shown in Figure 1.24. The concept of the EMA is identical with the exception that the power drive electronics drives a brushless DC motor operating a reduction gear that applies rotary motion allowing the jack ram to extend or retract to satisfy input demands. EMAs are therefore used to power the THS on civil aircraft and flap and slat drives and also find a use in helicopter flight control systems. A major concern regarding the EMA is the consideration of the actuator jamming case and this has negated their use in primary flight controls on conventional aircraft.

Actuator Matrix

Most of these actuation types are used in civil aircraft today. Table 1.1 lists how the various actuator types may be used for different actuation tasks on a typical civil airliner.

Table 1.1 Typical applications of flight control actuators

Actuator type	Power source	Primary flight control	Spoilers	Tailplane horizontal stabilator	Flaps and slats
Conventional Linear Actuator	Aircraft Hydraulic Systems: B/Y/G or L/C/R [1]	X	X		
Conventional Screw-jack Actuator	Aircraft Hydraulic or Electrical Systems [2]			X	X
Integrated Actuator Package (IAP)	Aircraft Electrical System (115VAC)	X	X		
Electrically Signalled Hydraulic Actuator	Aircraft Hydraulic Systems	X	X		
Electro-Hydrostatic Actuator (EHA)	Aircraft Electrical System [3] [4]	X	X		
Electro-Mechanical Actuator (EMA)	Aircraft Electrical System [3]			X	X

Notes: (1) B/Y/G = Blue/Green/Yellow or L/C/R = Left/Centre/Right (Boeing)
      (2) For THS and Flaps & Slats both hydraulic and electrical supplies are often used for redundancy
      (3) 3-phase VAC to 270 VDC matrix converter used in civil
      (4) 270 VDC aircraft electrical system used on F-35/JSF

1.11 Civil System Implementations

The flight control and guidance of civil transport aircraft has steadily been getting more sophisticated in recent years. Whereas Concorde was the first civil aircraft to have a fly-by-wire system, Airbus introduced a fly-by-wire system on to the A320 family [11] and a similar system has been carried forward to the A330/340. Boeing's first fly-by-wire system on the Boeing 777 was widely believed to a response to the Airbus technology development. The key differences between the Airbus and Boeing philosophies and implementations are described below.

1.11.1 Top-Level Comparison

The importance and integrity aspects of flight control lead to some form of monitoring function to ensure the safe operation of the control loop. Also for integrity and availability reasons, some form of redundancy is usually required. Figure 1.25 shows a top-level comparison between the Boeing and Airbus FBW implementations.

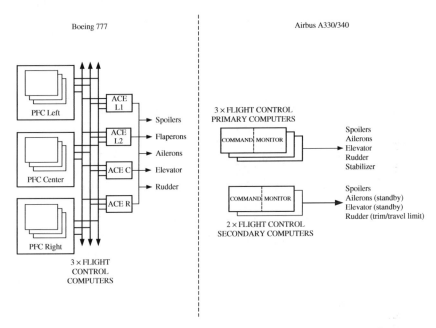

Figure 1.25   Top-level Boeing & Airbus FBW comparison

In the Boeing philosophy, shown in simplified form on the left of Figure 1.25, the system comprises three Primary Flight Computers (PFCs) each of which has three similar lanes with dissimilar hardware but the same software. Each lane has a separate role during an operating period and the roles are cycled after power up. Voting techniques are used to detect discrepancies or disagreements between lanes and the comparison techniques used vary for different types of data. Communication with the four Actuator Control Electronics (ACE) units is by multiple A629 flight control data buses. The ACE units directly drive the flight control actuators. A separate flight control DC system is provided to power the flight control system. The schemes used on the Boeing 777 will be described in more detail later in this Module.

The Airbus approach is shown on the right of Figure 1.25. Five main computers are used: three Flight Control Primary Computers (FCPCs) and two Flight Control Secondary Computers (FCSCs). Each computer comprises command and monitor elements with different software. The primary and secondary computers have different architectures and different hardware. Command outputs from the FCSCs to ailerons, elevators and the rudder are for standby use only. Power sources and signalling lanes are segregated.

1.11.2 Airbus Implementation

The Anglo-French Concorde apart, Airbus was the first aircraft manufacturer in recent years to introduce Fly-By-Wire (FBW) to civil transport aircraft. The original aircraft to utilise FBW was the A320 and the system has been used throughout the A319/320/321 family and more recently on the A330/340. The A320 philosophy will be described and A330/340 system briefly compared.

A320 FBW System

A schematic of the A320 flight control system is shown in Figure 1.26. The flight control surfaces are all hydraulically powered and are tabulated as follows:

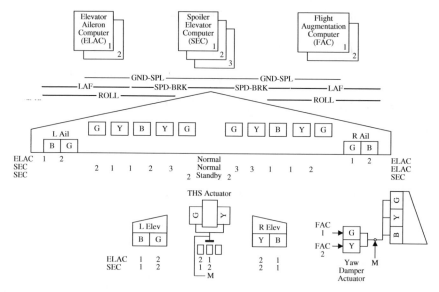

Figure 1.26 A320 flight control system

- *Electrical control:*
  Elevators 2
  Ailerons 2
  Roll spoilers 8
  Tailplane trim 1
  Slats 10
  Flaps 4
  Speedbrakes 6
  Lift dumpers 10
  Trims

- *Mechanical control*:
    Rudder
    Tailplane trim (reversionary mode)

The aircraft has three independent hydraulic power systems: blue (B), green (G) and yellow (Y). Figure 1.26 shows how these systems respectively power the hydraulic flight control actuators.

A total of seven computers undertake the flight control computation task as follows:

- Two Elevator/Aileron Computers (ELACs). The ELACs control the aileron and elevator actuators according to the notation in the figure
- Three Spoiler/Elevator Computers (SECs). The SECs control all of the spoilers and in addition provide secondary control to the elevator actuators. The various spoiler sections have different functions as shown namely:

    - ground spoiler mode: all spoilers
    - speed brake mode: inboard three spoiler sections
    - load alleviation mode: outboard two spoiler sections (plus ailerons); this function has recently been disabled and is no longer embodied in recent models
    - roll augmentation: outboard four spoiler sections

- Two Flight Augmentation Computers (FACs). These provide a conventional yaw damper function, interfacing only with the yaw damper actuators

The three aircraft hydraulic systems; blue, green and yellow provide hydraulic power to the flight control actuators according to the notation shown on the diagram.

In the very unlikely event of the failure of all computers it is still possible to fly and land the aircraft – this has been demonstrated during certification. In this case the Tailplane Horizontal Actuator (THS) and rudder sections are controlled directly by mechanical trim inputs – shown as M in the diagram – which allow pitch and lateral control of the aircraft to be maintained.

Another noteworthy feature of the Airbus FBW systems is that they do not use the conventional pitch and roll yoke. The pilot's pitch and roll inputs to the system are by means of a side-stick controller and this has been widely accepted by the international airline community.

In common with contemporary civil aircraft, the A320 is not an unstable aircraft like the EAP system briefly described earlier in this chapter. Instead the aircraft operates with a longitudinal stability margin of around 5% of aerodynamic mean chord or around half what would normally be expected for an aircraft of this type. This is sometimes termed relaxed stability. The A320 family can claim to be the widest application of civil FBW with over 3000 examples delivered.

A330/340 FBW System

The A330/340 FBW system bears many similarities to the A320 heritage as might expected.

The pilot's input to the Flight Control Primary Computers (FCPCs) and Flight Control Secondary Computers (FCSCs) is by means of the side-stick controller. The Flight Management Guidance and Envelope Computers (FMGECs) provide autopilot pitch commands to the FCPC. The normal method of commanding the elevator actuators is via the FCPC although they can be controlled by the FCSC in a standby mode. Three autotrim motors may be engaged via a clutch to drive the mechanical input to the THS.

For the pitch channel, the FCPCs provide primary control and the FCSCs the backup. Pilots' inputs are via the rudder pedals directly or, in the case of rudder trim, via the FCSC to the rudder trim motors.

The yaw damper function resides within the FCPCs rather than the separate Flight Augmentation Computers (FACs) used on the A320 family. Autopilot yaw demands are fed from the FMGECs to the FCPCs.

There is a variable travel limitation unit to limit the travel of the rudder input at various stages of flight. As before, the three hydraulic systems feed the rudder actuators and two yaw damper actuators as annotated on the figure.

Therefore although the implementation and notation of the flight control computers differs between the A320 and A330/340 a common philosophy can be identified between the two families.

The overall flight control system elements for the A330/340 are:

- Three Flight Control Primary Computers (FCPCs); the function of the FCPCs has been described
- Two Flight Control Secondary Computers (FCSCs); similarly, the function of the secondary computers has been explained
- Two Flight Control Data Concentrators (FCDCs); the FCDCs provide data from the primary and secondary flight computers for indication, recording and maintenance purposes
- Two Slat/Flap Control Computers (SFCCs); the FSCCs are each able to control the full-span leading-edge slats and trailing-edge flaps via the hydraulically driven slat and flap motors

Spoiler usage on the A330/340 differs from that on the A320. There is no load alleviation function and there are six pairs of spoilers versus the five pairs on the A320. Also the functions of the various spoiler pairs differ slightly from the A320 implementation. However, overall, the philosophy is the same.

A380 Implementation

The A380 flight control system represents the most advanced system flying today and follows the philosophy used by Airbus over the past 20 years.

Airbus Fly-By-Wire Evolution

The first airbus FBW aircraft was the A320 that was first certified in 1988.Since then the A320 family has expanded to include the A318, A319 and A321; the A330 and A340 aircraft have entered service and the A380 did so in October 2007. In that time the number of flight control actuators has increased with the size of the aircraft as may be seen in Table 1.2.

Table 1.2 Airbus family – roll effectors

Airbus model	Spoilers per wing	Ailerons/actuators per wing
A320 family	5	1/2
A330/340 family	6	2/4
A380	8	3/6

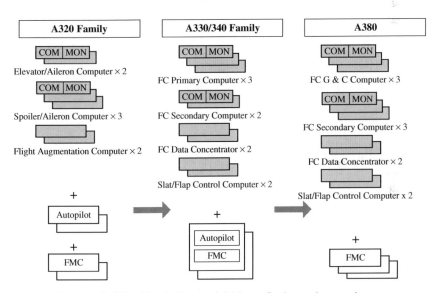

Figure 1.27 Evolution of Airbus fly-by-wire systems

The Airbus Family FBW has evolved historically from the A320 family through the A330/340 family to the latest A380 aircraft. Figure 1.27 clearly illustrates this progression. In this diagram the shaded portion represents the FBW or primary flight control system while the units shown below represent the associated autopilot and flight Management System (FMS) functions.

On the A320 family the autopilot and FMS functions are provided by standalone units. On the A330/340 flight guidance is provided by the Flight Management and Guidance Computers (FMGCs) that embody both autopilot and guidance functions. On the A380 integration has progressed with the autopilot function being subsumed into the FCS with the FMC as stand-alone.

Although the name of the computers has changed from application to application, a clear lineage may be seen with the A380 complement being:

- 3 x Flight Control Primary Computers (FCPCs)
- 3 x Flight Control Secondary Computers (FCSCs)
- 2 x Flight Control Data Concentrators (FCDCs)
- 2 x Flap/Slat Control Computers (FSCCs)

1.12 Fly-By-Wire Control Laws

While it is impossible to generalise, the approach to the application of control laws in a FBW system and the various reversionary modes does have a degree of similarity. The concept of having normal, direct and mechanical links has been outlined earlier in the chapter. The application of normal, alternate and direct control laws and, in the final analysis, mechanical reversion often follows the typical format outlined in Figure 1.28.

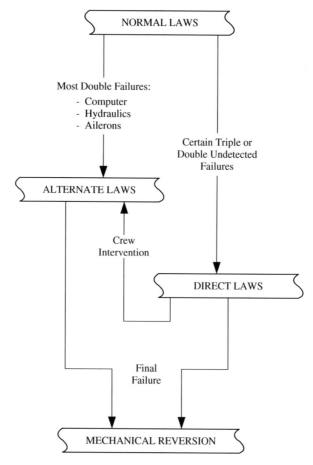

Figure 1.28 Typical interrelationship of FBW control laws

The authority of each of these levels may be summarised as follows:

- *Normal laws*: Provision of basic control laws with the addition of coordination algorithms to enhance the quality of handling and protection to avoid the exceedance of certain attitudes and attitude rates. Double failures in computing, sensors or actuation power channels will cause reversion to the Alternate mode
- *Alternate laws*: Provision of the basic control laws but without many of the additional handling enhancement features and protection offered by the Normal mode. Further failures cause reversion to the Mechanical mode
- *Direct laws*: Direct relationship from control stick to control surface, manual trimming, certain limitations depending upon aircraft CG and flight control system configuration. In certain specific cases crew intervention may enable re-engagement of the Alternate mode. Further failures result in reversion to Mechanical
- *Mechanical reversion*: Rudimentary manual control of the aircraft using pitch trim and rudder pedals to facilitate recovery of the aircraft electrical system or land the aircraft as soon as is practicable

1.13 A380 Flight Control Actuation

The electrical and hydraulic power derived for the A380 flight control actuators is summarised in Figure 1.29.

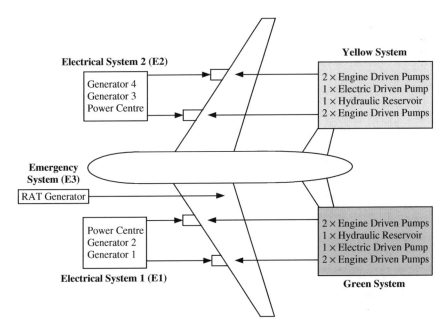

Figure 1.29   A380 hydraulic and electric power generation

Table 1.3 A380 Flight control system actuator matrix

LEFT WING				RIGHT WING			
AILERONS		SPOILERS		SPOILERS		AILERONS	
Inbd	G AC E2	1	Y	Y	1	G AC E2	Inbd
Mid	Y AC E1	2	G	G	2	Y AC E1	Mid
Outbd	Y G	3	Y	Y	3	Y G	Outbd
		4	G	G	4		
		5	Y + AC 2E	Y + AC 2E	5		
		6	G + AC 1E	G + AC 1E	6		
		7	Y	Y	7		
		8	G	G	8		

R ELEVATORS		THS	L ELEVATORS	
Inbd	AC E1 G	G	AC E1	Inbd
		Y	Y	
		AC E2		
Outbd	AC E2 G		AC E2	Outbd
			Y	

RUDDER

Upper
1 Y + AC E1
2 G + AC E2
Lower
1 G + AC E1
2 Y + AC E3

KEY:

G	Green Hydraulic System	AC E1	AC 1 Essential Side 1
Y	Yellow Hydraulic System	AC E2	AC 2 Essential Side 2
		AC E3	AC Essential (RAT)

The A380 flight control actuator configuration is shown in Table 1.3. Many of the actuators are powered only by the aircraft green (LH side powered by engines 1 and 2) and yellow (RH side powered by engines 3 and 4) hydraulic systems. However, many are powered by a combination of conventional hydraulic and electro-hydrostatic actuators (see Figure 1.30).

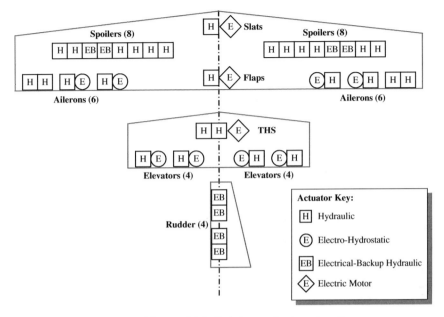

Figure 1.30 A 380 flight control actuation

The use of the various actuation types may be summarised as follows:

- The two outboard aileron surfaces and six spoiler surfaces on each wing are powered by conventional hydraulic actuators – yellow or green system
- The mid and inboard aileron surfaces and the inboard and outboard elevator surfaces are powered by both hydraulic and EHAs, each of which can drive the surface in the event of a failure of the other
- Two spoiler surfaces (five and six on each wing) and both rudder sections are powered by Electrical Backup Hydraulic Actuators (EBHAs) which combine the features of hydraulic actuators and EHAs
- The Tailplane Horizontal Stabilisor (THS) actuator is powered independently from green and yellow channels and from E2

For completeness, the diagram also shows the flap and slat drives. Slats may be powered by green or E1; flaps may be powered from green or yellow channels.

EBHAs receive a hydraulic input from the appropriate channel (green or yellow) and electrical channel (E1 or E2, or exceptionally E3 AC Essential (RAT)). In the case of the rudder, the upper surface is powered by green and yellow, E1 and E2 AC 2; the lower surface is powered by green and yellow, E1 and E3.

EBHAs are capable of two modes of operation:

- *Normal – hydraulic mode*: In the normal mode the actuator receives hydraulic power from the appropriate green or yellow hydraulic system and the SV moderates the supply to the actuator according to the FBW computer demand

- *Backup – EHA mode*: In the backup mode the actuator operates like an EHA. Electrical power is received from the aircraft AC electrical system and the FBW computer feeds demands to the EHA control package. The rotational direction and speed of the electrical motor determine the direction and rate of travel of the actuator ram

A top-level schematic of an EBHA is shown in Figure 1.31. The combination of multiple redundant FBW computing resources (three primary and three secondary flight control computers) and the actuator hydraulic and electrical power architectures described mean that the aircraft is not fitted with a mechanical reversion.

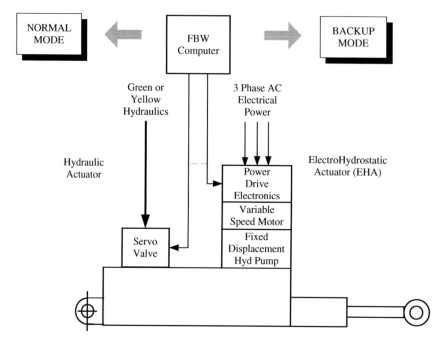

Figure 1.31   A380 EBHA modes of operation

1.14 Boeing 777 Implementation

Boeing ventured into the FBW field with the Boeing 777 partly, it has been said, to counter the technology lead established by Airbus with the A320. Whatever the reason, Boeing have approached the job with precision and professionalism and have developed a solution quite different to the Airbus philosophy. References [12] and [13] give a detailed description of the B777 FBW system.

The B777 PFCS is outlined at a system level in Figure 1.32. The drawing shows the three Primary Flight Control Computers (PFCS), four Actuator Control Electronics (ACEs) and three Autopilot Flight Director Computers (AFDCs) interfacing with the triple redundant A629 flight control buses. The AFDCs have terminals on both the flight control and A629 data buses. Attitude and information is provided by the ADIRU, and SAHRU and air data by the Air Data Modules (ADMs). The three Control and Display Units (CDUs) and the left and right Aircraft Information Management System (AIMS) cabinets provide the flight deck interface. In total there are 76 ARINC 629 couplers on the flight control buses.

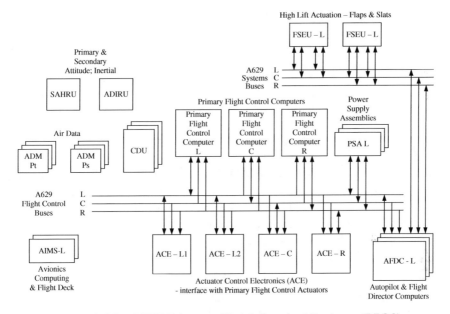

Figure 1.32 B777 Primary Flight Control System (PFCS)

The PFCS system comprises the following control surface actuators and feel actuators:

- Four elevators: left and right inboard and outboard
- Elevator feel: left and right
- Two rudders: upper and lower
- Four ailerons: left and right inboard and outboard
- Four flaperons: left and right inboard and outboard
- Fourteen spoilers: seven left and seven right

The flight control actuators are interfaced to the three A629 flight control data buses by means of four Actuator Control Electronics (ACE) units. These are:

- ACE Left 1
- ACE Left 2
- ACE Centre
- ACE Right

These units interface in turn with the flight control and feel actuators in accordance with the scheme shown in Table 1.4.

Table 1.4 ACE to PCU interface

ACE L1	ACE L2	ACE C	ACE R
ROB Aileron	LOB Aileron	LIB Aileron	RIB Aileron
LOB Aileron	RIB Aileron	ROB Aileron	LIB Aileron
		Upper Rudder	Lower Rudder
LIB Elevator	LOB Elevator	ROB Elevator	RIB Elevator
	L Elevator Feel	R Elevator Feel	
Spoiler 2	Spoiler 5	Spoiler 1	Spoiler 3
Spoiler 13	Spoiler 4	Spoiler 7	Spoiler 6
	Spoiler 11	Spoiler 8	Spoiler 9
	Spoiler 10	Spoiler 14	Spoiler 12

The Actuator Control Electronics (ACE) units contain the digital-to-analogue and analogue-to-digital elements of the system. A simplified schematic for an ACE is shown in Figure 1.33. Each ACE has a single interface with each of the A629 flight control data buses and the unit contains the signal conversion to interface the 'digital' and 'analogue' worlds.

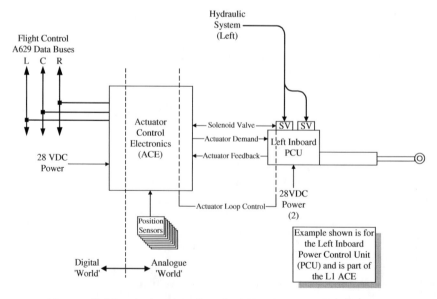

Figure 1.33 Actuator Control Electronics (ACE) Unit

The actuator control loop is shown in the centre-right of the diagram. The actuator demand is signalled to the Power Control Unit (PCU) which moves the actuator ram in accordance with the control demand and feeds back a ram position signal to the ACE, thereby closing the actuator control loop. The ACE also interfaces to the solenoid valve with a command to energise the solenoid valves to allow – in this example – the left hydraulic system to supply the actuator with motive power and at this point the control surface becomes 'live'.

The flight control computations are carried out in the Primary Flight Computers (PFCs) shown in Figure 1.34. The operation of the PFCs has been briefly described earlier in the chapter but will be recounted and amplified in this section.

Each PFC has three A629 interfaces with each of the A629 flight control buses, giving a total of nine data bus connections in all. These data bus interfaces and how they are connected and used form part of the overall Boeing 777 PFCS philosophy. The three active lanes within each PFC are embodied in dissimilar hardware. Each of the three lanes is allocated a different function as follows:

- *PFC command lane*: The command lane is effectively the channel in control. This lane will output the flight control commands on the appropriate A629 bus; e.g. PFC left will output commands on the left A629 bus
- *PFC standby lane*: The standby lane performs the same calculations as the command lane but does not output the commands on to the A629 bus. In effect the standby lane is a 'hot standby', ready to take command in the event that the command lane fails. The standby lane only transmits cross lane and cross-channel data on the A629 data bus
- *PFC monitor lane*: The monitor lane also performs the same calculations as the command lane. The monitor lane operates in this way for both the command lane and the standby lane. Like the standby lane, it only transmits cross lane and cross-channel data on the A629 data bus

Figure 1.34 shows that on the data bus, each PFC will only transmit aircraft control data on the appropriate left, centre or right A629 data bus. Within each PFC the command, standby and monitor lane operation will be in operation as previously described and only the command channel – shown as the upper channel in the figure – will actually transmit command data.

Within this PFC and A629 architecture:

- Cross lane comparisons are conducted via the like bus (in this case the left bus)
- Cross channel comparisons are conducted via the unlike buses (in this case the centre and right buses)

This use of standard A629 databuses to implement the flight control integration and to host the cross lane and cross-channel monitoring is believed to be unique in flight control. There are effectively nine lanes available to conduct the flight control function. In the event that a single lane fails, then only that lane will be

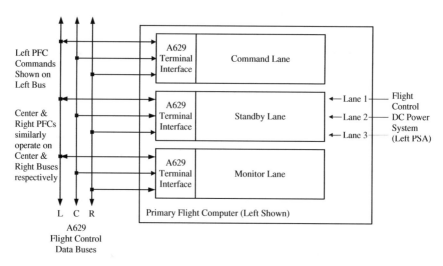

Figure 1.34 B777 Primary Flight Computer (PFC)

shut down. Subsequent loss of a second lane within that channel will cause that channel to shut down, as simplex control is not permitted. The aircraft may be operated indefinitely with one lane out of nine failed and the aircraft may be dispatched with two out of nine lanes failed for ten days. The aircraft may be operated for a day with one PFC channel inoperative.

The autopilot function of the B777 PFCS is undertaken by the three Autopilot Flight Director Computers (AFDCs): left, centre and right. The AFDCs have A629 interfaces on to the respective aircraft systems and flight control data buses. In other words, the left AFDC will interface on to the left A629 buses, the centre AFDC on to the centre buses and so on.

1.15 Interrelationship of Flight Control, Guidance and Flight Management

Figure 1.35 shows a generic example of the main control loops as they apply to aircraft flight control, flight guidance and flight management.

The inner loop provided by the FBW system and the pilot's controls effectively control the attitude of the aircraft.

The middle loop is that affected by the AFDS that controls the aircraft trajectory, that is, where the aircraft flies. Inputs to this loop are by means of the mode and datum selections on the FCU or equivalent control panel.

Finally, the FMS controls where the aircraft flies on the mission; for a civil transport aircraft this is the aircraft route. The MCDU controls the lateral demands of the aircraft by means of a series of waypoints within the route plan and executed by the FMS computer. Improved guidance required of 'free-flight' or DNS/ATM also requires accurate vertical or 3-Dimensional guidance, often with tight timing constraints upon arriving at a way-point or the entry to a terminal area.

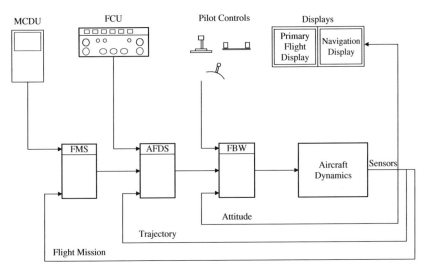

Figure 1.35 Definition of flight control, guidance and management

References

[1] British Aerospace (1990) *Hawk 200*. Marketing publication CO.095.0890.M5336.
[2] Farley, B. (1984) 'Electronic Control and Monitoring of Aircraft Secondary Flying Controls', *Aerospace*, March.
[3] Howard, R.W. (1973) 'Automatic Flight Controls in Fixed Wing Aircraft – The First 100 Years', *Aeronautical Journal*, November.
[4] Daley, E., Smith, R.B. (1982) 'Flight Clearance of the Jaguar Fly-By-Wire Aircraft', *Royal Aeronautical Society Symposium*, April.
[5] Yeo, C.J. (1984) 'Fly-by-Wire Jaguar', *Aerospace*, March.
[6] Kaul, H-J., Stella, F., Walker, M. (1984) 'The Flight Control System for the Experimental Aircraft Programme (EAP) Demonstrator Aircraft', *65th Flight Mechanics Panel Symposium, Toronto*, October.
[7] Young, B. (1987) 'Tornado/Jaguar/EAP Experience and Configuration of Design', *Royal Aeronautical Society Spring Convention*, May.
[8] Snelling, K.S., Corney, J.M. (1987) 'The Implementation of Active Control Systems', *Royal Aeronautical Society Spring Convention*, May.
[9] Anthony, M.J., Mattos, F. (1985) 'Advanced Flight Control Actuation Systems and their Interfaces with Digital Commands', *A-6 Symposium, San Diego*, October.
[10] White, J.A.P. (1978) 'The Development of Electromechanical Actuation for Aircraft Systems', *Aerospace*, November.
[11] Davies, C.R. (1987) 'Systems Aspects of Applying Active Control Technology to a Civil Transport Aircraft', *Royal Aeronautical Society Spring Convention*, May.
[12] B.G.S. Tucker (1993) 'Boeing 777 Primary Flight Control Computer System – Philosophy and Implementation', *RAeS Conference – Advanced Avionics on the A330/A340 and the Boeing 777*, November.
[13] James McWha (1995) '777 – Ready for Service', *RAeS Conference – The Design & Maintenance of Complex Systems on Modern Aircraft*, April.

2

Engine Control Systems

2.1 Introduction

The early jet engines based on a centrifugal compressor used a method of controlling fuel to the engine combustion chamber that used a fuel pump, a relief valve and a throttle valve. In series with these was a mechanical centrifugal governor. Barometric compensation of the relief valve was provided by a suitable bellows mechanism to maintain the full range of throttle movement at altitude. The design of such engines based upon Sir Frank Whittle's design was basically simple, using sound engineering practices and employing technology representing 'state of the art' of the day.

As gas turbine engine technology developed, demands for improved performance required substantial increases in pressure ratios and turbine inlet temperatures placing much more stress on the internal components. New developments such as the axial compressor and reheat (afterburning) created a demand for more complex methods of controlling airflow, fuel flow and exhaust gas flow.

Early gas turbine control systems were initially entirely hydro-mechanical. As engine and materials development continued a need arose to exercise greater control of turbine speeds and temperatures to suit prevailing atmospheric conditions and to achieve surge-free operation. The latter was particularly important in military engines where handling during rapid acceleration tended to place the engine under severe conditions of operation.

In support of the needed improvements, limited authority electronic trimmers sometimes referred to as 'supervisory controls' were developed to provide added functions such as temperature limiting and thrust management thus relieving the flight crew of this workload. This became important as new aircraft entering service eliminated the flight engineer position on the flight deck. Further developments in engine design led to the need to control more

Aircraft Systems: Mechanical, electrical, and avionics subsystems integration. 3rd Edition   Ian Moir and Allan Seabridge
© 2008 John Wiley & Sons, Ltd

parameters and eventually led to the use of full authority analogue control systems with electrical signalling from the throttle levers.

2.1.1 Engine/Airframe Interfaces

The engine is a major, high value item in any aircraft procurement programme. Often an engine is especially designed for a new aircraft – this is particularly true of military projects where a demanding set of requirements forces technology forward in propulsion and airframe areas. There is, however, a trend to make use of existing powerplant types or variant of types in an effort to reduce the development costs of a new project.

Whatever the case, control of the interfaces between the engine and the airframe is essential to allow the airframe contractor and the engine contractor to develop their products independently. The interface may be between the engine and a nacelle in the case of a podded, under-wing engine, as is common in commercial aircraft; or between the engine and the fuselage as is common in fast jet military aircraft types.

When full authority control systems were introduced in analogue form, semiconductor technology demanded that the electronic control units were mounted on the airframe. This led to a large number of wire harnesses and connectors at the engine–airframe interface. Together with the mechanical, fluid and power offtake interfaces, this was a measure of complexity that had the potential for interface errors that could compromise an aircraft development programme.

Although the emergence of rugged electronics, data buses and bleedless engines has simplified this interface, nevertheless it needs to be controlled. What often happens is that an Interface Control Document or ICD is generated that enables the major project contractors to declare and agree their interfaces.The nature of the interfaces and the potential for rework usually means that the ICD becomes an important contractual document.

Typical of the interfaces declared are the following.

Installation

- Engine mass, centre of mass and volume
- Engine space envelope
- Engine clearances under static and dynamic conditions
- Attachments
- Thrust bearings and fuselage loads
- Interface compatibility
- Turbine/disc containment measures
- Maintenance access points
- Drains and vents

- Engine change/winching points
- Ground crew intake and exhaust safety clearances
- Noise

System Connections

- Fuel connections
- Control system connections (throttles, reverse thrust command)
- Cockpit indications, alerts and warnings
- Air start interconnections
- Air data requirements
- Fire detection and protection
- Engine start/relight commands
- Engine health monitoring
- Ground equipment connections
- Inspection access

Power Offtakes

- Hydraulic power generation
- Electrical power generation
- Air bleeds

2.2 Engine Technology and Principles of Operation

The emergence of digital technology and serial data transmission systems, as well as higher performance electronic devices led to the introduction of the FADEC (Full Authority Digital Electronic Control). This, in turn led to the opportunity to integrate the control systems with the aircraft avionics and flight control systems, and to consider the mounting of complex electronic control units on the engine itself. When mounting these electronic controls on the engine, great care must be taken to isolate the units from the hostile environment by providing anti-vibration mounts and forced-air (or sometimes fuel) cooling.

Engine technology has advanced considerably with new materials and new manufacturing techniques leading to smaller, lighter and more efficient engines capable of delivering more thrust with considerable improvements in reliability and availability. The core of the gas turbine engine is the gas generator. Figure 2.1(a) shows single and two shaft versions of the typical gas generator. The single shaft version has limited capability because both the low pressure and high pressure stages of the compressor rotate at the same speed. In the two shaft design, the low pressure and high pressure spools can rotate at different speeds for improved performance and efficiency.

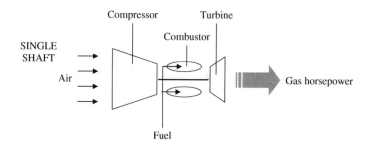

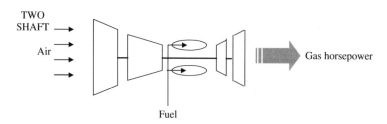

Figure 2.1a   Illustrations of single and multiple spool engines

In each case the output is 'gas horsepower', a stream of high energy gas that can be used to develop pure thrust as with a turbojet or to drive an additional turbine to develop torque that can be applied to a fan as in a turbofan or a propeller or rotor for turboprop and turbo-shaft applications respectively.

In turbofan applications the fan may be driven by the low pressure shaft of the gas generator with an additional low pressure turbine stage added to convert the exhaust gas energy into torque. This is typical of Pratt & Whitney and General Electric designs. Rolls Royce turbofan designs incorporate a third shaft allowing the fan and its power turbine to rotate at speed independent of either of the two gas generator shafts (see Figure 2.1(b)).

As implied by the figure, most of the thrust is generated by the fan since most of the high energy gas from the gas generator section is dissipated in the turbine connected to the fan. In the turbofan engine, thrust is generated by imparting a relatively small increase in velocity to a very large air mass flow through the fan while, in the older turbojet engines, the total air mass flow through the engine is much smaller and therefore, to achieve the same thrust, the velocity of the exhaust gasses must be much greater, i.e.

- Turbofan: Thrust $= \dot{M} \times \Delta v$ Large mass flow, small velocity change
- Turbojet: Thrust $= \dot{M} \times \Delta V$ Small mass flow, large velocity change

It is for this reason that today's large fan engines are much quieter than their turbojet or low by-pass ratio predecessors.

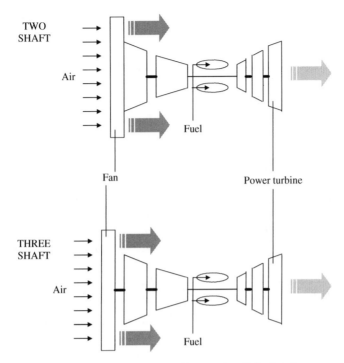

Figure 2.1b Two and three shaft turbofans

2.3 The Control Problem

The basic control action is to control a flow of fuel and air to the engine to allow it to operate at its optimum efficiency over a wide range of forward speeds, altitudes and temperatures while allowing the pilot to handle the engine without fear of malfunction. The degree of control required depends to a large extent upon the type of engine and the type of aircraft in which it is installed.

The military aircraft is usually specified to operate in worldwide conditions, and is expected to experience a wide range of operating temperatures. To be successful in combat the aircraft must be manoeuvrable. The pilot, therefore, expects to be able to demand minimum or maximum power with optimum acceleration rates, as well as to make small adjustments with equal ease, without fear of surge, stall, flame-out, over-speed or over-temperature. The pilot also needs a fairly linear relationship between throttle lever position and thrust.

The civil operator requires reliable, economical and long-term operation under clearly defined predictable conditions with minimum risk to passengers and schedules. For military engines the key to satisfactory performance is the ability to perform over large speed and altitude ranges as well as significant temperature variations.

To obtain these objectives, control can be exercised over the following aspects of engine control:

- Fuel flow – to allow varying engine speeds to be demanded and to allow the engine to be handled without damage by limiting rotating assembly speeds, rates of acceleration and temperatures
- Air flow – to allow the engine to be operated efficiently throughout the aircraft flight envelope and with adequate safety margins
- Exhaust gas flow – by burning the exhaust gases and varying the nozzle area to provide additional thrust

Electronic control has been applied in all these cases with varying degrees of complexity and control authority. Such control can take the form of simple limiter functions through to sophisticated multi-variable, full authority control systems closely integrated with other aircraft systems.

2.3.1 Fuel Flow Control

Control of power or thrust is achieved by regulating the fuel flow into the combustor. On turbo jet or turbo fan engines thrust can be controlled by setting an engine pressure ratio or, in the case of the larger commercial fan engines, by controlling fan speed, while on shaft power engines the speed of the gas generator is a measure of the power delivered to the propeller or to the rotor.

When changing the thrust or power setting the fuel control system must limit the rate of acceleration and deceleration of the engine rotating assemblies in order to prevent compressor surge or flame out. This control process is further complicated by the change in engine inlet conditions, i.e. inlet temperature, inlet pressure and Mach number that can occur as the aircraft moves around the flight envelope.

Airflow modulation through the compressor may also be necessary by the use of variable vanes and/or bleed valves to provide adequate surge margin under all operating conditions.

The control of power or thrust of the gas turbine engine is obtained by regulating the quantity of fuel injected into the combustion system.

When a higher thrust is required the throttle is opened and the fuel pressure to the burners increases due to the higher fuel flow. This has the effect of increasing the gas temperature which, in turn, increases the acceleration of the gases through the turbine to give a higher engine speed and correspondingly greater air flow, resulting in an increase in thrust.

The relationship between the air flow induced through the engine and the fuel supplied is, however, complicated by changes in altitude, air temperature and aircraft speed. These variables change the density of the air at the engine intake and consequently the mass of air flowing through the engine.

To meet this change in air flow a similar change in fuel flow must occur, otherwise the ratio of air to fuel will alter and the engine speed will increase or decrease from that originally selected by the pilot in setting the throttle lever position. Fuel flow must, therefore, be monitored to maintain the conditions demanded by the pilot whatever the changes in the outside world. Failure to do so would mean that the pilot would constantly need to make minor adjustments to throttle lever position, increasing his work load and distracting his attention from other aspects of aircraft operation.

The usual method of providing such control is by means of a fuel control unit (FCU) or fuel management units (FMU). The FCU/FMU is a hydro-mechanical device mounted on the engine. It is a complex engineering mech-anism containing valves to direct fuel and to restrict fuel flow, pneumatic capsules to modify flows according to prevailing atmospheric conditions, and dashpot/spring/damper combinations to control acceleration and deceleration rates. A complete description of an FCU is beyond the scope of this book. An excellent description can be found in *The Rolls-Royce Book of the Jet Engine* [1].

The engine speed must be controlled from idle to maximum rating. Over-speed must be avoided to reduce stresses in the rotating assemblies, and over-temperature must be avoided to prevent blade damage and to reduce thermal creep. The engine must be allowed to accelerate and decelerate smoothly with no risk of surge.

Such control influences are difficult to achieve manually. Therefore the FCU has, over the generations of jet engines, been designed to accommodate control inputs from external electronic devices. Electrical valves in the FCU can be connected to electronic control units to allow more precise and continuous automatic control of fuel flows in response to throttle demands, using measure-ments derived from the engine, to achieve steady state and transient control of the engine without fear of malfunction.

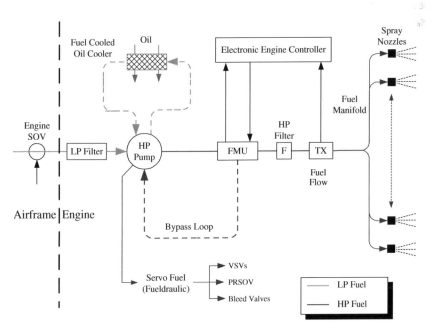

Figure 2.2 Fuel control system

A typical fuel control circuit is shown in Figure 2.2. This circuit also shows the way in which fuel is used as a cooling medium for oil by passing it through a Fuel Cooled Oil Cooler (FCOC).

On some military aircraft the fuel system receives a demand from the weapon release switch or gun trigger to preempt weapon release. This allows fuel flow to the engines to be modified to prevent an engine surge resulting from disturbance of the intake conditions from missile exhaust, shock from the gun muzzle or smoke from the gun breech. This facility is known as 'fuel dip'.

2.3.2 Air Flow Control

It is sometimes necessary to control the flow of air through to the engine to ensure efficient operation over a wide range of environmental and usage conditions to maintain a safe margin from the engine surge line. Most modern commercial engines have variable compressor vanes and/or bleed valves to provide optimum acceleration without surge though it is not a feature usually associated with military applications.

Figure 2.3 illustrates some aspects of air management at various stages in the compressor, showing movements of inlet guide vanes or stator vanes to achieve inlet air stability. In some high Mach number aircraft it was necessary to provide intake ramps and variable intake area control to maintain suitable air flow under all conditions of speed, altitude and manoeuvre. Concorde and Tornado are examples of aircraft with air intake control systems.

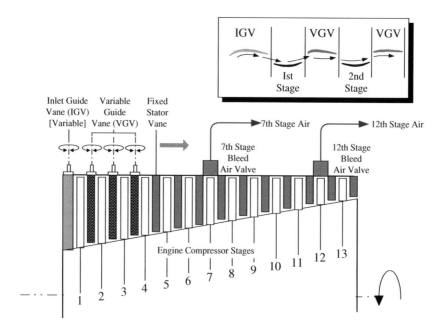

Figure 2.3   Engine air management

Figure 2.3 also shows where air is bled from the engine for various purposes, including engine stability reasons and also to provide a source of air for conditioning systems and bleed air systems such as wing leading edge anti-icing. Control of bleed air on the Trent 800 engine is shown in Figure 2.4.

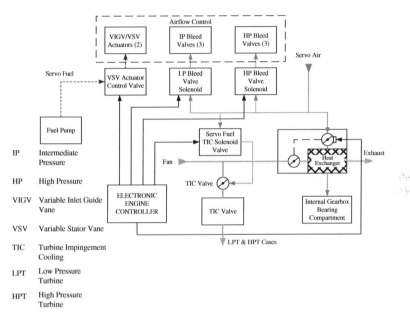

Figure 2.4 Bleed air control – RR Trent 800 example

2.3.3 Control Systems

The number of variables that affect engine performance is high and the nature of the variables is dynamic, so that the pilot cannot be expected constantly to adjust the throttle lever to compensate for changes, particularly in multi-engined aircraft. In the first gas turbine engined aircraft, however, the pilot was expected to do just that.

A throttle movement causes a change in the fuel flow to the combustion chamber spray nozzles. This, in turn, causes a change in engine speed and in exhaust gas temperature. Both of these parameters are measured; engine speed by means of a gearbox mounted speed probe and Exhaust Gas Temperature (EGT), or Turbine Gas Temperature (TGT), by means of thermocouples, and presented to the pilot as analogue readings on cockpit-mounted indicators. The pilot can monitor the readings and move the throttle to adjust the conditions to suit his own requirements or to meet the maximum settings recommended by the engine manufacturer. The FCU, with its internal capsules, looks after variations due to atmospheric changes.

In the dynamic conditions of an aircraft in flight at different altitudes, temperatures and speeds, continual adjustment by the pilot soon becomes impractical. He cannot be expected continuously to monitor the engine conditions safely for a flight of any significant duration. For this reason some form of automatic control is essential.

2.3.4 Control System Parameters

To perform any of the control functions electrically requires devices to sense engine operating conditions and to perform a controlling function. These can usually be conveniently subdivided into input and output devices producing input and output signals to the control system.

To put the control problem into perspective the control system can be regarded as a box on a block diagram receiving input signals from the aircraft and the engine and providing outputs to the engine and the aircraft systems. This system is shown diagramatically in Figure 2.5.

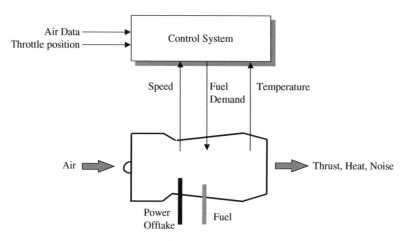

Figure 2.5 Engine control systems – basic inputs and outputs

The input signals provide information from the aircraft and the engine to be used in control algorithms, while the output signals provide the ability to perform a control function. Further signals derived from output devices provide feedback to allow loop closure and stable control. Typical inputs and outputs are described below.

2.3.5 Input Signals

- *Throttle position* – A transducer connected to the pilot's throttle lever allows thrust demand to be determined. The transducer may be connected directly to the throttle lever with electrical signalling to the control unit, or connected to the end of control rods to maintain mechanical operation as far as possible. The transducer may be a potentiometer providing a DC signal or a variable transformer to provide an AC signal. To provide suitable integrity of the signal a number of transducers will be used to ensure that a single failure does not lead to an uncommanded change in engine demand

- *Air data* – Airspeed and altitude can be obtained as electrical signals representing the pressure signals derived from airframe mounted capsule units. These can be obtained from the aircraft systems such as an air data computer (ADC) or from the flight control system air data sensors. The latter have the advantage that they are likely to be multiple redundant and safety monitored. In most applications, for reasons of autonomy, inlet pressure and temperature are measured using dedicated sensors located on the engine
- *Total temperature* – A total temperature probe mounted at the engine face provides the ideal signal. Temperature probes mounted on the airframe are usually provided, either in the intakes or on the aircraft structure
- *Engine speed* – The speed of rotation of the shafts of the engine is usually sensed by pulse probes located in such a way as to have their magnetic field interrupted by moving metallic parts of the engine or gearbox. The blades of the turbine or compressor, or gear box teeth, passing in front of a magnetic pole piece induce pulses into a coil or a number of coils wound around a magnet. The resulting pulses are detected and used in the control system as a measure of engine speed
- *Engine temperature* – The operating temperature of the engine cannot be measured directly since the conditions are too severe for any measuring device. The temperature can, however, be inferred from measurements taken elsewhere in the engine. The traditional method is to measure the temperature of the engine exhaust gas using thermocouples protruding into the gas stream. The thermocouples are usually arranged as a ring of parallel connected thermocouples to obtain a measurement of mean gas temperature and are usually of chromel-alumel junctions. A cold junction is provided to obtain a reference voltage. An alternative method is to measure the temperature of the turbine blades with an optical pyrometer. This takes the form of a fibre optic with a lens mounted on the engine casing and a semiconductor sensor mounted in a remote and cooler environment. Both of these temperatures can be used to determine an approximation of turbine entry temperature, which is the parameter on which the temperature control loop should ideally be closed
- *Nozzle position* – For those aircraft fitted with reheat (or afterburning) the position of the reheat nozzle may be measured using position sensors connected to the nozzle actuation mechanism or to the nozzle itself. An inductive pick-off is usually used since such types are relatively insensitive to temperature variations, an important point because of the harsh environment of the reheat exhaust
- *Fuel flow* – Fuel flow is measured by means of a turbine type flow meter installed in the fuel pipework to obtain a measure of fuel inlet flow as close to the engine as possible. Fuel flow measured by the turbine flow meter is for instrumentation and monitoring purposes and is not used as an input

to the engine control system. The dynamic response of this device is much too slow for this function. Instead the position of the fuel metering valve within the FCU is used as a measure of fuel flow

- *Pressure ratio* – The ratio of selected pressures between different stages of the engine can be measured by feeding pressure to both sides of a diaphragm operated device. The latest technology pressure ratio devices use two high accuracy pressure sensors and electronics to generate pressure ratio

2.3.6 Output Signals

- *Fuel flow control* – The fuel supply to the engine can be varied in a number of ways depending on the type of fuel control unit used. Solenoid operated devices, torque motor or stepper motor devices have all been employed on different engine types. Each device has its own particular failure modes and its own adherents
- *Air flow control* – The control of air flow at different stages of the engine can be applied by the use of guide vanes at the engine inlet, or by the use of bleed valves between engine stages. These are controlled automatically to preserve a controlled flow of air through the engine for varying flight conditions

2.4 Example Systems

Using various combinations of input and output devices to obtain information from the engine and the airframe environment, a control system can be designed to maintain the engine conditions stable throughout a range of operating conditions. The input signals and output servo demands an be combined in varying degrees of complexity to suit the type of engine, the type of aircraft, and the manner in which the aircraft is to be operated. Thus the systems of civil airliners, military trainers and high speed combat aircraft will differ significantly.

In a simple control system, such as may be used in a single engine trainer aircraft the primary pilot demand for thrust is made by movements of a throttle lever. Rods and levers connect the throttle lever to a fuel control unit (FCU) so that its position corresponds to a particular engine condition, say rpm or thrust. Under varying conditions of temperature and altitude this condition will not normally stay constant, but will increase or decrease according to air density, fuel temperature or demands for take-off power. To obtain a constant engine condition, the pilot would have continually to adjust the throttle lever, as was the case in the early days of jet engines. Such a system with the pilot in the loop is shown in Figure 2.6.

The flow of fuel to the combustion chambers can be modified by an electrical valve in the FCU that has either an infinitely variable characteristic, or moves in a large number of discrete steps to adjust fuel flow. This valve is situated in the engine fuel feed line so that flow is constricted, or is by-passed and returned to the fuel tanks, so that the amount of fuel entering the engine is different from that selected.

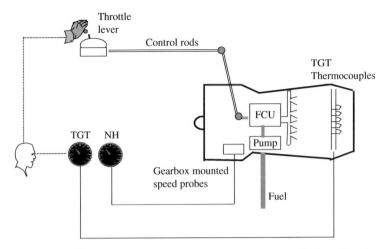

Figure 2.6 A simple engine control system – pilot in the loop

This valve forms part of a servo loop in the control system so that continuous small variations of fuel flow stabilise the engine condition around that demanded by the pilot. This will allow the system to compensate for varying atmospheric and barometric conditions, to ensure predictable acceleration and deceleration rates and to prevent over-temperature or over-speed conditions occurring over the available range – acting as a range speed governor; Figure 2.7 illustrates such a control system. It can be seen that the pilot shown in Figure 2.6 now acts in a supervisory role, relying on the control system to maintain basic control conditions while he monitors the indicators for signs of over-speed or over-temperature.

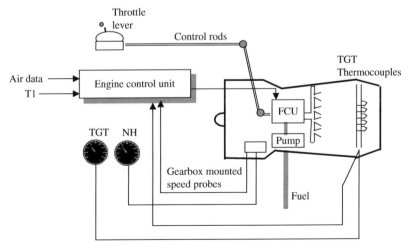

Figure 2.7 A simple limited authority engine control system

Even this task can be reduced considerably by incorporating an automatic means of signalling an over-speed or over-temperature. This can be performed in the control unit by setting a datum related to a particular engine type, or by setting a variable 'bug' on the cockpit indicator. If either preset datum is exceeded a signal is sent to the aircraft warning system to warn the pilot by means of a red light and signal tone (see Chapter 9). This principle is illustrated in Figure 2.8 which shows warning systems for both over-temperature and over-speed conditions.

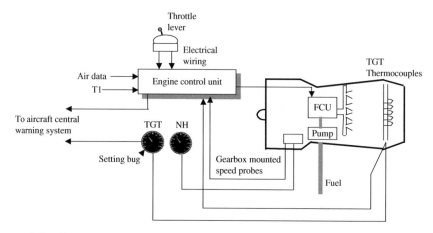

Figure 2.8 Engine control system with NH and TGT exceedence warnings

In this diagram the over-speed warning is provided by a mechanism in the turbine gas temperature (TGT) indicator. A knob on the indicator allows the pilot to set a 'bug' to a particular temperature. When the indicator pointer exceeds that setting, a pair of contacts in the indicator close and provide a signal to the aircraft central warning system. The over-speed warning is provided by a pair of contacts in the engine control unit. In practice either one method or the other is used in one aircraft type, rather than a mix of methods.

In many modern aircraft the simple throttle signalling system is retained, but with the replacement of rods and levers by electrical signalling from the throttle levers. This reduces friction and eliminates the possibility of jamming in the control rod circuit. An example of a system with electrical throttle signalling is illustrated in Figure 2.9. The removal of any mechanical links between the pilot and the engine means that the control unit has full authority control. There is nothing the pilot can do to correct an engine malfunction other than to shut down the engine. Because of this the throttle signalling circuit (like the rest of the control system) is designed with great care to ensure that all failures are detected and taken care of by the control system. For example, additional windings on the Tornado throttle position transducer enable the control system to detect open circuits and short circuits and to take corrective action.

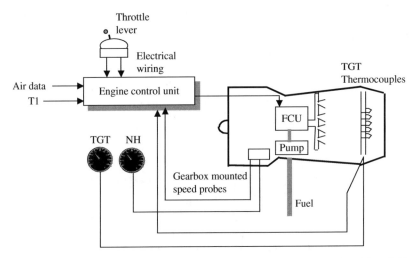

Figure 2.9 Full authority engine control system with electrical throttle signalling

For multiple engine types of similar complexity, the system is duplicated with no cross connection between the systems to reduce the risk of common mode failures. More functions can be added to the system to enable the engine to operate in more demanding situations. For example, air bleed valves between engine stages can be opened or closed to stabilise the engine as a function of speed or acceleration. The ignition system can be switched on during periods of heavy rain or icing; and all conditions can be signalled to the crew by cockpit instruments or warning lights.

The system illustrated in Figure 2.7 is typical of many systems engineered in the 1950s and 1960s. The BAE Canberra and Lightning aircraft contained engine control systems based on magnetic amplifiers used as an analogue control system. Developments in semiconductor technology led to the introduction of transistorised analogue amplifiers such as that used in the control unit for the Adour engine installed in the Sepecat Jaguar.

Jaguar was an early venture into European collaboration between British Aerospace (then British Aircraft Corporation) and Dassault (then Avions Louis Breguet). The engine control unit was manufactured by Elecma in France to control the Rolls-Royce/Turbomeca Adour twin engine combination. Each engine had its own control unit mounted on the airframe in a ventral bay between the two engines. Provision was made for the connection of test equipment and for adjustments to the unit to allow the engine to be set up correctly on engine ground runs.

Concorde made full use of electronic technology for the control of its four reheated Olympus 593 engines. The control system for each engine was designed as a full-authority self-monitoring system, completely independent of the others. The control units were mounted on the airframe and provided control for the main engine and reheat functions. This analogue system went into each of the production Concorde aircraft. A separate system provided control of the intake ramps to provide a suitable airflow to the engines under all flight conditions.

The Turbo Union RB 199 engines in the Panavia Tornado made full use of the experience gained on Concorde. Each engine was controlled by a single Main Engine Control Unit (MECU). Each MECU contained two independent lanes of dry engine control and a single reheat control lane. A single engine system is shown in Figure 2.10.

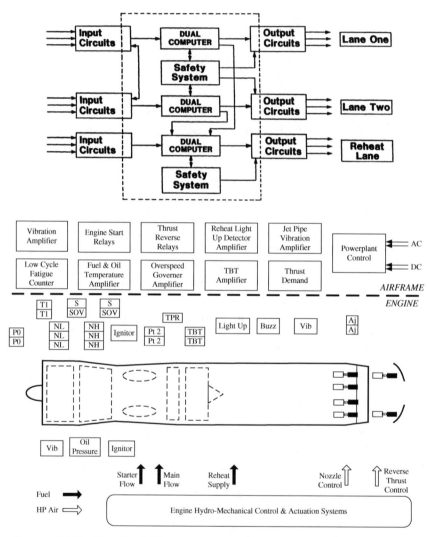

Figure 2.10 The RB 199 control system in the Panavia Tornado

The RB 199 is a complex engine, and a number of separate input conditioning units were required to provide the completed control and indication package. Instead of TGT, engine temperature was measured using an optical pyrometer monitoring the infra-red radiation of the turbine blades. This required a turbine blade temperature (TBT) amplifier which not only converted the pyrometer signal into a form suitable for connection to the MECU, but also provided a signal to the TBT indicator in the cockpit. The TBT indicator provided a signal to the aircraft central warning panel in the event of an over-temperature. This system is shown in Figure 2.11.

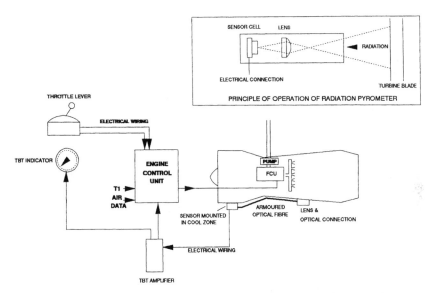

Figure 2.11 The RB199 turbine blade temperature system

Other individual electronic units were provided for monitoring vibration using piezo-electric transducers, for detecting the light-up of the reheat system using ultra-violet detectors, for providing an independent over-speed governor circuit for both HP and LP turbines, and for controlling reverse thrust. Throttle position was signalled using dual winding AC pick-offs.

All electronic units were airframe mounted in the aircraft front fuselage avionics bays. This required long lengths of multiple cable harnesses to run almost the full length of the aircraft. The harnesses had to be designed to allow physical separation, not only of each engine harness, but also each control lane, and for electro-magnetic health reasons. This resulted in a large weight of wiring in the aircraft and required a large number of connectors to allow the wiring to cross between the engine and the airframe. This was a heavy and costly arrangement, but one which was necessary because semiconductor technology was insufficiently advanced at that time to allow electronic control units to be mounted in the high temperature and vibration environment of the engine bay. There was an absolute limit on some devices that would be destroyed by high internal temperatures; the environment would lead to unacceptable low reliability for complex units.

Lucas Aerospace made considerable advances in technology in the development of integrated circuits mounted on ceramic multi-layer boards to provide a highly reliable engine control system. Roll-Royce, MTU and FIAT formed a joint engine company – Turbo Union, which designed and manufactured the engine, and acted as prime contractor for the engine control system.

In the early 1960s Rolls-Royce began to experiment with the use of digital control systems which led to a demonstration of such a system on a test rig. However, by the 1970s sufficient work had been done to enable them and Lucas Aerospace to design and build an experimental full-authority control system for use with multiple spool engines. Such a system was flown connected to a single engine of Concorde 002 in July 1976 [2]. This advance in technology went through several stages of design and approval before it became accepted as a suitable system for use in Tornado, and the MECU was replaced by the Digital Electronic Control Unit (DECU).

The concept of full authority digital control went a stage further in the BAE Experimental Aircraft Programme (EAP) in which the DECU became integrated with the aircraft avionics. A system had been installed in EAP to provide digital control and monitoring of all the aircraft utility systems. This system was known as utility systems management (USM) and was essentially a multi-computer system interconnected with a MIL-STD-1553B (Def Stan 00-18 Part 2) serial data transmission system. A simple and economic method of incorporating the RB 199 and its control system into the aircraft structure of data buses and multi-function cockpit displays was to provide a means of interconnection through USM [3]. This system is illustrated in Figure 2.12.

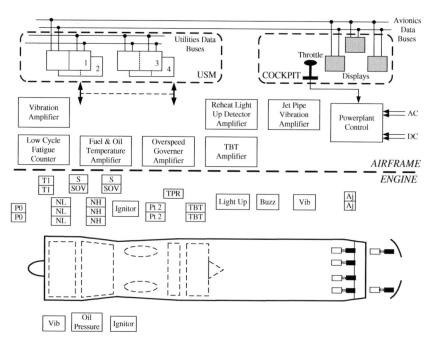

Figure 2.12   The RB199 control system in the BAE Systems EAP

The result of this gradual evolution of control systems has resulted in a control system in which the electronic control unit is now mounted on the engine with relatively few connections to airframe mounted signal sources. A typical modern engine control systems is illustrated in Figure 2.13.

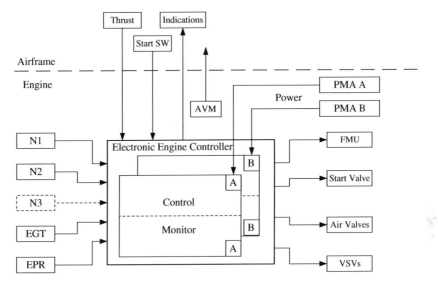

Figure 2.13 A modern simplified engine control system

Some modern examples of engines in various types of aircraft are shown below in Figures 2.14 to 2.16.

The EJ 200 engine manufactured by EUROJET is a high-technology engine that is smaller and lighter than other engines in its thrust class with low fuel consumption and high power to weight ratio. This makes it ideal for the

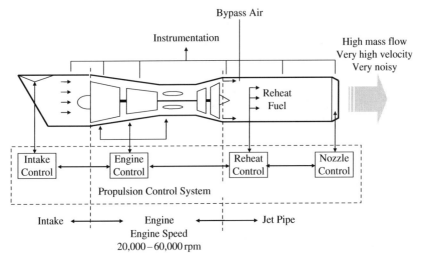

Figure 2.14 Turbojet Engine (EJ200 in Eurofighter Typhoon)

Typhoon in which two reheated engines are installed. The engine operates in a demanding flight envelope in Typhoon with its high manoeuvrability and high g capability. The control system maintains the engine operating with rapid throttle movements, or handling, high incidence and high turn rates without any malfunctions.

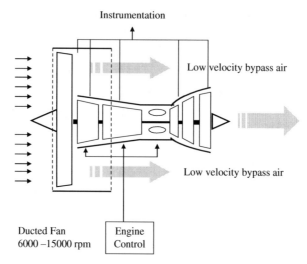

Figure 2.15   Turbofan (Trent 1000 in Boeing B787)

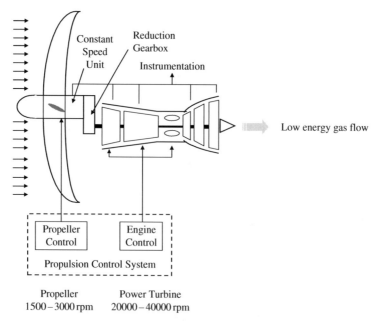

Figure 2.16   Turboprop (EPI TP400 in A400M)

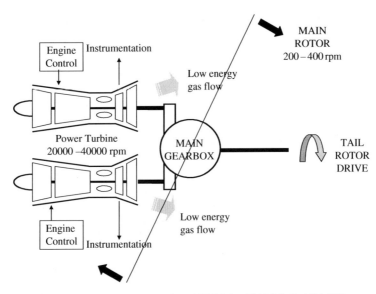

Figure 2.17   Helicopter (T800 in EH101 & NH-90)

2.5 Design Criteria

The engine and its control system are considered to be safety critical. That is to say that a failure may hazard the aircraft and the lives of the crew, passengers and people on the ground. For this reason the system is generally designed to eliminate common mode failures, to reduce the risk of single failures leading to engine failure and to contain the risk of failure within levels considered to be acceptable by engineering and certification authorities. As an example the Civil Aviation Authority set the integrity requirements for the Concorde engine control system [2]. These were:

- The in-flight shut down rate due to electronics failure must not exceed 2.3 x 10^{-6} per engine hour
- The upward runaway rate due to electronics failure must not exceed 1 x 10^{-6} per engine hour. The system also includes an independent over-speed trip to prevent catastrophic engine failure
- The downwards runaway rate due to electronics failure must not exceed 2.7 x 10^{-6} per engine hour

Similar design targets are set for every project and they are based upon what the certification authorities consider to be an acceptable failure rate. They are used by the engineer as targets that should never be exceeded, and are used as a budget from which individual control system components and modules can be allocated individual targets. The sum of all individual modules must never exceed the budget. A wise engineer will ensure that an adequate safety margin exists at the beginning of the design.

The design failure rate targets are based upon the well-known random failure properties of hardware. Every item of electronic hardware has a failure rate that can be obtained from a design handbook or from the component manufacturer's literature. This rate is based upon statistical evidence gathered from long-term tests under varying conditions, and may be factored by practical results from the use of components in service. The designer selects the correct components, ensures that they are not over-stressed in use and observes scrupulous quality control in design and manufacture. On the airframe side similar care is taken in the provision of cooling, freedom from vibration and by providing high-quality power supplies.

Nevertheless, failures will occur, albeit rarely. Techniques have been established to ensure that the effects of failure on system operation are minimised. A common method of reducing the effect of failures is to introduce redundancy into a system. Concorde, for example, had four engines, therefore a failure of at least one could be tolerated, even at take-off. Each engine had a separate control unit with no physical interconnections, each control unit has two independent lanes of control, and duplicated input signals were obtained from separate sources. The wiring harnesses were widely separated in the airframe to reduce the risk of mechanical damage or electromagnetic interference affecting more than one system.

In addition a separate over-speed governor was provided to ensure that the HP turbine was never allowed to over-speed and suffer catastrophic failure.

It is important that the entire system is designed to be fail-safe. For an aircraft, failsafe means that the system must be able to detect failures and to react to them by either failing to a condition of existing demand (fail frozen) or to a condition of maximum demand. This is to ensure that a failure in a critical regime of flight, such as take-off, will enable the pilot to continue with the take-off safely. For this reason fuel valves generally fail to the open position.

These techniques are used on many multiple engine combinations with electronic control systems. The technique are well established, well understood and can be analysed numerically to provide evidence of sound design.

Difficulties began to occur when digital control was introduced. Software does not have a numerical failure rate. If failures are present they will be caused by inadequate design, and not discovered during testing. The design process for software used in safety critical applications, such as engine control, should ensure that there are no incipient design faults.

In a multiple engine aircraft, as explained above, each engine will have its own independent control unit. For an analogue system the random failure characteristics of electronic components means that failures will generally be detectable and will be contained within one engine control system The possibility of two identical failures occurring in the same flight is extremely unlikely to occur on a second engine.

It is argued that, since software in independent digital control units is identical (hence a common mode failure potential), then it is possible for undetected design faults to manifest themselves with particular combinations of data and instruction. As the software and control systems are identical, then in theory the same set of conditions could occur on the same flight and may result in multiple engine shutdown.

To counteract this effect a number a number of techniques were used in some systems. Dissimilar redundancy was one such technique in which different teams of engineers designed and coded the software in different control lanes or control units. This was an extremely costly method, requiring two design teams, two test programmes, and two certification programmes. An alternative was to provide a mechanical reversionary mode that allowed the pilot to effect rudimentary control over the flow of fuel to the engine by means of a switch and solenoid valve.

However, the best method of producing sound software is to establish sound design principles. For this reason modern techniques of software design include structured methods of requirements analysis, software design, modular coding and thorough testing, as well as such techniques as static code analysis. Modern engine control systems are now well established and trusted and have achieved many trouble-free flying hours.

2.6 Engine Starting

To start the engines a sequence of events is required to allow fuel flow, to rotate the engine and to provide ignition energy. For a particular type of aircraft this sequence is unvarying, and can be performed manually with the pilot referring to a manual to ensure correct operation, or automatically by the engine control unit. Before describing a typical sequence of events, an explanation of some of the controls will be given.

2.6.1 Fuel Control

Fuel from the tanks to the engine feed line is interrupted by two shut-off cocks. The first is in the low pressure feed lines, at which fuel pressure is determined by the fuel boost pumps (see Chapter 3 – Fuel Systems). The valve, known as the LP cock or firewall shut-off cock, is situated close to the engine firewall. Its primary purpose is to isolate the engine in the event of a fire. It is usually a motor-driven valve controlled by a switch in the cockpit and, once opened, cannot be shut except by means of the switch. The switch is usually covered by a guard so that two actions are needed to select the switch to either open or close the cock. This helps to prevent inadvertent actions that may lead to accidental engine shutdown.

The second valve is in the high-pressure fuel line, in which the fuel pressure is determined by an engine-driven pump. The function of this valve is to open and close the fuel feed close to the engine inlet at the fuel control unit. It is opened manually by the pilot, or automatically by the engine control unit at an appropriate stage in the engine start cycle. The location of these valves is shown in Figure 2.18.

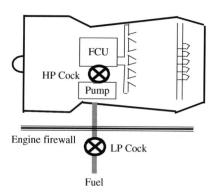

Figure 2.18 Typical location of LP and HP Cocks

2.6.2 Ignition Control

The ignition system consists of a high energy ignitor which is switched on for a period during the start cycle. The ignitors initiate combustion of the fuel vapour in the combustion chamber. An ignitor plug is supplied with electrical energy by an ignition exciter that produces stored energy from 1 to 6 joules depending on the type required. High energy systems are used for starting, and low energy systems can be provided to maintain engine ignition during aircraft operations in heavy rain, slushy runways or icing conditions. A typical ignition circuit is shown in Figure 2.19 and some examples of typical ignition equipment are shown in Figure 2.20.

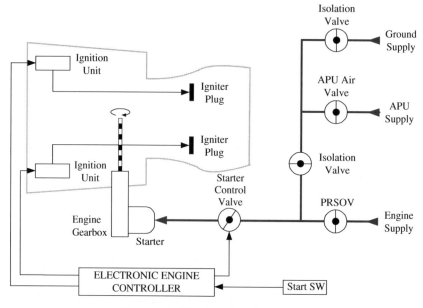

Figure 2.19 Engine ignition

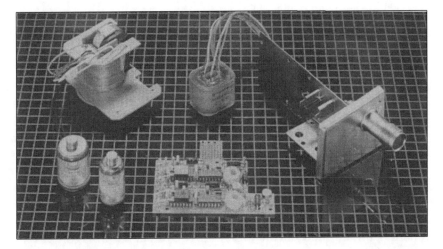

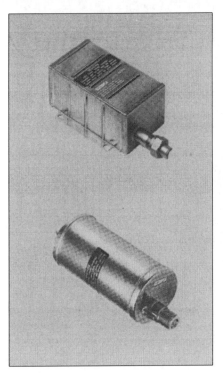
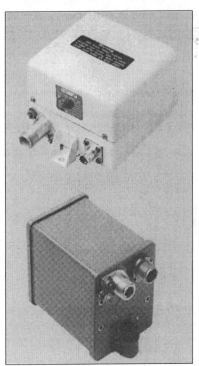

Figure 2.20 Some examples of high energy ignition equipment (Courtesy of BF Goodrich)

2.6.3 Engine Rotation

During the starting cycle the engine needs to be rotated until the fuel has ignited and the temperature of combustion is sufficient for the engine to rotate without assistance. At this point the engine is said to be self-sustaining. A number of methods are in current use for providing assistance by means of air, electrical energy or chemical energy. The most common method in modern use is to use an external air source or an internal auxiliary power unit to start the first engine, and to cross drive start the remaining engines. Some smaller engines and the More-Electric B787 use electrical engine start.

Air at high pressure can be provided by an external air compressor trolley connected to the engine by ground crew, or by air supplied by an onboard Auxiliary Power Unit (APU). This is a small gas turbine that is started prior to engine start. It has the advantage of making the aircraft independent of ground support and is useful at remote airfields. It is also used to provide electrical and hydraulic energy for other aircraft services. An example APU is shown in Figure 2.21, this has its own intake concealed beneath an opening hatch and its own exhaust positioned so as not to present a hot gas hazard to ground crew.

A functional diagram of a conventional APU is shown in Figure 2.22.

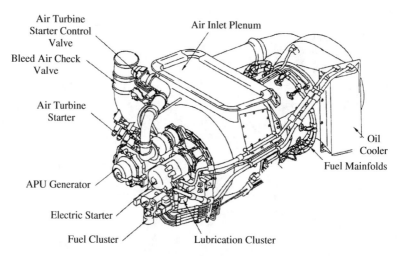

Figure 2.21   APU overview

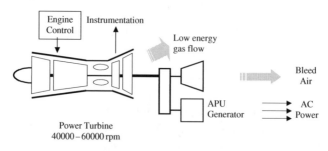

Figure 2.22   APU functional diagram

It is now common to see APUs used in flight to relight engines that may have failed or to provide energy for hydraulics or air during a relight procedure. To do this the APU may be left running throughput the flight, or it may be started prior to start of a failed engine. There are limitations in altitude and attitude that may limit the envelope in which either the APU or engine can be started. Figure 2.23 provides an example of the limitation that may apply.

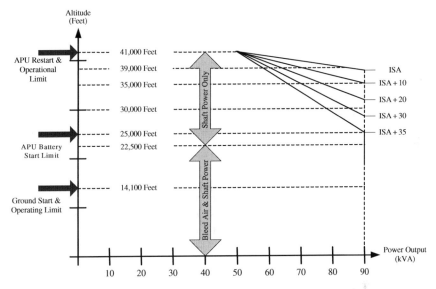

Figure 2.23   APU – airborne rating example

A DC motor mounted on the engine can be supplied with energy from an external battery truck or from the aircraft internal battery.

Chemical energy can be provided by the use of cartridges or a mono-fuel such as iso-propyl-nitrate (IPN) to rotate a small turbine connected to the engine.

2.6.4 Throttle Levers

The throttle lever assembly is often designed to incorporate HP cock switches so that the pilot has instinctive control of the fuel supply to the engine. Microswitches are located in the throttle box so that the throttle levers actuate the switches to shut the valves when the levers are at their aft end of travel. Pushing the levers forward automatically operates the switches to open the fuel cocks, which remain open during the normal operating range of the levers. Two distinct actions are required to actuate the switches again. The throttle lever must be pulled back to its aft position and a mechanical latch operated to allow the lever to travel further and shut off the fuel valve.

2.6.5 Starting Sequence

A typical start sequence is:

- Open LP cocks
- Rotate engine
- Supply ignition energy
- Set throttle levers to idle – open HP cocks
- When self-sustaining – switch off ignition
- Switch off or disconnect rotation power source

Together with status and warning lights to indicate 'start in progress', 'failed start' and 'engine fire' the pilot is provided with information on indicators of engine speeds, temperatures and pressures that he can use to monitor the engine start cycle.

In many modern aircraft the start cycle is automated so that the pilot has only to select START for the complete sequence to be conducted with no further intervention. This may be performed by an aircraft system such as Vehicle Management, or by the FADEC control unit. In future this sequence may be initiated by an automated pre-flight check list.

2.7 Engine Indications

Despite the fact that engine control systems have become very comprehensive in maintaining operating conditions at the most economic or highest performance, depending on the application, there is still a need to provide the pilot with an indication of certain engine parameters.

Under normal conditions the pilot is interested in engine condition only at the start and when something goes wrong. The engine control system, with its monitoring and warning capability, should inform the pilot when something untoward does happen. However, there may be circumstances when human intuition wins the day.

During engine start the pilot monitors (and checks with his co-pilot in a multi-crew aircraft) that start progresses satisfactorily with no observed sluggish accelerations, no low oil pressures or over-temperatures. Much of this monitoring involves pilot familiarity with the aircraft type and engine type, incurred over many starts. The crew may accept certain criteria that an automatic system would not.

During normal operation the control system should provide sufficient high integrity observation by self-monitoring and by checking certain parameters against preset values. In this way the system can monitor accelerations, rates of change, value exceedance and changes of state and issue the necessary warning.

Control systems are good at detecting sudden changes of level or state. However, slow, gradual but persistent drift and transient or intermittent changes of state are a designer's nightmare. The first may be due to degradation in performance of a component, e.g. a component becoming temperature sensitive, a gradually blocking filter or the partial occlusion of a pipe or duct. The second may be due to a loose connection some where in the system.

The pilot can observe the effects of these circumstances. In a four-engine aircraft, for example, one indicator reading differently to three others can be easily seen because the indicators are grouped with just such a purpose in mind.

Until recently all aircraft had at least one panel dedicated to engine instruments. These were in view at all times and took the form of circular pointer instruments, or occasionally vertical strip scales, reading such parameters as:

- Engine speed – NH and NL
- Engine temperature
- Pressure ratio
- Engine vibration
- Thrust (or torque)

In modern aircraft cockpits the individual indicator has largely given way to the Multi-Function Display Unit (MFDU). With a MFDU any information can be shown in any format, in full colour, at any time. This facility is often exploited to ensure that the pilot is only given the information that is essential for a particular phase of flight. This means that engine displays may occur on a single screen or page that is automatically presented to the pilot at certain times, say starting, take-off and landing, but is hidden at all other times. Provision is made for the pilot to select any page so that he can check from time to time, and an engine warning may automatically trigger the engine page to appear.

Engine indications are obtained from the same type of sensors and transducers that provide the inputs to the control system, as described earlier. However, for integrity reasons at least two sources of signal are required – one (or more) for control, another for the indicator. For example the engine rpm signal will be obtained from two separate coils of a speed sensor. This guards against a common mode failure that would otherwise affect both the control system and the indication system.

Such systems are the Engine Indication and Crew Alerting System (EICAS) used on Boeing and other aircraft and the Electronic Crew Alerting and Monitoring (ECAM) on Airbus aircraft.

Some examples of engine synoptic displays are shown in Figure 2.24 and the Trent 800 indication circuit is shown in Figure 2.25.

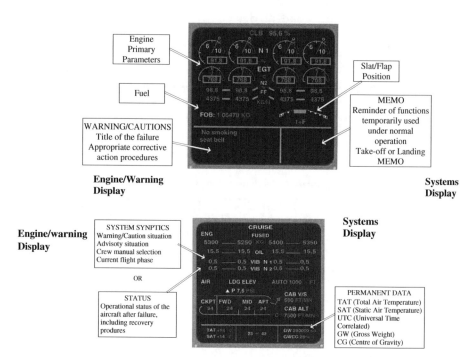

Figure 2.24 Some examples of engine synoptic displays

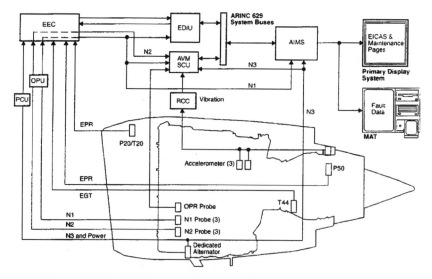

Figure 2.25 Rolls Royce Trent 800 – engine indications

2.8 Engine Oil Systems

Lubrication is vital for continued operation of the engine and all its high speed rotating mechanisms. Pressure, temperature and cleanliness can be monitored. Sensors are included in the oil system to provide warnings of low oil pressure and high oil temperature. Monitoring of particles in the oil can be performed by regular inspection of a magnetic plug in the oil reservoir or by counting particles in the fluid in a chip/particle detector. Any unusual particle density implies a failure somewhere in the rotating machinery and detection can be used to trigger a service or inspection. More recent technology advances have yielded methods of monitoring both ferrous and nonferrous debris. A typical oil circuit is shown in Figure 2.26.

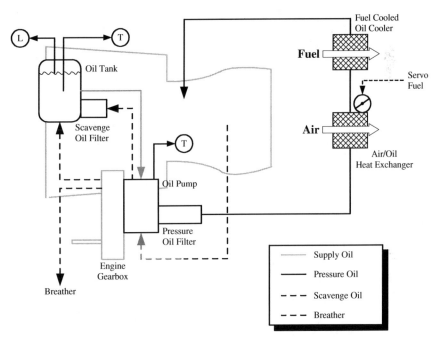

Figure 2.26 Engine oil system

2.9 Engine Offtakes

The engine is the prime mover for the majority of sources of power on the aircraft. An accessory gearbox enables accessories to be connected to the engine HP shaft and allows a starter connection so that the engine can be started from an external supply or from the Auxiliary Power Unit (APU). It is also a convenient place to obtain measurement of engine rotational speed by measuring the speed of rotation of the gearbox using a tachometer or pulse probe. An example accessory gearbox is shown in Figure 2.27.

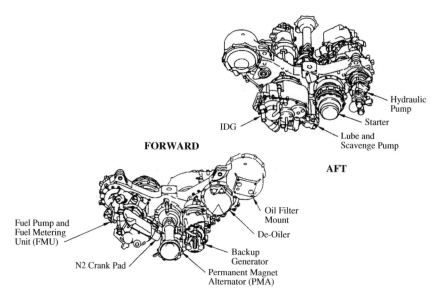

Figure 2.27 Typical accessory gearbox

Typical services, shown in Figure 2.28, include:

- Electrical power from generators
- Hydraulic power from hydraulic pumps
- Cabin and equipment conditioning system air from engine bleed
- Pneumatic power
- Anti and/or de-icing system air

It can be seen that many of the drives off the accessory gearbox are for the use of the engine:

- LP and HP fuel pumps
- Oil scavenge pumps; oil is used to cool the electrical generator as well as lubricate the engine
- PMAs to supply 28 VDC power for the dual channel FADEC
- Oil breather

Interfaces with the aircraft include:

- Supply of three-phase 115 VAC, 400 Hz electrical power – rated in the range from 40 to 90 kVA per channel on most civil transport aircraft; 120 kVA per channel on B777 and B767-400
- Supply of 3000 psi hydraulic power
- Engine tachometer and other engine indications

- Input of bleed air from a suitable air source to start the engine. This can be a ground power cart, the APU or air from the other engine if that has already been started

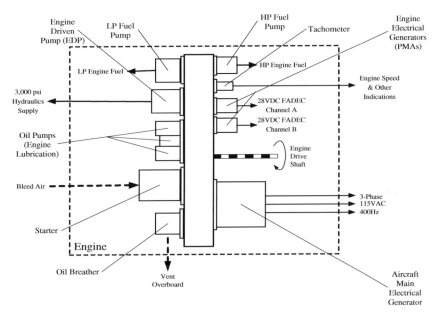

Figure 2.28   Engine power offtakes – typical

2.10 Reverse Thrust

A mechanism is provided on most engines to assist in decelerating the aircraft. On a turbo-prop engine this mechanism is to apply reverse pitch to the propeller blades. On a turbo-fan engine the usual mechanism is to deploy spoilers or buckets into the exhaust gas stream. Both of these methods have the effect of reversing the thrust provided by the engines to assist the brakes and shorten the landing distance.

Reverse thrust is commanded by the crew by a mechanism in the throttle levers, usually by pulling the levers back to idle, selecting reverse thrust and then increasing the throttle lever position towards maximum to achieve the required braking effect. The effect is often combined with lift dumping, in which air brakes and spoilers are deployed at the same time to provide a combined deceleration effect.

The thrust reverser circuit must be designed to prevent inadvertent operation in the air, and usually combined interlocks between throttle position, reverser selection and main wheel oleo weight on wheel switches. A typical circuit is shown in Figure 2.29.

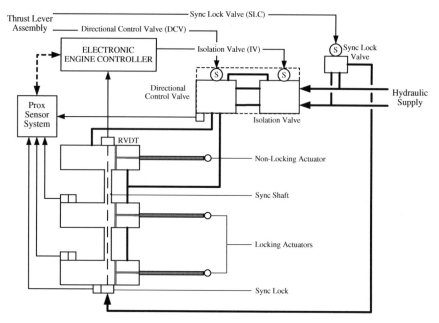

Figure 2.29   Reverse thrust circuit

2.11 Engine Control on Modern Civil Aircraft

Most commercial aircraft engines are twin shaft engines with LP and HP shafts. Some Rolls-Royce engines such as the RB211 and Trent family are triple shaft engine with LP, IP and HP shafts. A high proportion of air by-passes the engine core on a modern gas turbine engine; the ratio of bypass air to engine core air is called the by-pass ratio. The by-pass ratio for most civil engines is in the ratio of 4:1 to 5:1. The Rolls-Royce Trent engine is shown in Figure 2.30 as an example of a modern high bypass ratio engine for the modern generation of commercial airliners. Further views of the engine are shown in Figures 2.31 and 2.32.

Most modern civil engines use a Full Authority Digital Engine Control System (FADEC), mounted on the fan casing to perform all the functions of powerplant management and control.

The key areas of monitoring and control are:

- Various speed probes (N1, N2); temperature and pressure sensors (P2/T2, P2.5/T2.5, and T3); Exhaust Gas Temperature (EGT) and oil temperature and pressure sensors are shown
- The turbine case cooling loops – High Pressure (HP) and Low Pressure (LP)
- Engine start
- Fuel control for control of engine speed and, therefore, thrust

- The engine Permanent Magnet Alternators (PMAs) are small dedicated generators that supply primary power on the engine for critical control functions
- Various turbine blade cooling, Inlet Guide Vanes (IGVs), Variable Stator Vanes (VSVs) and bleed air controls

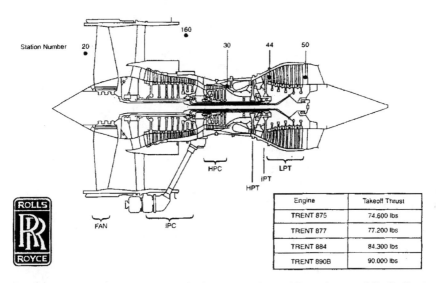

Engine	Takeoff Thrust
TRENT 875	74.600 lbs
TRENT 877	77.200 lbs
TRENT 884	84.300 lbs
TRENT 890B	90.000 lbs

Figure 2.30 Rolls Royce Trent 800 – overview (Courtesy of Rolls Royce)

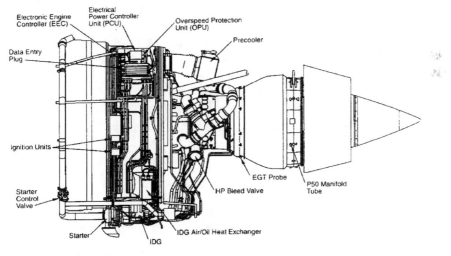

Figure 2.31 Rolls Royce Trent 800 – left side (Courtesy of Rolls Royce)

The engine supplies bleed air for a variety of functions as described in Chapter 7 – Pneumatic Systems. Bleed air provides the actuator motive power for some of the controls on the engine as well as supplying medium pressure air to the airframe for a variety of functions such as anti-icing, cabin pressurisation and cabin temperature control among other functions.

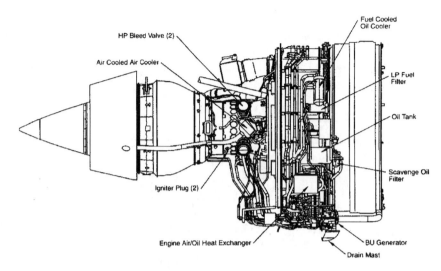

Figure 2.32   Rolls Royce Trent 800 – right side (Courtesy of Rolls Royce)

An important feature of commercial aircraft operations is the increasing use of two engine aircraft flying Extended Range Twin Operations (ETOPS) routes across trans-oceanic or wilderness terrain. The majority of trans-Atlantic flights today are ETOPS operations. The integrity of the engines and related systems is clearly vital for these operations and the engine in-flight shut down (IFSD) rate is central to determining whether 120 minute or 180 minute ETOPS approval may be granted by the certification authorities. Reference [5] is consulted for ETOPS clearance. It mandates that the engine IFSD needed for ETOPS approval is <50 per million flight hours and <20 per million flight hours for 120 minutes and 180 minutes respectively and the actual rate achieved in service is well below these minima.

Recently efforts have been made by Boeing to extend this to 208 minutes to take full account of the extended range of later versions of the B777. For further information refer to Chapter 11.12 (ETOPS).

References

[1] *The Rolls-Royce Book of the Jet Engine* (2005) Rolls-Royce Ltd.
[2] McNamara, J., Legge, C.J. and Roberts, E. (1979) 'Experimental Full-Authority Digital Engine Control on Concorde', AGARD Conference on Advanced Control Systems for Aircraft Powerplants (CP 274), October.
[3] McNamara, J. and Seabridge, A.G. (1982) 'Integrated Aircraft Avionics and Powerplant Control and Management Systems', ASME International Gas Turbine Conference, April.
[4] Bernstein, M.B. (1995), Monitoring of New Generation Turbofan Engines. RAeS Conference – The Design and Maintenance of Complex Systems on Modern Aircraft.
[5] FAA Advisory Circular AC 120–42A, Dec. 1998

3

Fuel Systems

3.1 Introduction

At the onset of aviation aircraft fuel systems were remarkably simple affairs. Fuel was gravity fed to the engine in most cases though higher performance engines would have an engine-mounted fuel pump. Tank configurations were extremely simple and fuel contents were visible float driven indications. In the case of the Tiger Moth, fuel indication was by means of a simple sight glass located on top of the fuel tank between the two upper wing sections.

Higher performance gave rise to more complexity within the fuel system. The need for transfer and booster pumps accompanied the arrival of high performance aircraft. More complex tank configurations introduced the need for multi-valve systems such that the flight crew could move fuel around the fuel tanks according to the needs at the time.

The arrival of jet turbine powered aircraft brought a range of engines that were much thirstier than their piston-engined predecessors: the early jet aircraft in general had a very short sortie length. More accurate fuel gauging systems were required to give the pilot advanced and accurate information regarding the aircraft fuel state in order that recovery to an airfield could be accomplished before running out of fuel. The higher performance jet engine also required considerably greater fuel delivery pressures to avoid cavitation and flame-out.

A further effect of the high fuel consumption was the use of under-wing or under-fuselage ventral tanks to enhance the range of the aircraft. These additional tanks further complicated the fuel system and tank pressurisation systems were developed to transfer the external fuel to the aircraft internal tanks. These systems brought the requirement for further valves to control tank pressurisation and ensure that the tanks could not be damaged by excessive pressure.

Aircraft Systems: Mechanical, electrical, and avionics subsystems integration. 3rd Edition   Ian Moir and Allan Seabridge
© 2008 John Wiley & Sons, Ltd

Fuel gauging systems became more complex as greater gauging accuracies were sought and achieved. Most systems are based upon capacitance measurement of the fuel level within the aircraft, using fuel probes placed at various locations within the fuel tanks. A large system may require some 30 or 40 probes or more to measure the contents accurately. Typical figures for the airliners of today are in the region of 1–2% accuracy, depending upon the sophistication of the systems, some of which can compensate for fuel temperature and density, aircraft attitude, fuel height and a variety of other variables.

Although not a new concept, the development of in-flight refuelling techniques has further extended the range of military aircraft and enhanced the flexibility of air power leading to a 'force-multiplier' effect. Military actions in the Falklands in early 1980s and in the Persian Gulf in 1991, as well as operations continuing today, have underlined the vital nature of in-flight refuelling (see Figures 3.1 and 3.2) and not just for fighter aircraft. In-flight refuelling has also been used to speed the pace of development programmes, especially in the US where the B-2, YF-22A and YF-23A flight test programmes all used the technique to extend sortie length soon after first flight.

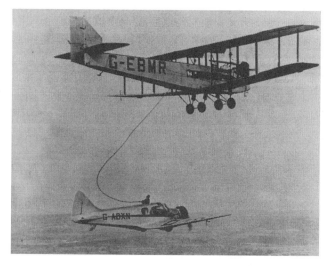

Figure 3.1 Handley Page W 10 tanker refuelling Sir Alan Cobham's Airspeed Courier in October 1934. S/Ldr W. Helmore had the draughty task of handling the fuel hose (Courtesy of Flight Refuelling Ltd/Cobham)

Modern aircraft fuel management and gauging systems are based upon a plethora of valves, pumps, probes, level sensors, switches etc. controlled by microprocessor based systems. This has led to more capable and more reliable systems needed for the aircraft to meet the exacting demands placed upon them.

Figure 3.2 Tornado GR 1s refuelling from a Vickers VC 10 tanker during the 1991 Gulf War (Courtesy of BAE Systems)

3.2 Characteristics of Fuel Systems

The purpose of an aircraft fuel system is primarily to provide a reliable supply of fuel to the engines. Without the motive power provided by them the aircraft is unable to sustain flight. Therefore the fuel system is an essential element in the overall suite of systems required to assure safe flight. Modern aircraft fuels are hydrocarbon fuels similar to those used in the automobile. Piston-engined aircraft use a higher octane fuel called AVGAS in aviation parlance. Jet engines use a cruder fuel with a wider distillation cut and with a lower flashpoint. AVTAG and AVTUR are typical jet engine fuels. The specific gravity of aviation fuels is around 0.8, that is about eight-tenths of the density of water. Therefore fuel may be quantified by reference to either volume (gallons or litres) or weight (pounds or kilograms). As the density of fuel varies according to temperature both may be used. The volume of an aircraft fuel tankage is fixed and therefore it will not be able to accommodate the same weight of fuel at high temperature when the fuel density is lower. For most practical purposes a gallon of fuel may be assumed to weigh around 8 lb (as opposed to 10 lb for a gallon of water).

The essential characteristics of a modern aircraft fuel management system may embrace some or all of the following modes of operation:

- Fuel pressurisation
- Engine feed
- Fuel transfer
- Refuel/defuel

- Fuel storage – there are many issues related to the storage and assured supply of fuel during aircraft flight; these issues vary from aircraft to aircraft and form the kernel of the overall aircraft fuel system requirements
- Vent systems
- Use of fuel as heat sink
- Fuel jettison
- In-flight refuelling

Before describing the operation of these typical modes of operation it is worth examining one and outlining the primary components that comprise such a system. It should also be stated that this represents the briefest introduction of issues addressed in a companion volume dedicated to aircraft fuel systems.

3.3 Description of Fuel System Components

3.3.1 Fuel Transfer Pumps

Fuel transfer pumps perform the task of transferring fuel between the aircraft fuel tanks to ensure that the engine fuel feed requirement is satisfied. On most aircraft this will require the supply of fuel to collector tanks which carry out the obvious task of collecting or consolidating fuel before engine feed; thereby assuring a guaranteed (short-term) supply to each engine. Transfer pumps may also be required to transfer fuel around the aircraft to maintain pitch or lateral trim. In the case of pitch trim this requirement is becoming more critical for unstable control configured aircraft where the task of active CG control may be placed upon the fuel management system.

On civil aircraft there is a requirement to transfer fuel from the fuselage centre wing tanks to tanks where fuel may typically be consolidated before engine feed. However there are FAR/JAR regulations which require independent engine feed systems. On more recent civil aircraft such as the Airbus A340 the horizontal stabiliser may contain up to 7 tonnes of fuel which has to be transferred to maintain the aircraft CG within acceptable limits during the cruise phase. Typically this schedule will be invoked when the aircraft has exceeded an altitude of FL250.

Older aircraft such as the Vickers VC10 also contain fuel in the empennage, in this case the fin, to increase fuel capacity. In these cases pumps are also required to transfer fuel forward to a centre tank for consolidation. A typical aircraft system will have a number of transfer pumps for the purposes of redundancy, as will be seen in the examples given later in this chapter.

An example of a fuel transfer pump is shown in Figure 3.3, this particular example being used on the Anglo-French Jaguar fighter. This is a fuel-lubricated pump; a feature shared by most aircraft fuel pumps. The pump has the capability of safely running dry in the event that no fuel should remain in the tank for any reason. Thermal protection is also incorporated to prevent over-heating. This particular pump is designed to supply in the region of 400 lb/minute at a pressure of 10 psi.

Figure 3.3 Jaguar fuel transfer pump (Courtesy of BAE Systems)

3.3.2 Fuel Booster Pumps

Fuel booster pumps, sometimes called engine feed pumps, are used to boost the fuel flow from the aircraft fuel system to the engine. One of the reasons for this is to prevent aeration (i.e. air in the fuel lines that could cause an engine 'flame-out' with consequent loss of power). Another reason in the case of military aircraft is to prevent 'cavitation' at high altitudes. Cavitation is a process in which the combination of high altitude, relatively high fuel temperature and high engine demand produce a set of circumstances where the fuel is inclined to vaporise. Vaporisation is a result of the combination of low fuel vapour pressure and high temperature. The effect is drastically to reduce the flow of fuel to the engine that can cause a flameout in the same way as aeration (as may be caused by air in the fuel). An aircraft system will possess a number of transfer pumps as will be illustrated later in the chapter.

The engine manufacturer usually imposes a requirement that fuel feed pressure must remain at least 5 psi above true vapour pressure at all times.

Booster pumps are usually electrically driven; for smaller aircraft such as the BAE Systems Jet Provost and the Harrier the pump is driven from the aircraft 28 VDC system with delivery pressures in the range 10–15 psi and flow rates up to 2.5 kg/sec of fuel. The higher fuel consumption of larger, high performance aircraft booster pumps are powered by three-phase AC motors; in the case of Tornado delivering 5 kg/sec. Booster pumps are cooled and lubricated by the fuel in which they are located in a similar way to transfer pumps, and may be specified to run for several hours in a 'dry' environment. Fuel pumps can also be hydraulically driven or, in certain cases, ram air turbine driven, such as the VC10 tanker in-flight refuelling pump.

While most of the larger aircraft use electric motor-driven pumps, ejector pumps are in common use for both fuel feed and transfer in some applications.

The example of a booster pump shown in Figure 3.4 is the double-ended pump used in the Tornado to provide uninterrupted fuel supply during normal and inverted flight/negative-g manoeuvres.

Figure 3.4 Tornado double-ended booster pump (Courtesy of BAE Systems)

3.3.3 Fuel Transfer Valves

A variety of fuel valves will typically be utilised in an aircraft fuel system. Shut-off valves perform the obvious function of shutting off fuel flow when required. This might involve stemming the flow of fuel to an engine, or it may involve the prevention of fuel transfer from one tank to another. Refuel/defuel valves are used during aircraft fuel replenishment to allow flow from the refuelling gallery to the fuel tanks. These valves will be controlled so that they shut off once the desired fuel load has been taken on board. Similarly, during defuelling the valves will be used so that the load may be reduced to the desired level – almost entirely used for maintenance purpose. Cross-feed valves are used when the fuel is required to be fed from one side of the aircraft to the other.

Fuel dump valves perform the critical function of dumping excess fuel from the aircraft tanks in an emergency. These valves are critical in operation in the sense that they are required to operate and dump fuel to reduce the fuel contents to the required levels during an in-flight emergency. Conversely, the valves are not required to operate and inadvertently dump fuel during normal flight.

The majority of the functions described are performed by motorised valves that are driven from position to position by small electric motors. Other valves with a discrete on/off function may be switched by electrically operated solenoids. Figure 3.5 shows an example of a transfer valve driven by a DC powered rotary actuator. An actuator of this type may be two-position (90°) or three-position (270°) or continually modulating over 90°.

Figure 3.5 Transfer Valve driven by a rotary actuator (Courtesy of High Temp Engineers/Cobham)

Fuel vent valves are used to vent the aircraft fuel tanks of air during the refuelling process; they may also be used to vent excess fuel from the tanks in flight. An example of such a valve is shown in Figure 3.6. This valve permits inward or outward venting of around 20–25 lb of air per minute during flight/pressure refuelling as appropriate. Venting fuel in flight must only be related to in-flight refuelling. The valve also permits venting of fuel (in the event of a refuelling valve failing to shut off) of about 800 lb/minute or 100 gallons/minute.

3.3.4 Non-Return Valves (NRVs)

A variety of non-return valves or check valves are required in an aircraft fuel system to preserve the fluid logic of the system. Non-return valves as the name suggests prevent the flow of fuel in the reverse sense. The use of non-return valves together with the various transfer and shut off valves utilised around the system ensure correct system operation in the system modes listed above and which will be described in more detail later in the chapter.

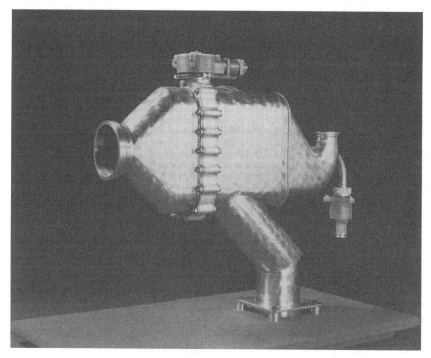

Figure 3.6 Typical fuel vent valve (Courtesy of High Temp Engineers/ Cobham)

3.4 Fuel Quantity Measurement

3.4.1 Level Sensors

Level sensors measure the fuel level in a particular tank and thereby influence fuel management system decisions. Level sensors are used to prevent fuel tank overfill during refuelling. Level sensors are also used for the critical low level sensing and display function to ensure that fuel levels do not drop below flight critical levels where the aircraft has insufficient fuel to return to a suitable airfield. Level sensors may be one of a number of types: Float operated; optical; sound or zener diodes – two of which are described below.

Float Level Sensors

Float level sensors act in a similar way to a domestic toilet cistern connected to the water supply shut-off valve that is closed as the float rises. The refuelling valve, operating in the same way, is a simple but effective way of measuring the fuel level but it has the disadvantage that, having moving parts, the float arm may stick or jam.

Zener Diode Level Sensors

By using simple solid state techniques it is possible to determine fluid levels accurately. The principle is based upon a positive temperature coefficient directly heated Zener diode. The response time when sensing from air to liquid is less than 2 seconds (refuelling valve) and from liquid to air less than 7 seconds (low level warning). Fluid level may be sensed to an accuracy of about plus/minus 2 mm and the power required is around 27 mA per channel at 28 V DC. The sensor operates in conjunction with an amplifier within a control unit and can accommodate multi-channel requirements. A typical fluid sensor of this type is shown in Figure 3.7. The advantage of this method of level sensing is accuracy and the fact that there are no moving parts. In more recent times this technique is disfavoured for safety reasons.

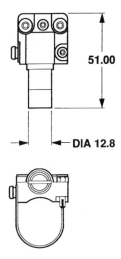

51.00

DIA 12.8

Figure 3.7   Solid state level sensor (Courtesy of Smiths Group – now GE Aviation)

Capacitance Sensors

Capacitance sensors were used on A340 and A380 for sensing fuel level. The advantage is that there is a measurable signal from the sensor under both states.

Ultrasonic Sensors

Ultrasonic point sensors are becoming favoured as point level sensors within the fuel sensing system.

3.4.2 Fuel Gauging Probes

Many of the aircraft functions relating to fuel are concerned with the measurement of fuel quantity on board the aircraft. For example, the attainment of a particular fuel level could result in a number of differing actions depending upon the circumstances: opening or closing fuel valves or turning on/off fuel pumps in order to achieve the desired system state. Quantity measurement is usually accomplished by a number of probes based upon the principle of fuel capacitance measurement at various locations throughout the tanks. Air and fuel have different dielectric values and by measuring the capacitance of a probe the fuel level may be inferred. The locations of the fuel probes are carefully chosen such that the effects of aircraft pitch and roll attitude changes are minimised as far as quantity measurement is concerned. Additional probes may cater for differences in fuel density and permittivity when uplifting fuel at differing airfields around the world as well as for fuel at different temperatures. Fuel gauging, or Fuel Quantity Indication Systems (FQIS) as they are sometimes known, are therefore an essential element in providing the flight and ground crews with adequate information relating to the amount of fuel contained within the aircraft tanks.

3.4.3 Fuel Quantity Measurement Basics

The underlying difficulties in accurately measuring aircraft fuel contents; also referred to as Fuel On Board (FOB) lie in the very nature of the agility and mobility of the air vehicle. The most obvious factors are:

- The difficulty in measuring a fluid level within a body in motion
- The fact that aircraft tanks are virtually never regular shapes
- The fact that aircraft fuel demonstrates diverse properties and has different composition when uplifted in different locations

Fuel quantity may be expressed as kilograms (1000 kilogram = a metric tonne), pounds (lb), or gallons – either Imperial or US gallons. A US gallon is 0.8 × an Imperial gallon (1 Imperial gallon = 8 × 20 = 160 fluid ounces). The Specific Gravity (SG) of fuel is around 0.8, therefore an Imperial gallon is roughly equivalent to 160/16 or 10 lbs whereas a US gallon equates to around 8 lbs. Since the contents of aircraft tanks are characterised by tank volume the amount of energy contained within a fuel load is therefore determined by the weight (mass) of the FOB; itself a function of fuel density and fuel temperature.

Fluid Motion

Measuring fuel level in flight is analogous to trying to run while carrying a bucket of water; the fluid appears to take on a mind of its own and the 'inertia' of the fluid has to be anticipated both when starting out and when stopping. This fluid can be ameliorated to a degree by natural boundaries such as wing ribs or fuselage frames that may protrude into the tank. The insertion of baffles may also prevent undue 'sloshing' of the fuel. This sloshing action can be modelled using 3D computer aided design tools together with fluid dynamic modelling tools such as Flowmaster. This enables a simulation of the fuel system, in whole or in part, to be modelled and subjected to aircraft manoeuvres to observe the effects on the fuel. Baffles can then be inserted into the model to allow observation of their effect on fuel slosh, and to optimise their location in a tank.

Tookey, Spicer and Diston (2002) describe an aircraft system model of this type [1].

3.4.4 Tank Shapes

Aircraft tank shapes vary greatly and are difficult to determine, particularly at an early stage in the aircraft design. Large, regular volumes are at a premium within an aircraft and the volumes available to the fuel system designer are usually those remaining when the structures and propulsion designers have had their day. Therefore not only are the tank shapes irregular but their boundaries may not be fixed until fairly late in the design. Once the tank boundaries are frozen, the tank designer has to characterise the volumetric shape of the tank to understand what the fluid level means for a variety of tank attitudes.

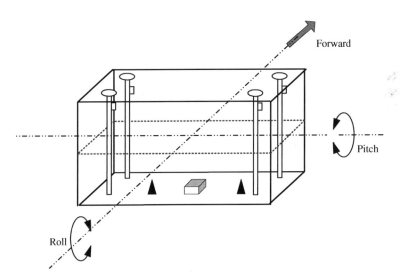

Figure 3.8 Simplified aircraft centre fuel tank

The problem may be better understood by referring to Figure 3.8. This is a representation of a simple rectangular tank that might approximate to the centre tank on many typical civil aircraft. While the shape is regular the tank will be rotating as the aircraft pitch and roll attitude alter. Aircraft accelerations will also occur as speed changes are made. The fluid contents of this tank, or more correctly, the fluid level may be determined by placing quantity probes in each corner of the tank. This may be acceptable for a basic configuration but to permit necessary levels of accuracy following a probe failure, additional probes may need to be added. In a sophisticated long range aircraft the probes may need to be replicated to provide dual redundant sensing.

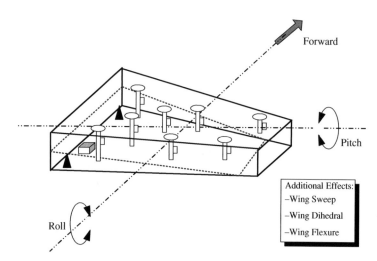

Figure 3.9   Simplified aircraft wing fuel tank

It may be seen from Figure 3.9 that this situation is significantly compounded compared to that for a wing tank as the tank is irregular and long and shallow. To compound the problem the wing may also be flexing during flight and therefore the tank shape may not be regular at certain stages. The accurate gauging of the wing tank fuel level under these circumstances is very difficult. To illustrate the problem the layout of the F-35 fuel tanks is shown in Figure 3.10.

3.4.5 Fuel Properties

As mentioned earlier, aircraft fuel is not a uniform commodity with readily repeatable characteristics. Fuel characteristics will depend upon the oilfield from which the original was extracted and the subsequent refinery process. Aviation fuel has a significant variation in density with temperature. For a typical commercial aircraft with a capacity of ~ 8000 gallons, the fuel load will vary from about 26 tonnes on a hot day to around 28 tonnes on a cold day. The measurement of fuel temperature is important as is density to establish the Fuel On Board (FOB). FOB is always defined in mass terms since this defines the amount of stored energy within the fuel and hence aircraft range.

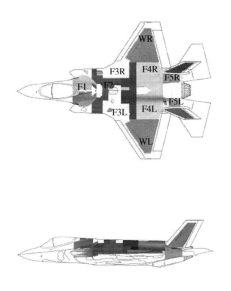

Seven Tanks: Nominal 18,000 lb of fuel:

F1 Centre Fuselage

F2 Centre Fuselage

F3 Engine Feed Tank (F3L & F3R)

F4 Wing Carry-Through (F4L & F4R)

F5 Aft Fuselage (F5L & F5R)

WL Left Wing Box

WR Right Wing Box

Figure 3.10 Joint strike fighter fuel tank layout (Courtesy of BAE Systems) (See Colour Plate 2)

Fuels will contain a variety of additives to aid their use, typical additives are:

- Antioxidants to prevent gumming
- Antistatic agents to dissipate static electricity
- Corrosion inhibitors
- Icing inhibitor agents to prevent suspended water in the fuel from freezing

The most typical and widely used jet fuel is a kerosene-paraffin mix called JET A-1, also produced as JET A in the US. JET B is a naphtha-kerosene fuel used for cold-weather applications though it is more difficult to handle. The military categorise their fuel by JP (Jet Propulsion) numbers depending upon the application. JP-4 is used by the US Air Force while JP-5 is used by the US Navy, the latter being less volatile for use on aircraft carriers. Specialised vehicles like the SR-71 strategic reconnaissance aircraft used JP-7 which was suited to its high altitude 75 000–80 000 ft and high speed (Mach 3 plus) flight envelope.

Typical JET A-1 characteristics are as follows:

Flash point	38 °C
Auto-ignition temperature	210 °C
Freezing point[a]	−47° [−40 °C for JET A]
Open air burning temperature	260–315 °C
Maximum burning temperature	980 °C

[a] With the advent of longer range flights over polar regions, the operation of aircraft under cold fuel soak conditions is becoming increasingly important as is described later in the chapter.

Fuel gauging probes are concentric cylindrical tubes with a diameter of about 1 inch. Despite experiments with glass-fibre probes, metal ones have been found to be the most reliable for minimum weight. Plastic, non-conducting cross-pins maintain the concentricity of the tubes while providing the necessary electrical insulation. Tank units may be either internally or externally mounted on straight or angled flanges, for both rigid and flexible tanks.

A number of factors may affect fuel measurement accuracy:

- *Tank geometry*. The optimum number of probes for a given tank is established by means of computerised techniques to model the tank and probe geometry. Each probe may then be 'characterised' to achieve a linear characteristic of the gauging system. This may be done by mechanical profiling to account for tank shape and provide a linear output. This is an expensive and repetitive manufacturing process which may be more effectively achieved by using 'linear' probes with the correction being derived in computer software for some of the more advanced microprocessor driven fuel gauging systems
- *Attitude envelope*. The most significant factor driving the probe array design versus accuracy is the attitude envelope. This will be different on the ground where higher accuracy may be desirable for refuelling. Also inertial reference data may not always be available on the ground
- *Permittivity variations*. Variations in the permittivity of the fuel may adversely affect gauging accuracy. Reference units may be used to compensate for the varying temperature within the fuel. These may be separate standalone units or may be incorporated into the probe itself

Examples of particular tank probes are shown in Figure 3.11.

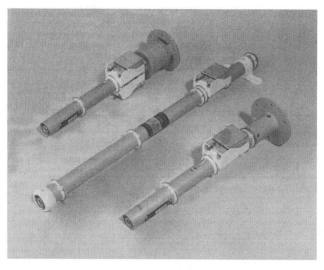

Figure 3.11 Examples of fuel probe units (Courtesy of Smiths Group – now GE Aviation)

3.4.6 Fuel Quantity Measurement Systems

Fuel quantity measurement systems using capacitance probes of the type already described may be implemented in one of two ways. These relate to the signalling techniques used to convey the fuel tank capacitance (and therefore tank contents) to the fuel indicator or computer:

- AC system
- DC system

The preference for capacitance gauging technology has shifted to AC from DC systems. Almost all aircraft new starts in the past 15 years have used AC gauging; namely: A340-500/600, A380, EMB 170/190, Global Express, A350XWB, B787, F-22 and F-35.

- *AC systems*
  In an AC system the tank unit information is conveyed by means of an AC voltage modulated by the measured tank capacitance and therefore fuel quantity. The problem with the AC signalling technique is that there is a greater risk of electro-magnetic interference (EMI) so that coaxial cables and connectors are required making the installation more complex, expensive and difficult to maintain. Therefore although individual AC tank units may be lighter, cheaper and more reliable (being simpler in construction) than the DC tank unit equivalent, the overall system penalties in terms of weight and cost may be greater
- *DC systems*
  In the DC system the probes are fed by a constant voltage/frequency probe drive and utilise automatic fuel probe diode temperature compensation. Fuel probe signals are rectified by the diodes and the resulting signal proportional to fuel contents returned to the processor as a DC analogue signal. The more complex coaxial cables and connectors of the AC system are not required. The overall system weight and cost of the DC system is therefore usually less than an AC system whose overall system reliability is usually better than for the DC system. There is an increasing tendency for modern systems to adopt the AC system due to the inherent benefits. A disadvantage of a DC system is the need for additional components within the fuel tank

In reality the choice between AC and DC systems will be heavily biased by the experience accrued by a specific airframe manufacturer.

Two examples of DC systems which have been in service for a considerable time are the systems used on board the Fokker F50/F100 and the Airbus A320.

3.4.7 Fokker F50/F100 System

The diagrammatic layout of this system and the system architecture are shown in Figures 3.12a and 3.12b respectively.

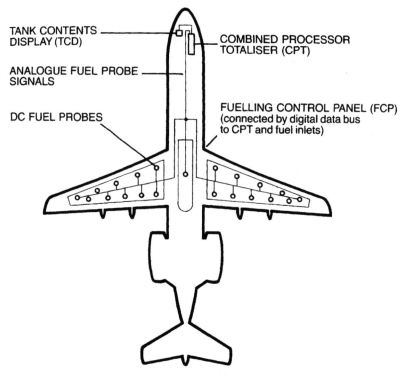

Figure 3.12a Fokker 100 diagrammatic layout (Courtesy of Smiths Group – now GE Aviation)

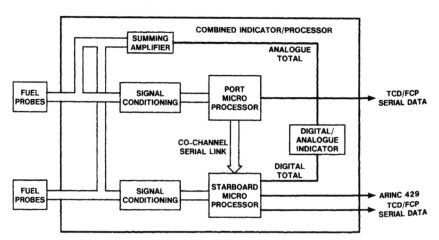

Figure 3.12b Fokker F50/F100 system architecture (Courtesy of Smiths Group – now GE Aviation)

Data from the DC fuel probes in the wing and fuselage tanks are summed and conditioned in the Combined Processor Totaliser (CPT) and fed to the fuel indicator portion of the unit. Dual 8-bit microprocessors process the information into serial digital form for transmission on ARINC 429 data buses to the

Total Contents Display (TCD) in the cockpit and the Fuel Control Panel (FCP) in the right wing root. The system displays individual tank contents to the crew. The FCP enables the aircraft to be automatically refuelled to preset fuel quantities without operator intervention, The accuracy of this type of system is of the order of 2%. The system is designed so that no single failure will cause total loss of all fuel gauging information.

3.4.8 Airbus A320 System

The DC fuel system used on the Airbus A320 is shown in Figures 3.13a and 3.13b.

The A320 example is more complex than the Fokker F50/F100 system. Linear DC gauging probes are located in the two wing tanks, three fuselage tanks; later models such as the A340 also have a tank located in the tailplane. Densitometers are fitted in the wing and centre fuselage tanks. The system also uses attitude data supplied by the aircraft systems. The system is based upon a dual redundant computer architecture using Motorola 68000 microprocessors: each processor handles identical data and in the event of one processor failing the other automatically takes over the computation tasks without any loss of

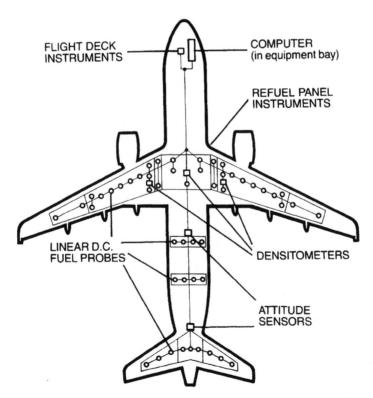

Figure 3.13a Airbus diagrammatic layout (Courtesy of Smiths Group – now GE Aviation)

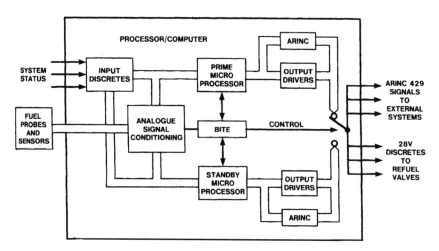

Figure 3.13b Airbus system architecture (Courtesy of Smiths Group – now GE Aviation)

continuity. The system is designed to fail with 'graceful degradation', that is to degrade gently in accuracy while informing the crew.

In this system data relating to the tank geometry is stored in memory together with the computed fuel density, permittivity, fuel temperatures, aircraft attitude and other relevant aircraft information. The computers then use various algorithms to calculate the true mass of fuel. Multiple ARINC 429 serial data buses provide data to the flight management computer and the various displays. In this system discrete signal outputs are used to control the operation of refuelling valves or transfer valves. The overall accuracy of this system is in the order of 1%.

Further information regarding these systems is given in [2] and [3].

3.4.9 'Smart' Probes

A further variation on the theme of capacitance probes is the 'smart' probe used on the Eurofighter Typhoon and BAE Systems Nimrod aircraft. The probes are active or 'smart' in that each probe has dedicated electronics associated with the probe. Each is supplied with a regulated and protected DC voltage supply to power the local electronics. The local electronics process the capacitance value to produce a pulse train the period of which is proportional to the capacitance sensed and therefore the fuel level measured by the probe. The benefit of this type of system is to provide a means of reducing the EMI susceptibility of the fuel probe transmission system. Twisted, screened three-wire signal lines are used which are simpler than coaxial cables but nonetheless expensive in wiring terms. A disadvantage is the need to provide electronics for each individual probe in a relatively hostile environment within the airframe.

3.4.10 Ultrasonic Probes

All of the previously mentioned systems use capacitive measurement techniques to sense fuel level. Ultrasonic techniques have been developed which utilise ultrasonic transducers to measure fuel level instead of the conventional capacitive means. The sensor is located at the bottom of the waveguide. The waveguide arrangement at the base of the tank directs the ultrasonic transmission back to the transducer. To measure height with ultrasonics the speed of sound in the fuel medium is required. This is generally measured using a fixed reference in the waveguide. A portion of the ultrasonic wave is reflected directly back to the transducer and serves as a reference signal. The time taken for the signal to be reflected back from the fuel surface is measured and by using a simple ratiometric calculation the fuel height may be determined. Fuel level may be measured by comparing the time of propagation for the reference signal with that for the fuel level reflected signal. This type of quantity measuring system was introduced on the Boeing 777 airliner which first flew on revenue service in June 1995 and of which there are several hundred examples in service today.

3.5 Fuel System Operating Modes

The modes of operation described in the following paragraphs are typical of many aircraft fuel systems. Each is described as an example in a particular fuel system. Any system may exhibit many but probably not all of these modes. In an aircraft the fuel tanks and components have to compete with other systems, notably structure and engines for the useful volume contained within the aircraft profile. Therefore fuel tanks are irregular shapes and the layman would be surprised by how many tanks there are, particularly within the fuselage where competition for usable volume is more intense. The proliferation of tanks increases the complexity of the interconnecting pipes and certainly does not ease the task of accurate fuel measurement. As an example of a typical fighter aircraft fuel tank configuration see Figure 3.14 that shows the internal fuel tank configuration for EAP.

   This is a simplified diagram showing only the main fuel transfer lines; refuelling and vent lines have been omitted for clarity. Whereas the wing fuel tanks are fairly straightforward in shape, the fuselage tanks are more numerous and of more complex geometry than might be supposed. The segregation of fuel tanks into smaller tanks longitudinally (fore and aft) is due to the need to avoid aircraft structural members. The shape of most of the fuselage tanks also shows clearly the impositions caused by the engine intakes. Furthermore as an experimental aircraft EAP was not equipped for in-flight refuelling nor was any external under-wing or ventral tanks fitted. It can be seen that a fully operational fighter would have a correspondingly more complicated fuel system than the one shown.

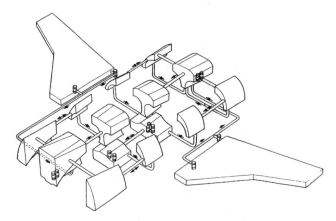

Figure 3.14   Simplified EAP fuel system (Courtesy of BAE Systems)

3.5.1 Pressurisation

Fuel pressurisation is sometimes required to assist in forcing the fuel under relatively low pressure from certain tanks to others that are more strategically placed within the system. On some aircraft there may be no need for a pressurisation system at all; it may be sufficient to gravity feed the fuel or rely on transfer pumps to move it around the system. On other aircraft ram air pressure may be utilised to give a low but positive pressure differential. Some fighter aircraft have a dedicated pressurisation system using high pressure air derived from the engine bleed system.

The engine bleed air pressure in this case would be reduced by means of pressure reducing valves (PRVs) to a more acceptable level. For a combat aircraft which may have a number of external fuel tanks fitted the relative regulating pressure settings of the PRVs may be used to effectively sequence the transfer of fuel from the external and internal tanks in the desired manner. For example, on an aircraft fitted with under-wing and under-fuselage (ventral) tanks it may be required to feed from under-wing, then the ventral and finally the internal wing/fuselage tanks. The PRVs may be set to ensure that this sequence is preserved, by applying a higher differential pressure to those tanks required to transfer fuel first.

In some aircraft such as the F-22, inert gas is used to pressurise the fuel tanks. Inert gas for this purpose can be obtained from an On-Board Inert Gas Generating System (OBIGGS).

3.5.2 Engine Feed

The supply of fuel to the engines is by far the most critical element of the fuel system. Fuel is usually collected or consolidated before being fed into the engine feed lines. The example in Figure 3.15 shows a typical combat aircraft, the fuel is consolidated in two collector tanks; one for each engine.

This schematic diagram may be reconciled with the EAP example depicted in Figure 3.14. The fuel transfer from the aircraft fuel tanks into the collector tanks is fully described in the fuel transfer section.

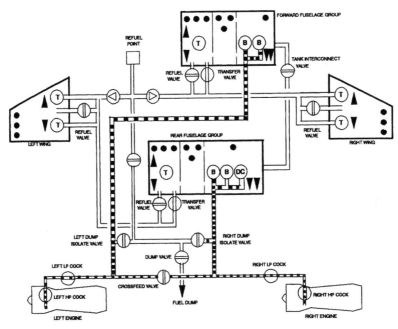

Figure 3.15 Typical fighter aircraft engine feed

The collector tanks may hold sufficient fuel for several minutes of flying, depending upon the engine throttle settings at the time. The contents of these tanks will be gauged as part of the overall fuel contents measuring system; however, due to the criticality of the engine of the engine fuel feed function additional measurement sensors are added. It is usual to provide low-level sensors that measure and indicate when the collector tanks are almost empty. These low-level sensors generate critical warnings to inform the pilot that he is about to run out of fuel and that the engine will subsequently flame out. The low-level warnings are a last ditch indication that the pilot should be preparing to evacuate the aircraft if he is not already doing so.

The collector tanks contain the booster pumps that are pressurising the flow of fuel to the engines. It is usual for two booster pumps to be provided so that one is always available in the event that the other should fail. Booster pumps are immersed in the fuel and for a combat aircraft the scavenge pipes feeding fuel to the pump inlets will have a provision such that a feed is maintained during inverted or negative-g flight. Note that the booster pump example shown in Figure 3.4 had such a facility. Booster pumps are usually powered by 115V AC three-phase motors of the type described in Chapter 5 – Electrical Systems. However the motor itself is controlled by a three-phase relay, the relay coil being energised by a 28V DC supply. An auxiliary contact will provide a status signal back to the fuel management system, alternatively a pressure switch or measuring sensor may be located in the delivery outlet of the pump which can indicate that the pump is supplying normal delivery pressure. Booster pumps are fuel lubricated and also have the capability of running dry should that be necessary.

Downstream of the booster pump is the engine high pressure (HP) pump which is driven by the engine accessory gearbox. Engine HP pumps are two-stage pumps; the first stage provides pressure to pass the fuel through heat exchangers and filters and to provide a positive inlet pressure to the second stage. The second stage supplies high pressure fuel (around 1500 to 2000 psi) to the engine fuel control system.

A number of shut-off valves are associated with the control of fuel to the engine. A pilot operated low pressure (LP) cock provides the means of isolating the fuel supply between the booster pump and the HP engine driven pump. This valve may also be associated with a firewall shut-off function which isolates the supply of fuel to the engine compartment in the event of an engine fire. A cross-feed valve located upstream of the LP cocks provides the capability of feeding both engines from one collector tank if necessary; in most cases the cross-feed valve would be closed as shown in Figure 3.15. The pilot may also operate a high pressure (HP) cock that has the ability to isolate the fuel supply on the engine itself. In normal operation both the LP and HP cocks re open allowing an unimpeded supply of fuel to the engine. The cocks are only closed in the case of normal engine shut-down or in flight following an engine fire.

3.5.3 Fuel Transfer

The task of fuel transfer is to move fuel from the main wing and fuselage tanks to the collector tanks. In commercial transport there tend to be fewer tanks of more regular shape and transfer pumps may merely be used for redistributing fuel around the tanks. In the example given in Figure 3.16 the fuselage and wing tanks for the Experimental Aircraft Programme (EAP) are shown. The main tankage comprises left and right wing tanks and forward and rear fuselage tanks.

Two transfer pumps are provided in each wing tank and two in each of the fuselage groups. Transfer pumps are usually activated by the level of fuel in the tank that they supply. Once the fuel has reached a certain level measured by the fuel gauging system, or possibly by the use of level sensors, the pumps will run and transfer fuel until the tank level is restored to the desired level. In the EAP this means that the forward and rear groups are replenished from the left and right wing tanks respectively in normal operation. The fuselage groups in turn top up the collector tanks with the aid of further transfer pumps. The tank interconnect valve also provides for fuel crossfeed from one fuel system (left/forward) to the other (right/rear) which allows fuel to be balanced between left and right or permits one system to feed both engines if the need arises. Transfer pumps operate in a similar fashion to booster pumps; they are also electrically operated by 115 VAC three-phase electrical power driving an induction motor. The duty cycle of the transfer pumps is not continuous like the booster pumps, rather their operation is a periodic on-off cycle as they are required to top up the relevant aircraft tanks subject to fuel demand.

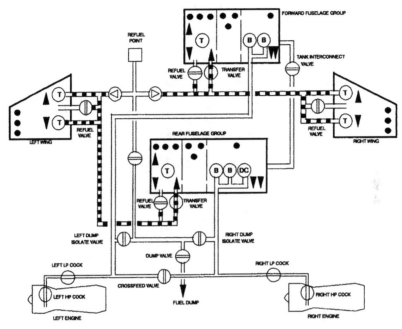

Figure 3.16   EAP fuel transfer operation

It should also be noted that fuel transfer in some aircraft may be performed in order to modify the fuel CG so that the aircraft longitudinal and lateral CG are kept within strict limits. This may be for economy reasons, to maintain an optimum trim, or it may be ensure that the Flight Control System (FCS) is able to interpret pilot inputs to obtain optimum performance without damaging the aircraft. This means that the fuel system and FCS must exchange information with appropriate integrity and this can significantly affect the design of each system. Examples of where this is implemented are highly agile aircraft such as Eurofighter Typhoon and F-35.

3.5.4 Refuel/Defuel

Aircraft refuelling and defuelling is controlled by a separate subsystem within the overall fuel system (see Figure 3.17). The aircraft is fuelled by means of a refuelling receptacle that connects to the refuelling tanker. From the receptacle it enters a refueling gallery which distributes the incoming fuel to the various aircraft tanks. The control of fuel entry into each tank is undertaken by valves that are under the control of the fuel management system. In the crudest sense fuel will enter the tanks until they are full, whereupon the refuelling valve will be shut off preventing the entry of any more.

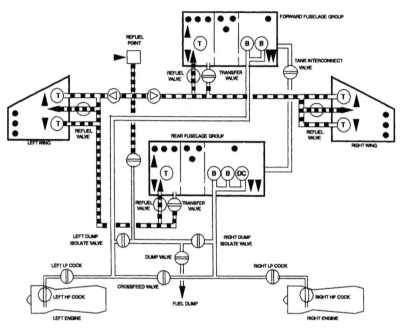

Figure 3.17 Refuelling operation schematic

In a very simple system this shut-off may be accomplished by means of a simple float operated mechanical valve. In more sophisticated systems the fuel management system has control over the operation of the refuelling valve, usually by electrical means such as a solenoid operated or motorised valve. A typical system may comprise a mixture of both types. In most cases the aircraft is not filled to capacity, rather the maintenance crew select a fuel load and set the appropriate levels at the refuel/defuel panel adjacent to the refuelling receptacle – often located under the aircraft wing in an accessible position.

The defuelling process is almost the reverse of that for refuelling. It may be necessary to defuel the aircraft for maintenance reasons. In general defuelling is carried out relatively infrequently compared to refuelling. When it is performed the fuel in the tanks must be completely emptied out and the tank volume purged with air to make the tank space safe to operate in, i.e. to reduce fuel vapour to allow maintenance crew to work in the tanks, and to reduce the risk of an explosive atmosphere leading to a fire or explosion.

In some simpler aircraft it is possible to carry out over-wing refuelling. This is undertaken at remote airstrips where there may not be any dedicated refuelling machinery such as a fuel bowser and the fuel is provided in drums.

In this situation an over-wing panel is removed and fuel is poured manually into the wing tanks.

Certain aircraft, usually commuter and commercial types, have devices called magnetic level indicators (MLIs) which are equivalent to a fluid level dipstick. The MLIs are mounted under the wing and when a simple catch is released the indicator drops until the upper portion is level with the fuel surface. The extended portion of the MLI is graduated so that the amount by which the device extends can be measured. And hence the level of fuel in the tanks can be deduced and cross-checked with the level indicated by the aircraft fuel gauges. For an example the BAE ATP has a total of eight MLIs fitted, four for each wing tank

3.5.5 Vent Systems

Commercial aircraft use what is termed an 'open vent system' to connect the ullage space above the fuel in each tank to the outside air. The provision of adequate fuel tank venting throughout the aircraft operational flight envelope is critical in that it allows the tanks to 'breathe' as the aircraft climbs and descends. Without this provision large pressure differences could develop between the ullage and outside air resulting in very large forces on the tank structure. It is impractical to accommodate these forces via the wing structural design because of the resultant weight penalty; therefore the design of the vent system plays a critical role in protecting the tank structure from structural failure as the aircraft transitions between ground and cruise altitudes.

During the refuel process, the uplifted fuel displaces air in the fuel tanks. For safety and environmental reasons, spillage of fuel to the outside must be avoided. To accomplish this consistently and reliably, a vent box (sometimes referred to as a surge tank) is provided to capture any fuel that may enter the vent lines which connect to the various fuel tanks.

Since pressure refuelling involves the application of a relatively high positive pressure (typically 50 psi) to speed the refuelling process it becomes necessary to protect against a failed open refuel valve. To do this a pressure relief valve usually installed on the upper wing surface prevents the build-up of internal tank pressure to a level that could damage the aircraft structure. During maximum rates of descent a pressure difference in the opposite direction must be avoided by adequate sizing of the vent lines and/or by designing the relief valve to be double-acting.

In military aircraft where operation at extremely high altitudes is required, a closed vent system is employed to prevent excess vaporisation or boiling of the fuel. Here the tanks are slightly pressurised typically using bleed air from the engines. A climb and dive valve must now be employed to maintain a safe pressure differential between the ullage and the outside air.

The reference to fuel/no-air valves at the end of this paragraph is usually associated with pressure transfer of fuel from an external tank and not relevant to the vent discussion.

3.5.6 Use of Fuel as a Heat Sink

In certain aircraft such as high performance jet fighters and Concorde the aircraft fuel performs the very important function of acting as a heat sink for heat generated within the aircraft during flight. For Concorde the kinetic heat is generated by air friction during prolonged flight at very high speeds (Mach 2) in the cruise. In the case of fighter aircraft prolonged operation at high speeds is not likely because of the punitive fuel consumption. The aircraft will generate a lot of heat, particularly from the hydraulic and environmental control system, which needs to be 'sunk' in the fuel.

3.5.7 External Fuel Tanks

Combat aircraft increase range by the use of external fuel tanks. These are usually mounted underwing but have also been belly mounted (ventral tanks) and overwing mounted. The BAE Lightning Mk 6 had a ventral tank fitted for normal operation and over-wing long range ferry tanks as shown in Figure 3.18. The ventral tank had a capacity of 609 gallons/4872 lb while the overwing ferry tanks had a capacity of 540 gallons/4320 lb each. This compares to the aircraft internal fuel capacity of 716 gallons/5728 lb.

The McDonnell Douglas F-15 Eagle fighter usually carries underwing tanks but can also carry close-fitting ventral tanks called conformal tanks to further extend range. In this case the underwing tanks add a capacity of 1484 gallons/11 869 lb and the conformal tanks add 1216 gallons/9728 lb. The internal fuel capacity of the F-15 is 1637 gallons/13 094 lb. Figure 3.19 shows a F-15 with a centre line and conformal tanks fitted.

Figure 3.18   Lightning F6 with over-wing tanks (Courtesy of BAE Systems)

Figure 3.19 F-15E Eagle (Courtesy of Boeing)

External fuel tanks have a disadvantage in that they cause significant additional drag, thereby reducing range and the benefits of the extra fuel they provide. Some fuel tanks are not stressed for supersonic flight and an aircraft operating with external tanks may be subject to a 'q' or airspeed limitation as well as a 'g' limit due to the higher weight and accompanying higher structural loading. It is common for an aircraft to jettison underwing tanks before combat though this is clearly expensive and may cause logistic difficulties during a prolonged conflict.

3.5.8 Fuel Jettison

Fuel constitutes a large portion of overall aircraft weight, particularly at the beginning of a flight. Therefore if an aircraft suffers an emergency or malfunction shortly after take-off it may prove necessary to jettison a large proportion of the fuel in order to reduce weight rapidly. This may be to reduce the aircraft weight from close to maximum All-Up Weight (AUW) to a level that is acceptable for landing; many aircraft are not stressed to land with a full fuel load. Alternatively if an engine has failed the fuel may need to be jettisoned merely to remain airborne. On an aircraft such as EAP the fuel jettison valves are tapped off from the engine feed lines with left and right jettison valves feeding fuel from the left and right engine feed lines respectively (see Figure 3.20).

A fuel jettison master valve is provided downstream to prevent inadvertent fuel jettison which could itself present a flight safety hazard. Only when both left and right and master valves are opened will fuel be jettisoned overboard.

On a civil transport fuel dumping is likely to be achieved by different means with the fuel being ejected from jettison masts situated at the rear of each wingtip. On an aircraft such as EAP the jettison valves are electrically operated motorised valves as are many of the valves in the fuel system.

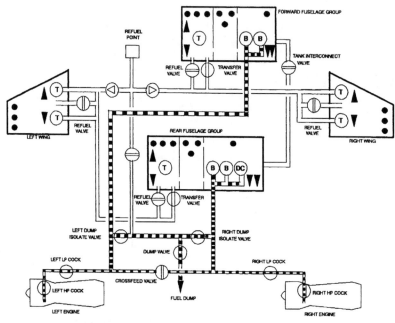

Figure 3.20   EAP fuel jettison system

3.5.9 In-Flight Refuelling

For many years the principle of in-flight refuelling has been known. In fact the first demonstration of in-flight refuelling occurred in April 1934 (Figure 3.1). Today it is an important and inherent method of operating military aircraft. The use of the principle was first widely applied to fighter aircraft because of their high rates of fuel consumption and short sortie length. However, more recently, and particularly during the Falklands campaign the use of in-flight refuelling was extended to transports (Hercules and VC10), maritime patrol aircraft (Nimrod), and tankers (Tristar and VC10). The ability to refuel an aircraft in the air greatly adds to the flexibility of air power giving what is termed in military parlance a 'force multiplier' effect. In the Falklands campaign it was the sheer distance between Ascension Island and the Falklands themselves with virtually no diversions in between that required extensive use of in-flight refuelling. For fighter aircraft maintaining a combat air patrol over a specific objective the operational advantage is gained by keeping armed aircraft in the air, around the clock if necessary.

There are two methods of in-flight refuelling widely in use today. One – the probe and drogue method – is that generally favoured by the Royal Air Force, US Navy and others. The other – the boom and receptacle – is used almost exclusively by the US Air Force. In the former the tanker aircraft trails a refuelling hose with a large drogue attached, behind the aircraft. The recipient is fitted with a fuel probe that may be either fixed or retractable when not in use. The pilot of the receiving aircraft has the responsibility of inserting the refuelling probe into the tanker drogue. When positive pressure is exerted on the drogue by the refuelling probe fuel is able to pass to the receiving aircraft. The transfer of fuel is monitored by the tanker and by the gauging system of the recipient. Contact is broken when the receiving aircraft drops back and the positive pressure between probe and drogue is lost. At this point the refuelling operation is complete. Royal Air Force tankers usually operate with one drogue from the aircraft centre line and one from under-wing refuelling pods, so a total of three stations is available. It is possible to refuel more than one aircraft at a time using this method. See Figure 3.21 for an example of probe and drogue in-flight refuelling.

Figure 3.21 Probe and drogue in-flight refuelling (Courtesy of BAE Systems)

In the boom method, sometimes called the flying boom, the technique is different. The responsibility for making contact is that of the boom operator in the tanker who flies the boom so that the recipient makes contact in a similar manner to the drogue method. The receiving aircraft has a receptacle on its upper surface into which the refuelling boom is inserted. A tanker has one boom mounted on the centre line from the rear of the aircraft and therefore the number of aircraft refuelling using this method may be limited. See Figure 3.22.

Figure 3.22   Boom in-flight refuelling (Courtesy of US Air Force)

Air-to-air refuelling is now extensively used during aircraft flight test programmes in the UK where it is possible to extend the duration of flight tests and effectively accelerate programme completion. The Northrop B-2 stealth bomber, the competing Lockheed/Boeing/General Dynamics YF-22A and the Northrop/McDonnell Douglas YF-23A Advanced Tactical Fighter proto-types used this technique during their respective development programmes. This is a graphic illustration of how commonplace this activity has now become.

In terms of interfacing with the normal refuelling system, the air-to-air refuelling probe feeds into the refuelling lines via a non-return valve (NRV) which only permits flow from the probe into the system and not vice-versa. Therefore once probe contact has been made and is maintained, air-to-air refuelling continues in an identical fashion to the normal refuelling operation except that the aircrew determine when to halt the process.

3.6 Integrated Civil Aircraft Systems

The integration of aircraft civil fuel systems has become more prevalent over the last decade or so using digital data buses and the supply of hardware from one or manufacturers. Most civil aircraft have a fuel tank configuration as shown in Figure 3.23.

This configuration comprises left and right wing tanks and a centre tank. However, it is also possible for aircraft to have an aft or trim tank. The major system functions are:

- Engine and APU feed
- Fuel transfer
- Refuel/defuel
- Fuel jettison

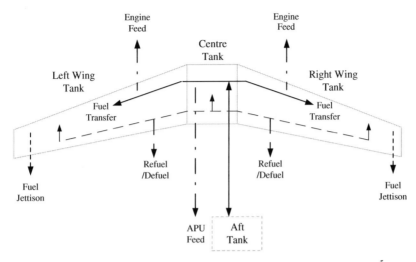

Figure 3.23 Typical civil aircraft fuel tank configuration

Depending upon the aircraft configuration and the degree of control, the aft or trim tank may be used as a means of controlling the aircraft centre of gravity (CG). Altering the contents of a trim tank can reduce trim drag and improve aircraft range; it is also possible to reduce the structural weight of the tailplane. Most aircraft have variations on this basic topology although the number of wing tanks may also be dictated by the wing structure, the number of engines, or the need to partition fuel to cater for engine turbine disc burst zones. This section addresses three examples:

- Bombardier Global Express
- Boeing 777
- Airbus A340-500/600

3.6.1 Bombardier Global Express

The Fuel Management & Quantity Gauging System (FMQGS) developed by Parker Aerospace for the Bombardier Global Express is typical of a family of systems which may be found fitted to regional aircraft and business jets. The Global Express has a true intercontinental range capability exceeding 6500 miles and is cleared to 51 000 ft.

The system has interfaces to:

- Engine Indication and Crew Alerting System (EICAS) and ground crew via A429 data buses
- Cockpit control panel for APU and engine selector switches and fire handles
- Cockpit fuel panel for fuel system mode selections

- Electrical load management system for supplying power to the electrically power pumps and valves the system receives status discretes from fuel pumps and valves
- Cockpit and wing Refuel/Defuel Control Panels (RDCPs)

See Figure 3.24.

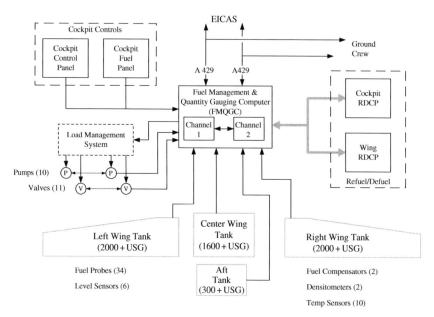

Figure 3.24 Simplified Global Express fuel system (Courtesy of Parker Aerospace)

The heart of the system is the dual channel Fuel Management and Quantity Gauging Computer (FMQCG) which embraces the following functions:

- *Fuel management*
  The fuel management function provides the following:

  - Control, status and Built-In Test (BIT) of all system pumps, valves and pressure sensors
  - Fuel transfer – burn sequence and lateral balance
  - Flight crew and ground crew interface
  - Automatic/manual refuel/defuel operation
  - BIT fault detection and annunciation

- *Optional thermal management*
  The operation of the aircraft for long periods at altitude provides extreme cold soak conditions. The system provides control of the return of warm fuel

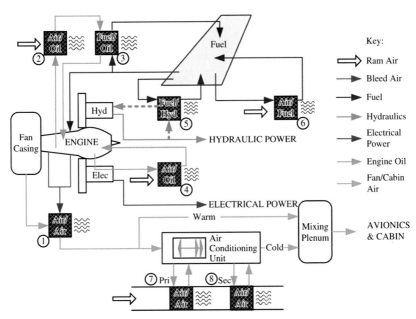

Plate 1 Typical civil systems interaction – heat exchange between system (See Figure 0.5)

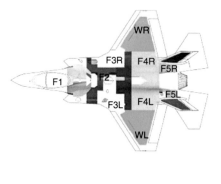

Seven tanks: Nominal 18,00lb of fuel

F1 Centre Fuselage

F2 Centre Fuselage

F3 Engina Feed Tank (F3L & F3R)

F4 Wing carry-Throught (F4L & F4R)

F5 Aft Fuselage (F5L & F5R)

WL Left Wing Box

WR Right Wing Box

Plate 2 Joint strike fighter fuel tank layout (Courtesy of BAE Systems) (See Figure 3.10)

from the engine oil coolers to the wing tanks when extreme low temperature operation might be encountered. Refer to the section on cold fuel management

- *Fuel quantity gauging*
  Fuel quantity gauging using the following sensors:

  - Linear AC capacitance fuel probes (34)
  - Level sensors – software adjustable (6)
  - Fuel compensators (2)
  - Self-calibrating densitometers (1)
  - Temperature sensors (10)

The FMQGS is an ARINC 600 LRU designed to meet the DO 160C environment. The unit contains a dual-channel microprocessor architecture hosting software to DO 178B Level B. On this system Parker Aerospace performed the role of systems integrator, taking responsibility for design and development, controlling configuration and certifying the system [4].

3.6.2 Boeing 777

The Boeing 777 in contrast uses an integrated architecture based upon A429 and A629 data buses as shown in Figure 3.25. This diagram emphasises the refuel function which is controlled via the Electrical Load Management System (ELMS) P310 Standby Power Management Panel in association with the integrated refuel panel and the Fuel Quantity Processor Unit (FQPU).

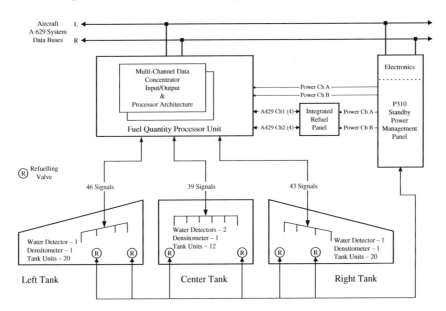

Figure 3.25 Simplified portrayal of B777 fuel gauging/fuel management (Courtesy of Smiths Group – now GE Aviation)

There are six refuelling valves, marked as R on the diagram, two in each of the left wing, centre and right wing tanks. The P310 panel provides power to the FQPU, integrated refuelling panel and controls the operation of the refuelling valves. The FQPU and refuelling panel communicate by means of dual A429 data links. The top level integration of the FQPU and ELMS P310 panel is via the aircraft systems left and right A629 data buses. This system permits the automatic refuelling of the aircraft to a preset value, as the FQPU senses the fuel tank quantities reaching their assigned value, messages are sent to the ELMS to shut off the refuelling valves until all three tanks have attained the correct level.

The function of the B777 ELMS is described in Chapter 5 – Electrical Systems. In this mode of operation the ELMS is able to power up the necessary components of the fuel system to accomplish refuelling during ground maintenance operations without the need to power the entire aircraft.

The FQPU is a multi-channel multi-processor controller which processes the fuel quantity information provided by a total of 52 tank units (probes), 4 water detectors and 3 densitometers located in the three fuel tanks. The B777 uses ultrasonic fuel probes, the first civil airliner to do so.

The ELMS, FQPU and integrated refuelling panel are supplied by Smiths Group – now GE Aviation.

3.6.3 A340-500/600 Fuel System

The aircraft fuel system has the primary task of providing fuel to the engines in order that thrust may be generated and powered flight sustained. Fuel is also used extensively as a heat sink for the aircraft hydraulic, engine oil and other systems. On some aircraft, namely those from the Airbus wide-bodied stable use the transfer of fuel fore and aft to modify the aircraft's CG position to minimise trim drag. Fuel is also moved inboard and outboard within the wing tanks to alleviate structural loading.

Fuel Tank Configuration

The Airbus A340-600 fuel tank configuration is shown in Figure 3.26. The fuel tanks include:

- Four collector cells – one for each engine
- Centre tank
- Left wing inner 1, inner 2 and outer tanks
- Right wing inner 1, inner 2 and outer tanks
- Trim tank located in the tailplane

The wing tank configuration is largely determined by the need to retain sufficient fuel to enable continued flight following a turbine disc failure in one of the engines. In this analysis the wing/engine(s) geometry together with the internal wing tank partitioning determines the contents and shape of the

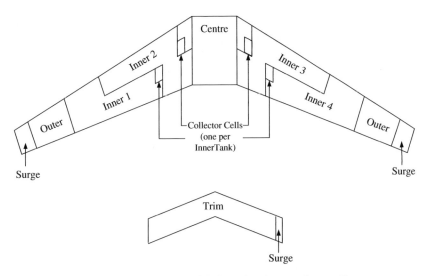

Figure 3.26   A340-600 fuel tank configuration

wing tanks and the length of time the aircraft can continue to fly following a catastrophic failure.

The A340-500 fuel systems is similar with the exception that there is provision of a Rear Centre Tank (RCT) provided at the forward end of the rear cargo hold to provide additional fuel for ultra-long-range routes.

A secondary gauging system using dissimilar (and less accurate) technology is an alternative to Magnetic Level Indicators (MLIs) – sometimes known as drip sticks.These may be used on very large aircraft such as the A380 where accessing the MLIs may be impractical. MLIs are effectively sealed float indicators that when released from the lower surface of the tank drop to indicate a calibrated scale that allow the fluid level in the fuel tank at that location to be unambiguously determined.

Fuel Transfers

Apart from fuel feed to the four engines and the APU, the A340-600 fuel system performs the following fuel transfers whose need is determined by factors external to the fuel system:

- Forward and aft fuel transfers to modify the aircraft's CG position
- Inboard and outboard transfers within the wing to reduce wing bending moments and hence structural fatigue

Forward and Aft Fuel Transfers

Figure 3.27 depicts the concept of forward and aft fuel transfers. As more fuel is transferred aft then the aircraft CG moves aft. Conversely as fuel is transferred forward so the CG moves forward. The use of fuel to modify the aircraft CG in this manner enables the aircraft to operate in a relaxed CG mode where the aircraft remains stable but the trim drag induced by the tailplane is reduced. This also has an impact upon the tailplane structure and it has been calculated that tailplane structural weight may be reduced by 1000 kg as a consequence.

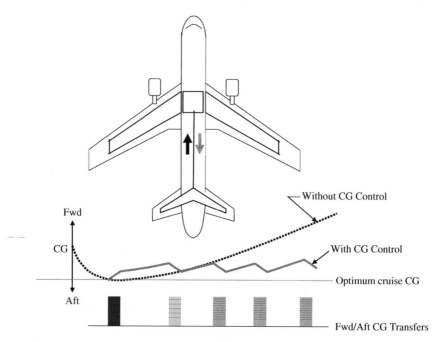

Figure 3.27 Effect of fore and aft fuel transfers on CG

The inboard engines are well forward of the outboard engine feed tanks so pose less of a problem for the fuel system following rotor burst. To accommodate an outboard engine rotor burst the inner feed tanks have a separate compartment (forward/inboard) holding about 8 tonnes of fuel. This same compartment contains the engine collector cell with 1.4 tonnes of fuel. During cruise this inner compartment is kept full via a transfer ejector which tops this compartment up from the main inner feed tank.

Following an outboard engine rotor burst which may result in penetration of the main inner feed tank; the 8 tonne compartment being outside the critical zone holds fuel good for four hours of flight.

The transfer of fuel fore and aft for CG control is the only transfer determined by factors outside the fuel system.

Figure 3.28 illustrates the CG control process that involves keeping the longitudinal CG essentially constant during cruise.

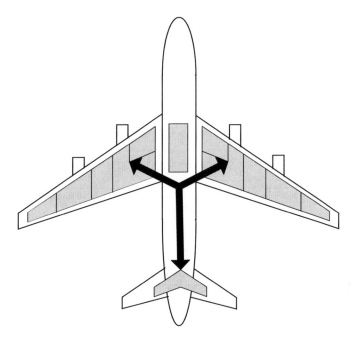

Figure 3.28 Generic forward and aft fuel transfers

Inboard and Outboard Fuel Transfers

Inboard and outboard transfers as shown in Figure 3.29 have an impact upon reducing the wing bending moment. When the aircraft is on the ground no lift forces are acting upwards and consequently the wing droops due to the weight of fuel inside. Therefore on the ground fuel is moved inboard to the centre or inner tanks to minimise this effect. Conversely when the aircraft becomes airborne the wings generate lift and the wings flex upwards. To counteract this upwards lift force, fuel is moved to the outboard tanks from the centre and inner wing tanks, thereby counteracting the effect.

For wing load alleviation, the outboard wing tanks are kept full as long as possible. Their contents are transferred to the feed tanks when all other auxiliary tank fuel has been consumed and the feed tanks have depleted to some predetermined value. This transfer is not external to the fuel system.

Fuel burn sequencing requires the centre tank to transfer to each feed tank to keep the feed tanks within 10% of full until the centre tank fuel is depleted. (While there is fuel in the centre tank all fore and aft transfers are to and from the centre tank. Subsequently, fore and aft transfers will take place between the feed tanks and the tail tank.) After the centre tank has been emptied, any lateral balance is accommodated by transferring fuel between left and right feed tanks. As already mentioned the outer wing tanks transfer their contents to the inner feed tanks near the end of the flight.

Emergency operation following loss of AC power: the cross-feed valves are opened and all engines are fed from one set of feed pumps until power is

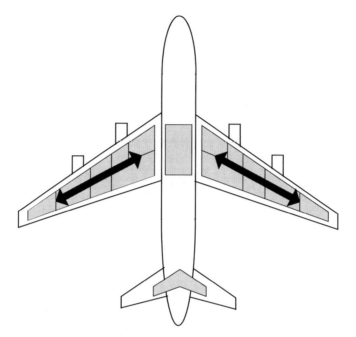

Figure 3.29   Generic inboard and outboard fuel transfers

restored. This configuration is selected automatically. Integrity of the transfer system is maintained by allowing the flight crew to monitor and manually override any transfer activities via the overhead panel.

Additional Fuel Transfers

As well as these structural load alleviation fuel transfers the aircraft fuel system also has to accommodate the normal fuel transfers:

- Engine and APU feed
- Intertank transfers
- Refuel and defuel
- During an emergency:fuel jettison

As well as this normal fuel system functionality the system also has to embody redundancy, such that the system may continue to operate safely after several failures.

Fuel System Components Complement

The outcome of all of these requirements is a fairly complex and sophisticated fuel system with an extensive array of fuel pumps, valves and fuel quantity sensors. Figure 3.20 shows the layout of the A340-600 fuel system.

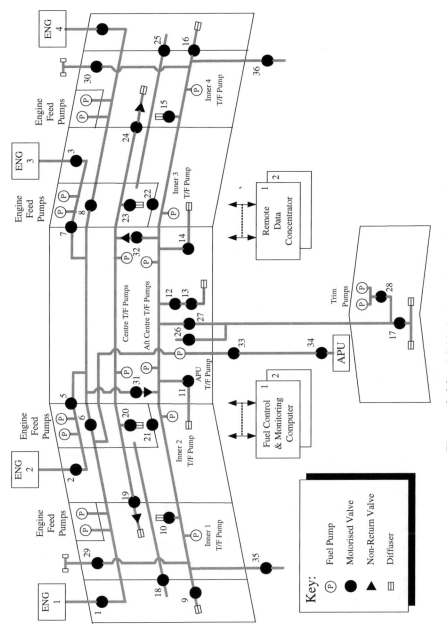

Figure 3.30 A340 fuel system schematic

The system comprises the following.

Fuel Pumps

2 boost pumps per engine.	8
2 centre transfer pumps.	2
2 aft centre transfer pumps.	2
Transfer pump – Inner 1 to Inner 4.	4
APU transfer pump.	1
Trim transfer pumps.	2
Total	19

Fuel Valves

The total fuel valve complement for the A340-600 fuel system is identified in Table 3.1. There are a total of 36 fuel valves in the system.

Fuel Quantity Sensors

In addition to the fuel pumps and valves already identified there are ~ 150 sensors in the Fuel Quantity Indication System (FQIS) plus densitometers and fuel temperature sensors.

Fuel System Control

The fuel system control is carried out by a total of four electronic controllers:

- Two Fuel Data Concentrators (FDCs)
- Two Fuel Control and Monitoring Computers (FCMCs)

Refer to Figure 3.31 which depicts a system control architecture.

The two independent FDCs provide excitation and return signal processing for the combined fuel quantity measurement and level sensing capacitive probes. Probes are segregated into two separate groups per tank and segregated such that any single failure will not cause a single failure. The wiring harnesses are a crucial element in the fuel quantity measurement system for reasons of sensitivity, electrical intrinsic safety and EMI/lightning strike reasons. For this reason the aircraft wiring harnesses are manufactured by the same company that provides the probes and the signal conditioning and control – in this case Parker Aerospace. The length of harnesses is minimised and they are segregated from all other wiring. To satisfy the minimum harness length requirement the FDCs are located in the centre fuselage adjacent to the wing break. Densitometers are included in the sensor set to take account of fuel density; the effect of which was described earlier in the chapter. Dual in-tank temperature sensors are fitted to the trim, outer and engine feed tanks to enable the flight crew to monitor fuel temperature throughout – see the section on fuel cold soak. Within the normal ground standing attitude of the aircraft the fuel quantity indication system accuracy will be ~0.4% of the full capacity at empty to ~1% at full.

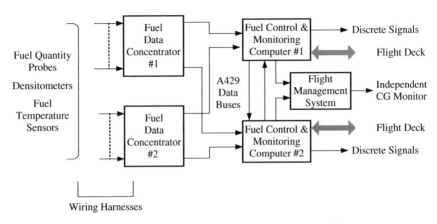

Figure 3.31 A340 fuel system control architecture

   The conditioned data from the FDCs is sent via A429 buses to the FCMCs. The FCMCs process the data to generate fuel quantity indication, fuel level indication, refuel control, CG calculations and control, and the control of other fuel transfers. Depending upon the aircraft's Zero Fuel Weight (ZFW) and the fuel on board the system will control the aircraft CG to within 2% of Mean Aerodynamic Chord (MAC). To assure the integrity of the CG control system an independent monitor is provided by the Flight Management System (FMS).
   For further detailed data refer to reference 5.

Table 3.1 A340 fuel system – fuel valve identification

1	Engine 1 LP	17	Trim Inlet	33	APU Isolation
2	Engine 2 LP	18	Left Outer Transfer	34	APU Aft Feed
3	Engine 3 LP	19	Inner 1 Transfer	35	Left Jettison
4	Engine 4 LP	20	Inner 2 Transfer	36	Right Jettison
5	Crossfeed 1	21	Transfer Control		
6	Crossfeed 2	22	Transfer Control		
7	Crossfeed 3	23	Inner 3 Transfer		
8	Crossfeed 4	24	Inner 4 Transfer		
9	Outer Inlet	25	Right Outer Transfer		
10	Inner 1 Inlet	26	Trim Forward Transfer		
11	Inner 2 Inlet	27	Trim Pipe Isolation		
12	Centre Inlet	28	Trim Isolation		
13	Centre Restrictor	29	Left Refuel Isolation		
14	Inner 3 inlet	30	Right Refuel Isolation		
15	Inner 4 Inlet	31	Defuel		
16	Outer Inlet	32	Auxiliary Refuel		

3.7 Fuel Tank Safety

Fuel tank safety based upon explosion suppression has been a significant issue in military aircraft for many years. The need to protect the airframe and fuel system from the effects of small arms fire or explosive fragments has been a consideration in battle damage alleviation during the design of many military platforms including the C-130, C-5 Galaxy, F-16, C-17, F-22 and many others. These systems use a variety of techniques:

- Reticulated foam
- Stored liquid nitrogen (C-5)
- Fire retarding additive mixing for short term protection (F-16, A-6, F-117)

OBIGGS using air separation technology was first used on C-17. Early separation technology was employed and there was a need for onboard storage of inert gas to cope with the descent case. This system was upgraded about two years ago. The F-22 uses OBIGGS and the A400M will have an OBIGGS.

Fuel tank safety embraces a number of issues relating to the electrical components and installation as well as providing an oxygen depleted environment in the ullage volume. These electrical and component issues include:

- *In-tank wiring*. The possibility of electrical energy entering the fuel tank due to normal operation, short circuits, and induced current/voltage on to fuel systems wiring that may potentially lead to ignition of flammable vapours. An earlier energy limit of 200 μJoules has been superseded by a lower limit of 20 μJoules for in-tank electrical design.* Allowable current limits are now 30 mAmps whereas previously no limits were specified. Advisory circular (AC) 25.981-1B refers [6]
- *Pump wiring*. Spark erosion and hot spots due to short circuits in the pump wiring
- *Pump dry-running*. Mechanical sparks generated due to component wear or Foreign Object Damage (FOD) inside the pumps
- *Bonding*. Electrical discharges occurring within the fuel tank due to lightning. High Intensity Radiation Fields (HIRF), static and/or fault currents
- *Adjacent systems*. Ignition sources adjacent to the fuel tanks

  - ignition of the fuel in the tank due to electrical arcing external to the fuel tank penetrating the tank wall and causing auto-ignition of the fuel due to heating of the tank wall
  - explosions within the adjacent area

- *Arc gaps*. Inadequate separation between components and structure that could allow electrical arcing due to lightning

See Airbus *FAST* magazine Issue 33 [7].

---

* Joule equals 1 watt per second; a typical car light bulb rated at 10 watts consumes 10 Joules per second. Therefore the specified energy limits in a fuel tank are ~500 000 times lower.

Particular emphasis has been placed on fuel tank safety in the commercial arena since the TWA flight 800 accident in July 1996. This occurrence was in addition to other previous ground-based incidents which had resulted in significant events leading to severe aircraft damage or loss. Extensive development and analysis illustrated that fuel tank inerting could have been effective if Air Separation Modules (ASMs), based on hollow-fibre membrane technology, could be used in an efficient way. To illustrate this, the Federal Aviation Administration (FAA), with the assistance of several aviation-oriented companies, developed an onboard inert gas generation system with ASMs that uses aircraft bleed air to generate nitrogen-enriched air at varying flow and purity (reduced oxygen concentration) during a commercial airplane flight cycle.

The FAA performed a series of ground and flight tests designed to prove the simplified inerting concept that is being proposed by the FAA. The FAA-developed system was mounted in the cargo bay of an A320 operated by Airbus for the purposes of research and development and used to inert the aircraft centre wing fuel tank during testing. The system and centre wing fuel tank were instrumented to allow for the analysis of the system performance as well as inerting capability. The FAA onboard oxygen analysis system was used to measure the oxygen concentration in the centre wing tank continuously during operation of the inerting system. Boeing undertook a trial installation on a Boeing 747 aircraft in a similar timeframe.

The results of the tests indicated that the concept of the simplified inerting system is valid and that the air separation module dynamic characteristics were as expected for the limited test plan performed. Both one and two ASM configuration tests gave the expected performance with ASM pressure having the expected effect on flow rate and the duel-flow performance being predictable. Bleed air consumption was greater than expected during the cruise phase of flight. Additional research is needed to determine what changes in system design or operational methodology would best reduce the bleed airflow and the associated cost.

3.7.1 Principles of Fuel Inerting

In April 2001 the FAA issued Special Federal Aviation Regulation (SFAR) 88, applicable to aircraft registered in the USA [8]. The JAA produced a similar document – JAA INT/POL 25/12 which was mandatory for all airbus aircraft [9]. These documents provided a methodology to categorise the hazards in fuel tanks. On a civil aircraft the main fuel tanks usually comprise left, centre and right wing tanks as shown in Figure 3.32. The centre wing fuel tank is categorised as hazardous; requiring fuel tank inerting due to the temperatures encountered and the proximity to external heat sources of which the air conditioning units represent a significant heat source. Left and right wing tanks are usually considered to be nonhazardous, primarily as the fuel contained within is much cooler and the fuel does not suffer from the proximity of hot aircraft components. Other tanks fitted to some aircraft types such as fuselage (long-range) tanks and tailplane trim tanks are similarly unaffected. It follows that aircraft without a centre tank may also avoid the need to fit an

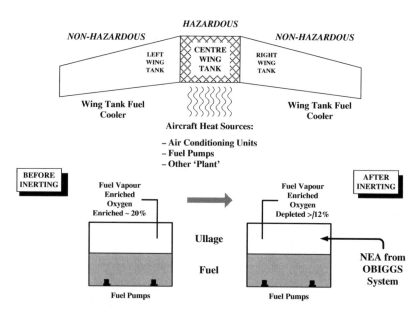

Figure 3.32 Principles of fuel tank inerting

inerting system if the hazard analysis provided that the remaining tanks meet the nonhazardous criteria.

In a normal fuel tank the air gap or ullage volume comprises a fuel-enriched vapour with a composition of around 20 % of gaseous oxygen. This mixture can provide an explosive mixture in certain circumstances when a heat source or a spark is present. The fuel tank inerting system – sometimes called a Flammability Reduction System (FRS) – receives nitrogen-enriched air from the On-Board Inert Gas Generating System (OBIGGS). This reduces the percentage of oxygen in the fuel-enriched vapour to not greater than 12 %: a figure that has been shown by experimentation to provide the limit of a safe fuel/air mixture. For military aircraft a more conservative figure of 9 % is used. The actual oxygen content achieved depends upon the ullage volume and the ability of the engines to supply bleed air and will usually be well below the 12 % limit. The most difficult design condition is during aircraft descent where the fuel tanks are almost empty (large ullage volume) and the engines are throttled back, limiting their ability to supply bleed air to the OBIGGS.

3.7.2 Air Separation Technology

The separation technology employs the use of fibre bundles contained within a cylinder as shown in Figure 3.33. There fibres are specially treated with a proprietary process that encourage those gas molecules including oxygen molecules – oxygen(O_2), carbon dioxide (CO_2) and water vapour (H_2O) to migrate towards the vent while nitrogen (N_2) is encouraged to flow straight through the unit. To facilitate this process the air is usually maintained at a temperature typically around 80 °C.

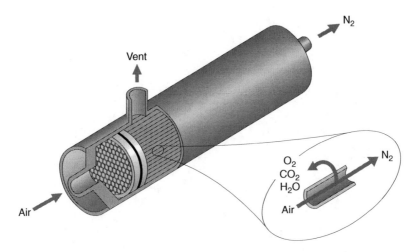

Figure 3.33 Air separation module (Courtesy of Parker Aerospace)

The aircraft installation combines several separation modules on to a pallet which is designed to add separator modules in parallel to match the capacity needs of the application. The 747 flight test demonstrator for the NGS programme had five separator modules as shown in Figure 3.34.

The production design is expected to have three separation modules implying the greater margin than expected performance margin obtained during the flight evaluation program. The 737 NGS pallet will have only one separator module as this is the smallest of the Boeing aircraft fleet.

The big unknown in OBIGGS so far is the useful life of the fibre bundles used in the separation process. Every effort is being made to control the quality of the air entering the separator modules since contaminants are known to degrade fibre performance. Measurement of oxygen concentration in an intrinsically safe manner is a current technology challenge. To date operational systems operate open loop and only by sampling the Nitrogen Enriched Air (NEA) discharged from the separator module(s) during ground maintenance can the degree of inertness be assessed.

3.7.3 Typical Fuel Inerting System

A typical fuel tank inerting system is shown in Figure 3.35.

The source of air to feed the system is bleed air extracted from the engines. After passing through a control and Shut-Off Valve (SOV) the air is fed through an air/air heat exchanger to reduce the temperature to the optimum 80 °C for the air separation module operation. After being passed through a filter to remove liquid droplets and particulates, the air enters a series of Air Separation Modules (ASMs) – usually three to five depending upon the instal- lation. The ASMs separate out the nitrogen and oxygen components within the air. Oxygen Enriched Air (OEA) is collected in a manifold and dumped overboard. Nitrogen Enriched Air (NEA) is controlled via a series of valves

Figure 3.34 Boeing 747 trial installation (Courtesy of Parker Aerospace)

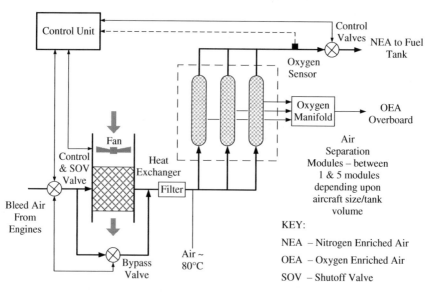

Figure 3.35 Typical fuel tank inerting system

before being fed to the tank ullage volume to reduce the oxygen content to safe levels.

The Boeing 787, which as a More-Electric aircraft has no bleed air extracted from the engines, uses a different method to provide air to the fuel tank inerting system. Air is extracted internally from the aircraft by means of a long tube that runs the length of the fuselage. This air is then compressed using an electrically powered compressor and fed through the ASMs in a similar manner to the conventional bleed air solution.

New build Boeing commercial aircraft are presently being fitted with a fuel tank inerting system or Nitrogen Generating System (NGS) and a considerable portion of existing aircraft fleets will need to be retrofitted by 2014. There are presently no plans to fit such systems to existing Airbus aircraft though some US registered Airbus aircraft may be affected.

It is understood that the A350XWB will have an OBIGGS supplying NEA to all wing tanks. This larger system will necessitate the use of between 7 and 12 ASMs; the precise number to be determined during the design and evaluation phase.

3.8 Polar Operations – Cold Fuel Management

The official opening of cross-polar routes in February 2001 marked an important new development in long-range air transport. These four new routes: Polar 1, Polar 2, Polar 3 and Polar 4 provide more direct great circle routes from the continental US to major Asian cities flying over the North Pole. These routes offer savings in time and fuel as well as having environmental advantages. Due to the inherent risks in flying such long distances over extremely inhospitable terrain, airlines have to address special considerations when embarking upon polar routes. These are:

- Regulatory guidance
- En-route alternate airports
- Cold fuel management
- Communication and navigation

The cold fuel management requirements have an impact upon the design and operation of the aircraft fuel system over and above normal operation.

3.8.1 Minimum Equipment List (MEL)

In order to operate these routes the following additions are made to the aircraft MEL:

- A fuel quantity indication system that includes a temperature indication
- For two-engine aircraft an APU that includes electrical and pneumatic (bleed air) supply to the designated capability
- An auto-throttle system
- Acceptable communications fit

In addition, extra training is required for flight crew and maintenance personnel to address operational and maintenance procedures for cold fuel management

3.8.2 Cold Fuel Characteristics

Aircraft fuel is a complex mixture of different hydro-carbons that do not all solidify at the same temperature. When fuel is cooled, an increasing proportion of wax crystals form in the fuel as certain of the constituents begin to freeze. The danger on an aircraft is that wax crystals block fuel lines and filters causing fuel starvation in the engine resulting in power fluctuations, power loss and eventually flame out. There are recognised standard procedures to measure the freezing point of fuel, the critical point is for the fuel to remain in a sufficiently fluid state so as to be able to flow or be pumped. The pour point, defined as the lowest point at which the fuel can flow, is the lowest temperature at which the fuel flows before attaining a semi-rigid state. There are several stages in this process as the fuel progressively cools, typically:

- At 3 °C above freezing the fuel will appear as a clear homogenous fluid
- At or around freezing point wax crystals will begin to form in the fuel as some of the hydro-carbons begin to solidify. The fuel effectively becomes a slush of wax crystals and liquid fuel
- At the pouring point the fuel begins to solidify and finally becomes a near solid block of wax. As the freezing point is defined as the temperature at which the last wax crystal melts, the pouring point is usually $\sim 6\,°C$ below the freezing point

Different fuels have different freezing points as shown in Table 3.2.

Table 3.2 Specified freezing point of different fuels

Fuel	Freezing point
JET A (US)	−40 °C
JP-5 (US Navy)	−46 °C
JET A1	−47 °C
RT/TS-1 (Russia)	−50 °C
JET B	−50 °C
TH (Russia)	−53 °C
JP-4 (US Air Force)	−58 °C
JP-8 (US Military)	[−47 °C]

The variability in measuring fuel freezing point means there is a reproducibility of $\sim 2.5\,°C$ in the test.

JP-8 is the military equivalent of JET A-1 with the addition of corrosion inhibitor and anti-icing additives

JET A is the fuel used predominantly in the US and also that with the highest freezing point and has attracted the most attention. Some US operators perform fuel freezing measurement tests and it has been found that there is variation of the freezing point of JET A when uplifted from different locations around the US – see Table 3.3. The airline is not necessarily allowed to accept the freezing point of the uplifted fuel, more likely the default or specification value shown in Table 3.2 will be applied. The reason is that uplifting fuel at different locations results in a blend of fuel in the tanks, each with its unique freezing point and consequently the freezing point of the fuel on board may vary considerably.

Table 3.3 Variation in JET A freezing point at selected airports

Airport location	Average freezing point (°C)	Range of freezing points (°C)
Atlanta	−43	−41.6 to −46–6
Chicago	−43	−42.4 to −44.7
Dallas – Fort Worth	−43	−41.4 to −45.9
Los Angeles	−50	−46.8 to −58.2
Miami	−47	−41.0 to −53.1
New York	−45	−44.0 to −46.4
San Francisco	−45	−44.2 to −56.1

To obviate this difficulty operational procedural may be applied as follows:

- The highest freezing point of the fuel load in the last three uplifts is applied
- Transferring fuel into the centre tank before commencing uplift may help in establishing the freezing point of the fuel uplifted for the leg. This is because it is the wing fuel temperature that is being measured; the fuselage/centre tank fuel will generally be warmer
- Transfer of fuel between tanks in flight to ensure that colder (wing) fuel is interchanged with warmer (fuselage) fuel

3.8.3 Fuel Temperature Indication

On modern aircraft the fuel temperature is measured and displayed on the EICAS or ECAM as appropriate. This display alters to an alert colour, typically amber, when the fuel drops below the cold fuel threshold and annunciates the low fuel temperature to the crew. This cold threshold is usually set at an appropriate level for JET A(−37 °C) or JET A-1 (−44 °C) though the setting may be customised if the actual freezing point of the fuel on board is known. Both Boeing and Airbus have designed software to aid the flight crews in addressing the cold fuel issue at the flight planning stage.

For further information upon this subject see references [10] and [11].

References

[1] Tookey, R. Spicer, M and Diston, D. (2002) Integrated design and analysis of an aircraft fuel system. NATO AVT Symposium on Reduction of Time and Cost through Advanced Modelling and Virtual Simulation.
[2] Smiths Industries Marketing Publication SAV 247X Issue 2.
[3] Smiths Industries Marketing Publication SIA 663.
[4] Parker Aerospace Marketing Publication GPDS9709-FMCG.
[5] *FAST* No. 26, September 2000.
[6] Advisory Circular (AC) 25.981-1B, Fuel Tank Ignition Source Prevention Guidelines, 14 April 2001.
[7] Preventing Ignition Sources inside Fuel Tanks, *FAST* No. 33, December 2003.
[8] Federal Aviation Regulation (SFAR) 88, Fuel Tank System Fault Tolerance Evaluation Requirements, 12 December 2002.
[9] JAA INT/POL 25/12.
[10] Polar Route Operations, *Boeing Aero* Magazine No. 16, October 2001.
[11] Low Fuel Temperatures, *FAST* No. 36, July 2005.

4

Hydraulic Systems

4.1 Introduction

Hydraulic systems made their appearance on aircraft in the early 1930s when the retractable undercarriage was introduced. Since that time an increasing number of tasks have been performed by the application of hydraulic power and the power demand has consequently increased greatly. Hydraulic power was seen as an efficient means of transferring power from small low energy movements in the cockpit to high energy demands in the aircraft. Hydraulic systems now have an important role to play in all modern aircraft, both military and civil.

The introduction of powered flying controls was an obvious application for hydraulic power by which the pilot was able to move the control surfaces with every increasing speeds and demands for manoeuvrability. This application brought hydraulics in the area of safety critical systems in which single failures could not be allowed to hazard the aircraft. The system developed to take account of this using multiple pumps, accumulators to store energy and methods of isolating leaks.

The hydraulic system today remains a most effective source of power for both primary and secondary flying controls, and for undercarriage, braking and anti-skid systems. However, it will become apparent later in the book that more-electric systems are being considered to replace hydraulically powered systems in some areas.

From the beginning the use of hydraulics as a means of transmitting power has not gone unchallenged. Of the various alternatives considered the chief contender has been the use of the electrical systems. The lure of the all-electric aeroplane has been a tempting prize, and numerous technical papers have evaluated the relative merits over at least the last thirty years. Hydraulics power has nevertheless maintained its position due to a unique combination of

Aircraft Systems: Mechanical, electrical, and avionics subsystems integration. 3rd Edition   Ian Moir and Allan Seabridge
© 2008 John Wiley & Sons, Ltd

desirable features, not least of which is low weight per unit power. Even with the advent of rare earth magnetic materials, the electric motor cannot yet match the power to weight ratio of a hydraulic actuator, particularly above 3 kW.

In choosing any type of system certain general characteristics, often conflicting, are sought. The principal requirements are low weight, low volume, low initial cost, high reliability and low maintenance. The latter two are the crucial constituents of low cost of ownership. Hydraulic systems meet all these requirements reasonably well, and have additional attractions. The small pipe diameters lend themselves to flexibility of installation, the use of oil as the working fluid provides a degree of lubrication, and the system overloads can be withstood without damage. Within the limits of their structural strength, actuators can stall and in some cases actually reverse direction. They will return to working condition perfectly normally on removal of the overload. Many mechanical engineers consider that these attractions make the hydraulic system more flexible and more robust than an electrical actuation system with the same power demand.

The last decade has seen the ever-accelerating introduction of microprocessors, both for monitoring system performance and to perform control functions. This has proved to be a major step forward, permitting some previous shortcomings to be overcome and opening the way to so-called 'smart' pumps and valves.

4.2 Hydraulic Circuit Design

The majority of aircraft in use today need hydraulic power for a number of tasks. Many of the functions to be performed affect the safe operation of the aircraft and must not operate incorrectly, i.e. must operate when commanded, must not operate when not commanded and must not fail totally under single failure conditions.

These requirements together with the type of aircraft, determine the design of a hydraulic system. When starting the design of any new hydraulic system the engineer must first determine the functions to be performed, and secondly he must assess their importance to flight safety. Thus a list of functions as illustrated in Figure 4.1 may appear as:

Primary flight controls: Elevators
                                     Rudders
                                     Ailerons
                                     Canards

Secondary flight controls: Flaps
                                     Slats
                                     Spoilers
                                     Airbrakes

Utility systems: Undercarriage – gear and doors
Wheelbrakes and anti-skid
Parking brake
Nosewheel steering
In-flight refuelling probe
Cargo doors
Loading ramp
Passenger stairs
Bomb bay doors
Gun purging scoop
Canopy Actuation

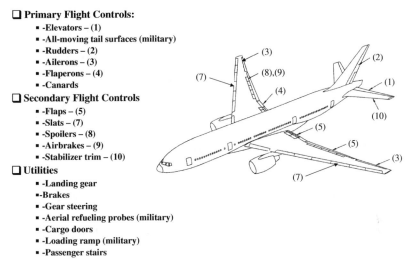

☐ **Primary Flight Controls:**
- -**Elevators – (1)**
- -**All-moving tail surfaces (military)**
- -**Rudders – (2)**
- -**Ailerons – (3)**
- -**Flaperons – (4)**
- -**Canards**

☐ **Secondary Flight Controls**
- -**Flaps – (5)**
- -**Slats – (7)**
- -**Spoilers – (8)**
- -**Airbrakes – (9)**
- -**Stabilizer trim – (10)**

☐ **Utilities**
- -**Landing gear**
- -**Brakes**
- -**Gear steering**
- -**Aerial refueling probes (military)**
- -**Cargo doors**
- -**Loading ramp (military)**
- -**Passenger stairs**

Figure 4.1 Hydraulic system loads

Many other functions are carried out on various aircraft by hydraulics, but those listed above may be used as a typical example of modern aircraft systems. The wise designer will always allow for the addition of further functions during the development of an aircraft.

From the above list the designer may conclude that all primary flight controls are critical to flight safety and consequently no single failures must be allowed to prevent, or even momentarily interrupt their operation. This does not necessarily mean that their performance cannot be allowed to degrade to some predetermined level, but that the degradation must always be controlled systematically and the pilot must be made aware of the state of the system. The same reasoning may apply to some secondary flight controls, for example, flaps and slats.

Other functions, commonly known as 'services' or 'utilities', may be considered expendable after a failure, or may needed to operate in just one direction after a positive emergency selection by the pilot. In this case the designer must provide for the emergency movement to take place in the correct direction, for example, undercarriages must go down when selected and flight refuelling probes must go out when selected. It is not essential for them to return to their previous position in an emergency, since the aircraft can land and take on fuel – both safe conditions.

Wheelbrakes tend to be a special case where power is frequently provided automatically or on selection, from three sources. One of these is a stored energy source which also allows a parking brake function to be provided.

The scope and scale of a hydraulic system must be determined by analysing the requirements of the users of the system. These users will have different demands of integrity and power depending on their application.

Some of these hydraulic demands will be continuous closed loop servo control systems, while others will be a demand to move from one position to another – a discrete or 'bang-bang' demand. All will contribute in their own way to the peak and continuous demand for hydraulic power and to the system architecture. In order to understand each requirement the following parameters need to be quantified:

- *Pressure* – What will be the primary pressure of the system? This will be determined by the appropriate standards and the technology of the system
- *Integrity* – Is the system flight safety critical or can its loss or degradation be tolerated? This determines the number of independent sources of hydraulic power that must be provided, and determines the need for a reversionary source of power
- *Flow rate* – What is the rate of the demand, in angular or linear motion per second, or in litres per second in order to achieve the desired action?
- *Duty cycle* – What is the ratio of demand for energy compared to quiescent conditions. This will be high for continuously variable demands such as primary flight control actuation on an unstable aircraft (throughout the flight), whereas it will be low for use as a source of energy for undercarriage lowering and retraction (twice per flight)
- *Emergency or reversionary use* – Are there any elements of the system that are intended to provide a source of power under emergency conditions for other power generation systems? An example of this is a hydraulic powered electrical generator. Is there a need for a source of power in the event of main engine loss to provide hydraulic power which will demand the use of reversionary devices?
- *Heat load and dissipation* – The amount of energy or heat load that the components of the system contribute to hydraulic fluid temperature

Analysis of these aspects enables decisions to be made on the number and type of components required for the complete system. These components include the following:

- A source of energy – engine, auxiliary power unit or ram air turbine
- A reservoir
- A filter to maintain clean hydraulic fluid

- A multiple redundant distribution system – pipes, valves, shut-off cocks
- Pressure and temperature sensors
- A mechanism for hydraulic oil cooling
- A means of exercising demand – actuators, motors, pumps
- A means of storing energy such as an accumulator

A simple hydraulic system is portrayed in Figure 4.2.

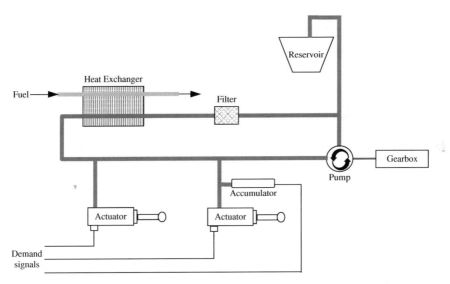

Figure 4.2   A simple hydraulic system

The primary source of power on an aircraft is the engine, and the hydraulic pump is connected to the engine gearbox. The pump causes a flow of fluid at a certain pressure, through stainless steel pipes to various actuating devices. A reservoir ensures that sufficient fluid is available under all conditions of demand.

This simple system is unlikely to satisfy the condition stated above, and in practice most aircraft contain multiple pumps and connections of pipes to ensure that single failures and leaks do not deplete the whole system of power. A more complex system, although still not adequate in practice is shown in Figure 4.3 as a simple example to describe the various components of a hydraulic system before going on to show some real-life examples.

To achieve the levels of safety described above requires at least two hydraulic circuits as shown in Figure 4.3. The degree of redundancy necessary is very largely controlled by specifications and mandatory regulations issued by the national and international bodies charged with air safety. The require-ments differ considerably between military and civil aircraft. Military aircraft frequently have two independent circuits, large civil transports and passenger aircraft invariably have three or more. In both types additional auxiliary power units and means of transferring power from one system to another are usually provided.

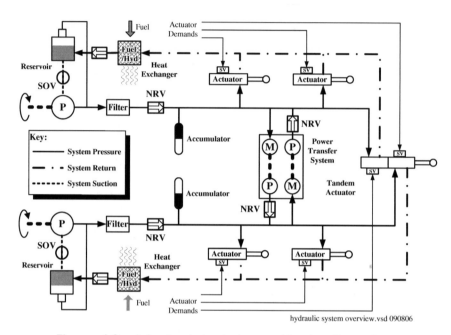

Figure 4.3 A typical dual channel hydraulic system

4.3 Hydraulic Actuation

On military aircraft the primary flight control actuator normally consists of two pistons in tandem on a common ram as illustrated in Figure 4.4.

Each piston acts within its own cylinder and is connected to a different hydraulic system. The ram is connected at a single point to a control. The philosophy is different on civil aircraft where each control surface is split into two or more independent parts. Each part has its own control actuator, each of which is connected to a different hydraulic system as shown in Figure 4.5.

The majority of actuators remain in a quiescent state, either fully extended or fully closed. They control devices which have two discrete positions, for example air brakes or in-flight refuelling probes that are either IN or OUT, or undercarriages which are either UP or DOWN. Although there is obviously a finite time during which these devices are travelling, it is usually undesirable that they should stop while in transition. They are essentially two state devices.

The actuator can be commanded to one or other of its states by a mechanical or electrical demand. This demand moves a valve that allows the hydraulic fluid at pressure to enter the actuator and move the ram in either direction.

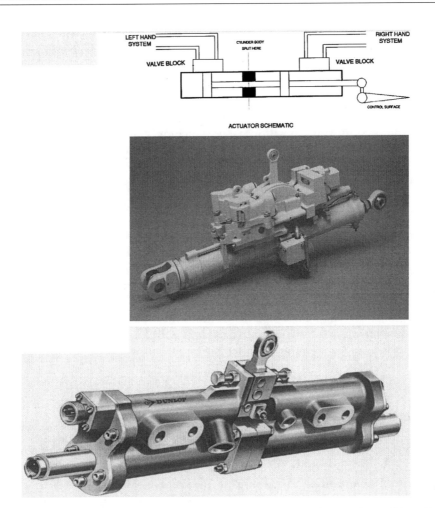

Figure 4.4 A flight control actuator (Courtesy of Claverham/Hamilton Sundstrand)

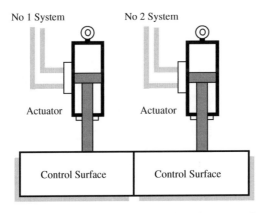

Figure 4.5 Civil aircraft control surface actuation

A mechanical system can be commanded by direct rod, lever or cable connection from a pilot control lever to the actuator. An electrical system can be connected by means of a solenoid or motor that is operated by a pilot or by a computer output.

In some instances it is necessary to signal the position of the actuator, and hence the device it moves, back to the pilot. This can be achieved by connecting a continuous position sensor such as a potentiometer, or by using microswitches at each end of travel to power a lamp or magnetic indicator. Some devices, however, are not simply two state, but are continuously variable. Examples are active primary flight control surfaces or engine reheat nozzles. These devices need to be variable and are usually controlled electrically by computers, which drive torque motors or stepper motors connected to a variable valve on the actuator. This allows the actuator to be driven to any point in its range, stopped, advanced or reversed as often and as rapidly as required.

A continuous position sensor connected into the computer servo loop allows the computer to drive the actuator accurately in accordance with the demands of the control system. Like the computers driving the actuator, the motors and position sensors must be multiple redundant. In the case of a quadruplex flight control system, an actuator will be equipped with four torque motors and four position sensors, each connected to a different computer (refer to Chapter 1 – Flight Control).

4.4 Hydraulic Fluid

The working fluid will be considered as a physical medium for transmitting power, and the conditions under which it is expected to work, for example maximum temperature and maximum flow rate are described.

Safety regulations bring about some differences between military and civil aircraft fluids. With very few exceptions modern military aircraft have, until recently, operated exclusively on a mineral based fluid known variously as:

- DTD 585   in the UK
- MIL-H-5606   in the USA
- AIR 320   in France
- H 515   NATO

This fluid has many advantages. It is freely available throughout the world, reasonably priced, and has a low rate of change of viscosity with respect to temperature compared to other fluids. Unfortunately, being a petroleum based fluid, it is flammable and is limited to a working temperature of about 130 °C. One of the rare departures from DTD 585 was made to overcome this upper temperature limit. This led to the use of DP 47, known also as Silcodyne, in the ill-fated TSR2.

Since the Vietnam War much industry research has been directed to the task of finding a fluid with reduced flammability, hence improving aircraft safety following accident or damage, particularly battle damage in combat aircraft. This work has resulted in the introduction of MIL-H-83282, an entirely synthetic fluid, now adopted for all US Navy aircraft. It is miscible with DTD 5858 and, although slightly more viscous below 20 °C, it compares well enough.

In real terms the designer of military aircraft hydraulic systems has little or no choice of fluid since defence ministries of the purchasing nations will specify the fluid to be used for their particular project. Most specifications now ask for systems to be compatible with both DTD 585 and MIL-H-5606.

Commercial aircraft make use of phosphate ester fluids which are fire resistant, e.g.

- Solutia Skydrol LD-4, Skydrol 500B-4 or Skydrol 5
- Exxon Type IV HJ4AP or Type V HJ5MP

These fluids are not fireproof – there are certain combinations of fluid spray and hot surfaces which will allow them to ignite and burn. Industry standard tests are conducted to demonstrate a level of confidence that ignition or fire will not occur and the hydraulic system design is influenced by these test results.

4.5 Fluid Pressure

Similarly little choice is available with respect to working pressure. Systems have become standardised at 3000 psi or 4000 psi. These have been chosen to keep weight to a minimum, while staying within the body of experience built up for pumping and containing the fluid. Many studies have been undertaken by industry to raise the standard working pressure. Pressure targets have varied from 5000 psi to 8000 psi, and all resulting systems studies claim to show reduced system component mass and volume. Interestingly DTD 585 cannot be used above 5000 psi because of shear breakdown within the fluid.

A detailed study would show that the optimum pressure will differ for every aircraft design. This is obviously impractical and would preclude the common use of well-proven components and test equipment.

4.6 Fluid Temperature

With fast jet aircraft capable of sustained operation above Mach 1, there are advantages in operating the system at high temperatures, but this is limited by the fluid used. For many years the use of DTD 585 has limited temperatures to about 130 °C, and components and seals have been qualified accordingly. The use of MIL-H-83282 has raised this limit to 200 °C and many other fluids have been used from time to time, for example on Concorde and TSR2, to allow high temperature systems to be used.

A disadvantage to operating at high temperatures is that phosphate ester based fluids can degrade as a result of hydrolysis and oxidation. As temperature increases, so the viscosity of the fluid falls. At some point lubricity will be reduced to the extent that connected actuators and motors may be damaged.

4.7 Fluid Flow Rate

Determination of the flow rate is a more difficult problem. When the nominal system pressure is chosen it must be remembered that this is, in effect, a stall pressure. That is to say, that apart from some very low quiescent leakage, no flow will be present in the circuit. The designer must allocate some realistic pressure drop that can be achieved in full flow conditions from pump outlet to reservoir. This is usually about 20–25% of nominal pressure.

Having established this, the pressure drop across each actuator will be known. The aerodynamic loads and flight control laws will determine the piston area and rate of movement. The designer must then decide which actuators will be required to act simultaneously and at what speed they will move. The sum of these will give the maximum flow rate demanded of the system. It is important also to know at what part of the flight this demand takes place.

It is normal to represent the flow demands at various phases of the flight – take-off, cruise etc. – graphically. The maximum flow rate does not necessarily size the pump to be used. It is frequently found that the flow required on approach provides the design case, when the engine rpm, and hence pump rpm, are low.

It may be found that the absolute maximum flow demand is of very short duration, involving very small volumes of oil at very high velocities. In this case sizing a pump to meet this demand may not be justified. An accumulator can be used to augment the flow available, but care must be taken. An accumulator contains a compressed gas cylinder, and the gas is used to provide energy to augment system pressure. Therefore, the fluid volume and pressure available will depend on the gas temperature. In a situation where the flow demanded will exceed the pump capabilities the system pressure is controlled by the accumulator, not the pump. This case will influence the circuit pressure drop calculations if the necessary pressure across the actuator piston is to be maintained.

The frequency of maximum demand must also be known, and time must be available for the pump to recharge the accumulator if it is not eventually to empty by repeated use.

4.8 Hydraulic Piping

When the system architecture is defined for all aircraft systems using hydraulic power, then it is possible to design the pipe layout in the aircraft. This layout will take into account the need to separate pipes to avoid common mode failures as a result of accidental damage or the effect of battle damage in a military aircraft. Once this layout has been obtained it is possible to measure the lengths of pipe and to calculate the flow rate in each section and branch of pipe. It is likely that the first attempts to define a layout will result in straight lines only, but this is adequate for a reasonably accurate initial calculation.

If an allowable pressure drop of 25% has been selected throughout the system, this may now be further divided between pressure pipes, return pipes and components. The designer will eventually control the specifications for the components, and in this sense he can allocate any value he chooses for pressure drop across each component. It must be appreciated, however, that these values must eventually be achieved without excessive penalties, being incurred by over-large porting or body sizes.

Once pipe lengths, flow rates and permissible pressure drops are known, pipe diameters can be calculated using the normal expression governing friction flow in pipes. It is normal to assume a fluid temperature of $0\,^{\circ}\mathrm{C}$ for calculations, and in most cases flow in aircraft hydraulic systems is turbulent. Pressure losses in the system piping can be significant and care should be taken to determine accurately pipe diameters. Theoretical sizes will be modified by the need to use standard pipe ranges, and this must be taken into account.

4.9 Hydraulic Pumps

A system will contain one or more hydraulic pumps depending on the type of aircraft and the conclusions reached after a thorough safety analysis and the consequent need for redundancy of hydraulic supply to the aircraft systems.

The pump is normally mounted on an engine-driven gearbox. In civil applications the pump is mounted on an accessory gearbox mounted on the engine casing. For military applications the pump is mounted on an Aircraft Mounted Accessory Drive (AMAD) mounted on the airframe. The pump speed is therefore directly related to engine speed, and must therefore be capable of working over a wide speed range. The degree of gearing between the pump and the engine varies between engine types, and is chosen from a specified range of preferred values. A typical maximum continuous speed for a modern military aircraft is 6000 rpm, but this is largely influenced by pump size, the smallest pumps running fastest.

The universally used pump type is known as variable delivery, constant pressure. Demand on the pump tends to be continuous throughout a flight, but frequently varying in magnitude. This type of pump makes it possible to meet this sort of demand pattern without too much wastage of power. Within the flow capabilities of these pumps the pressure can be maintained within 5% of nominal except during the short transitional stages from low flow to high flow. This also helps to optimise the overall efficiency of the system. A characteristic curve for a nominally constant pressure pump is shown in Figure 4.6.

The pumps are designed to sense outlet pressure and feed back this signal to a plate carrying the reciprocating pistons. The plate is free to move at an

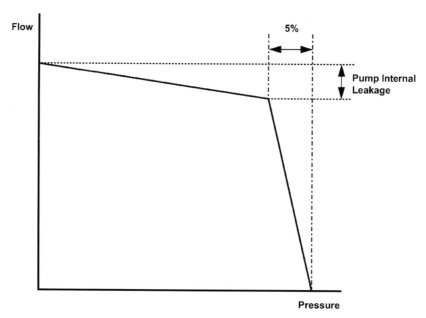

Figure 4.6 Characteristic curve for a 'constant pressure' pump

angle to the longitudinal axis of the rotating drive shaft. There are normally nine pistons arranged diametrically around the plate. The position of the plate therefore varies the amount of reciprocating movement of each piston.

Examples of different types of hydraulic pump are shown in Figures 4.7 and 4.8 together with their salient characteristics.

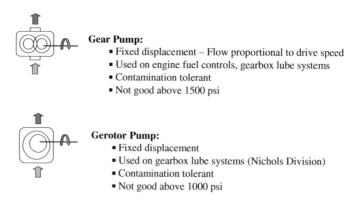

Figure 4.7 Examples of hydraulic pump technology

A more detailed diagrammatic representation of a variable displacement piston pump showing the working principle is shown in Figure 4.9. This is the preferred type of pump in use today and some commercial examples of

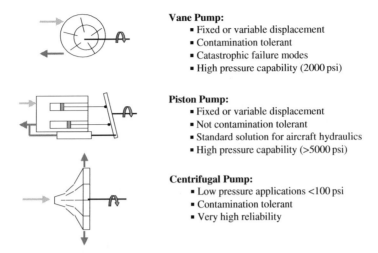

Vane Pump:
- Fixed or variable displacement
- Contamination tolerant
- Catastrophic failure modes
- High pressure capability (2000 psi)

Piston Pump:
- Fixed or variable displacement
- Not contamination tolerant
- Standard solution for aircraft hydraulics
- High pressure capability (>5000 psi)

Centrifugal Pump:
- Low pressure applications <100 psi
- Contamination tolerant
- Very high reliability

Figure 4.8   Examples of hydraulic pump technology

hydraulic pumps can be seen in Figure 4.10. A cross-section drawing of a piston pump is shown in Figure 4.10, together with an example pump.

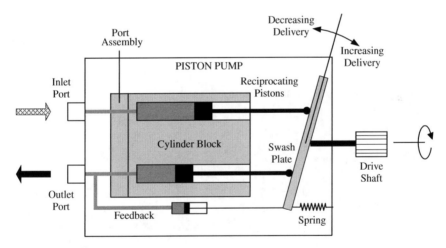

Figure 4.9   Working principle of a piston pump

When the plate is at 90° to the linear axis, there is no linear displacement of the pistons. Up to its maximum limit the plate will move to displace the volume needed to maintain nominal system pressure. When flow demands beyond maximum displacement are made the system pressure drops rapidly to zero. For short periods pressure can be maintained by means of an accumulator as described above. An example of an accumulator used in the Challenger and RJ Series aircraft can be seen in Figure 4.11. Also shown is a bootstrap reservoir from the Gulfstream V aircraft.

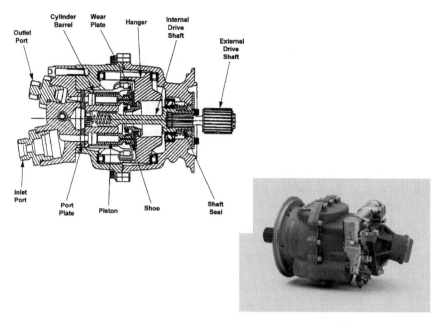

Figure 4.10   Cross section and external views of a piston pump

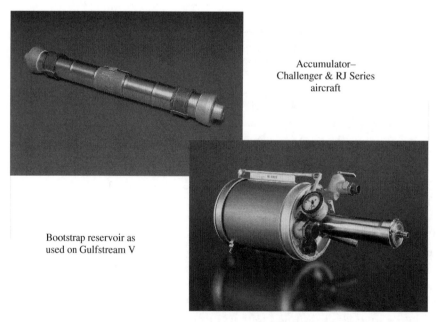

Figure 4.11   Examples of a hydraulic accumulator and a bootstrap reservoir

4.10 Fluid Conditioning

Under normal working conditions hydraulic fluid needs cooling and cleaning. Occasionally it is necessary to de-aerate by the connection of ground equipment, although increasingly modern systems are being produced with devices to bleed off any air accumulating in the reservoir.

For cooling purposes the fuel/hydraulic heat exchanger is used. This ensures that cooling on the ground is available. Further air/fluid cooling may be provided once the aircraft is in flight. Since heat exchangers are low pressure devices they are normally situated in the return line to the actuator/service.

When a pump is running off load, all the heat generated by its inefficiencies is carried away by the pump case drain line. The heat exchanger should therefore be positioned to cool this flow before its entry into the reservoir. Care must be taken to determine the maximum pressure experienced by the heat exchanger and to ensure that, not only is adequate strength present to prevent external burst, but in addition no failure occurs across the matrix between fuel and hydraulic fluid.

The introduction of servo-valves with very fine clearances emphasised the need for very clean fluids. The filter manufacturers responded to this by developing filter elements made of resin bonded paper supported by arrangements of metal tubes and wire mesh. This produces filter elements of high strength capable of withstanding differential pressures of one and a half times the system pressure.

These filters are capable, under carefully designed test conditions, of stopping all particles of contaminant above five microns in size, and a high percentage of particles below this size. This characteristic has led to filter elements becoming known by an absolute rating, the two examples above being five micron absolute.

More recent work is based on the ratio of particles upstream and downstream of the filter unit. This is referred to as the 'beta' rating. When specifying and choosing filter elements it is most important to specify the test method to be used.

Several standards exist defining the cleanliness of the fluid and these are based on a number of particles in the series of size ranges. Typically these are: 5–15 microns, 15–25 microns, 25–50 microns, 50–100 microns and above 100 microns, to be found in 100 ml of liquid. Unfortunately there is no way of calculating the relationship between the element's absolute rating and the desired cleanliness level. The choice of elements rests entirely on past experience and test results. In most cases it has been found that an adequate level of cleanliness can be achieved and maintained by the use of a 5 micron absolute return line filter in combination with a 15 micron pressure line filter. This combination also gives acceptable element life. Filters are not used in the pump inlet line. Figure 4.12 shows various filter units.

A further consequence of the demand for clean fluid has been a need for a means of measuring the cleanliness levels achieved. Electronic automatic counters are now available that are capable of providing rapid counts with a repeatability to within 5% in a form suitable for rapid interrogation by ground servicing crews.

Figure 4.12 Some typical filter units (Courtesy of Claverham/Hamilton Sundstrand)

4.11 Hydraulic Reservoir

The requirements for this component vary depending on the type of aircraft involved. For most military aircraft the reservoir must be fully aerobatic. This means that the fluid must be fully contained, with no air/fluid interfaces, and a supply of fluid must be maintained in all aircraft attitudes and g conditions. In order to achieve a good volumetric efficiency from the pump, reservoir pressure must be sufficient to accelerate a full charge of fluid into each cylinder while it is open to the inlet port.

The need to meet pump response times may double the pressure required for stabilised flow conditions.

The volume of the reservoir is controlled by national specifications and includes all differential volumes in the system, allowance for thermal expansion and a generous emergency margin.

It is common practice to isolate certain parts of the system when the reservoir level falls below a predetermined point. This is an attempt to isolate leaks within the system and to provide further protection for flight safety critical subsystems. The cut-off point must ensure sufficient volume for the remaining systems under all conditions. The reservoir will be protected by a pressure relief valve which can dump fluid overboard.

4.12 Warnings and Status

Several instruments are normally situated in the hydraulic power generation system to monitor continuously its performance. Pressure transducers monitor system pressure and transmit this signal to gauges in the cockpit. Pressure switches are also incorporated to provide a warning of low pressure in the system on the central warning panel. Filter blockage indicators show the condition of the filter elements to ground servicing personnel, and a fluid temperature warning may be given to the aircrew. With increasing use of microprocessor based system management units, more in-depth health monitoring of all major components is possible with data displayed to ground crews on a maintenance data panel.

4.13 Emergency Power Sources

All hydraulic systems have some form of emergency power source. In its simplest form this will be an accumulator. It is mandatory for wheel-brake systems to have a standby accumulator capable of supplying power for a predetermined number of brake applications when all other sources of power are inoperative. Cockpit canopies are frequently opened and closed hydraulically and emergency opening can be achieved by the use of accumulator stored energy.

Accumulators may also be used to provide sufficient flight control actuator movement to recover the aircraft to straight and level flight so that the crew can eject safely in the event of total systems failure.

To supply emergency power for longer periods an electric motor driven pump may be provided. Battery size and weight are the main limitations in this case, and to minimise these factors, the flow available is usually kept as low as possible to operate only those devices considered indispensable. Frequently it is also possible to operate at some pressure below nominal system pressure, even so it is unlikely that an acceptable installation can be achieved which will provide power for more than five or six minutes.

Weight may be kept to a minimum by the use of a one-shot battery. This allows the latest battery technology to be exploited without any concessions being made to obtain recharge capabilities. Selection will be automatic from a pressure switch with additional cockpit selection also being available.

For continuous emergency supply a Ram Air Turbine (RAT) may be used. This carries with it several disadvantages. Space must be found to stow the turbine and carriage assembly, a small accumulator is needed to deploy the turbine in emergency, and because speed governing and blade feathering are employed the assembly is complicated. Hydraulic pumps and/or emergency electrical generators can be mounted immediately behind the turbine on the same shaft. It is, however, more common to mount them at the bottom of the carriage arm close to the deployment hinge axis. This involves the use of drive shafts and gears. To keep the turbine blade swept diameter at a reasonable figure, the power developed must be kept low and it may be difficult to mount the assembly on the airframe so that the airflow is not impeded by the fuselage at peculiar aircraft attitudes. Deployment of the RAT is as for the electric motor-driven pump

In spite of these drawbacks, ram air turbines have several times proved their worth, particularly on civil aircraft, providing the only means of hydraulic power until an emergency has been dealt with and the aircraft has been recovered to a safe attitude.

In some cases a wind-milling engine may in certain circumstances provide sufficient energy to power an emergency generator.However, as the rates of rotation are relatively low, $\sim 18\%$ for a military turbojet and $\sim 8\%$ rpm for a large turbofan, then special measures and generation techniques need to be taken to extract useable electric power under these situations. Nevertheless, the F/A-18 cyclo-converter already in service provides this capability and more-electric technologies are being developed to provide useable power for civil applications in terms of fan-shaft driven generation.

4.14 Proof of Design

All the effort put into designing an hydraulic system culminates in the testing to prove that the design works in the required way. All the systems in an aircraft must be qualified before the aircraft is approved for flight. The qualification is built up through a series of steps starting with demonstrations that each individual component meets its specification. This will include proof and burst pressure tests, fatigue, vibration, acceleration and functional tests. These may be complemented by accelerated life tests.

Satisfactory completion of the tests is formalised in a Declaration of Design and Performance Certificate signed by the specialist company responsible for design and manufacture of the component, and by the company designing the aircraft.

The entire hydraulic system is then built up into a test rig. The rig consists of a steel structure representing the aircraft into which the hydraulic piping and all components are mounted in their correct relationship to each other. The pipes will be the correct diameter, shape and length. Flight standard pumps will provide the correct flows and pressures. The rig will incorporate loading devices to simulate aerodynamic and other loads on the undercarriage and other surface actuators. Strain gauging and other load techniques are used to measure forces and stresses as required. It is normal to 'fly' the rig for several hundred hours in advance of actual flight hours on the prototype aircraft.

Ultimately, before a customer accepts the aircraft into service, the hydraulic system can be declared fully qualified on the basis of the evidence obtained from the rig plus flight testing.

The cost and effort involved is considerable, but a well-designed and operated hydraulic test rig is crucial to the process of formal qualification and certification of the aircraft. A typical test rig is shown in Figure 4.13.

The process of testing the hydraulic system is illustrated in Figure 4.14. The suppliers of hydraulic system components will design their equipment to meet specifications issued by hydraulic engineers. The components will be tested at individual component level or assembled subsystems. This includes testing to meet environmental conditions (see Chapter 13). Following these tests, equipment and subsystems will be assembled on a hydraulic test rig which allows testing to be conducted at realistic system pump pressures. If models have been used up to this point, then rig test results will be used to validate the model results.

The next stage is to combine all the equipment on to an 'iron bird' rig which will make use of actuators and landing gear loaded to simulate realistic flight conditions. This type of rig represents the most faithful representation of the system short of the aircraft itself, and it is safer and more economical to operate. This will enable test results to be obtained at realistic pressures, loads, rates and endurance. Evidence gained from these tests is used to gain approval to start aircraft testing. This approach is in common with other major aircraft subsystems.

Figure 4.13 Hydraulic systems rig (Courtesy of BAE Systems)

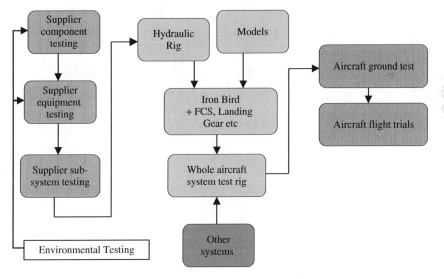

Figure 4.14 Hydraulic systems testing process

4.15 Aircraft System Applications

Since the range of hydraulic system design is dependent on the type of aircraft, it would not be sensible to give a single example. The following applications cover a range of single and multiple engine aircraft of both civil and military types.

4.15.1 The Avro RJ Hydraulic System

The Avro RJ family consists of the RJ70, RJ85 and RJ100 aircraft seating from 70 to 128 passengers. The RJ is a four-engine regional jet airliner designed for worldwide operations. Its hydraulic system has been designed to combine the lightness and simplicity of a two-engine design with the backup levels associated with a four-engine system.

Two independent systems each operate at a nominal 3000 psi. Hydraulic system controls and annunciations are located on the pilot's overhead panel. An amber caption on the master warning panel, plus a single audio chime draws attention to fault warnings on the overhead panel. Figure 4.15 shows the RJ family hydraulic system schematic.

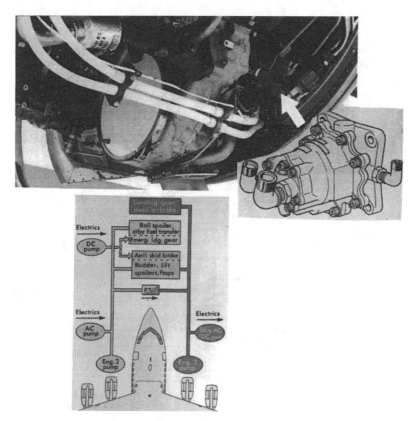

Figure 4.15   BAE Systems 146 regional jet hydraulic system (Courtesy of BAE Systems)

The systems are designated Yellow and Green and are normally pressurised by a self regulating engine driven pump on the inboard engines. Each system has an independent hydraulic reservoir, pressurised by regulated air bleed from its respective engine. Flareless pipe couplings with swaged fittings are used throughout for reliability and ease of repair.

Yellow and Green systems are geographically segregated as far as possible. The Yellow system is on the left of the aircraft and the Green on the right. Backup power for the Yellow system is provided by an AC electric pump, and backup for power for the Green system is provided by a power transfer unit (PTU) driven by the Yellow system.

An electrically operated DC pump, fed from a segregated hydraulic supply, provides emergency lowering of the landing gear and operation of the brakes in the event of failures in both the Yellow and Green systems.

The AC pump, PTU, hydraulic reservoirs etc, is housed in a pressurised and vented hydraulic equipment bay and are fully protected from foreign object damage.

The primary power generation components of the Yellow system are:

- Engine Driven Pump (EDP) on No. 2 engine
- Standby AC powered hydraulic pump
- Emergency DC powered hydraulic pump
- Accumulator
- Reservoir

All these components, except for the EDP, are located in the hydraulics equipment bay. The components are shown in Figure 4.16.

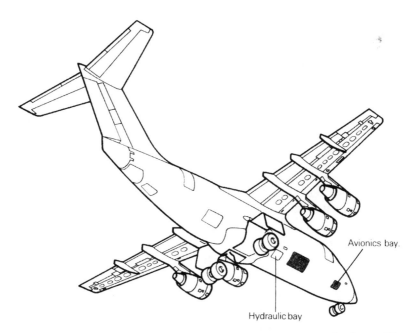

Figure 4.16 Yellow system components in the hydraulic bay (Courtesy of BAE Systems)

Yellow Hydraulic System

The Yellow system powers the following services:

- 1 flap motor
- Flap asymmetry brakes
- Roll spoilers
- 2 lift spoilers (inner spoilers on the left and right wing)
- 1 rudder servo control
- Standby fuel pumps (left and right)
- Landing gear emergency lock down
- Wheel brakes including park brake
- Airstairs through the AC pump
- Power transfer unit (PTU)

Yellow System Standby AC Pump

In the event of an EDP failure, the Yellow system is supported by a standby AC pump. The pump is continuously rated and is capable of maintaining the system pressure at 3000 psi. The AC pump is controlled by a three-position switch on the hydraulics overhead panel on the flight deck. This panel also includes the amber pump high temperature and failure annunciators.

The pump may be selected ON or OFF manually, but normally operates in automatic mode. In this mode a pressure switch in the Yellow and Green systems switches and latches the pump ON if the delivery pressure of either EDP falls below 1500 psi. The standby pump therefore supports the Yellow system directly and the Green system indirectly via the PTU.

Yellow System Backup DC Pump

In the event of a failure of both Yellow and Green systems the DC backup pump provides emergency lock-down of the main landing gear and operation of the Yellow system wheel brakes. On the ground it can provide brake pressure in the Yellow system for parking, starting or towing.

The system has its own DC powered hydraulic pump, fluid supply and an accumulator. The DC pump is controlled from the hydraulics overhead panel on the flight deck and is supplied from the emergency DC bus-bar. Hydraulic fluid is supplied from a segregated reservoir in the Yellow tank system.

The Yellow system accumulator is connected to the Yellow system wheel brakes and is protected from all other services by non-return valves. The accumulator stabilises the system and assists the DC pump. The accumulator is pressurised by the Yellow EDP, AC pump or DC pump.

Yellow System Reservoir

A 15.5 litre reservoir is provided for the Yellow system. It is pressurised by bleed air regulated to 50 psi from the engine HP compressor. The reservoir incorporates the following:

- A pressure gauge
- A sight glass
- An air low pressure switch
- Inward and outward relief valves
- A bursting disc to protect against manual failure of the outward relief valve
- A ground charge connection and manual pressure release lever
- A contents transmitter

Indications of tank contents are provided on the flight deck overhead panel that also includes amber low quantity and high temperature annunciators.

Engine Driven Pump

The Yellow system Engine Driven Pump (EDP) is mounted on the left inner engine auxiliary gearbox at the bottom of the engine to ensure easy maintenance access. The EDP has an associated motorised isolation valve. When the valve is closed it isolates the pump from the tank and provides an idling circuit to offload the pump. If the engine fire handle is pulled to its fullest extent the valve closes automatically, preventing more fluid reaching the pump.

A two-position switch on the overhead hydraulic panel controls the position of the EDP isolation valve. An amber annunciator on the overhead panel illuminates when the valve is travelling and remains ON until it reaches the selected position. The EDP also has an associated relief valve which opens to allow excess pressure back to the tank at 3500 psi.

Green Hydraulic System

The primary power generation components of the Green system are:

- Engine Driven Pump (EDP) on No. 3 engine
- Power Transfer Unit (PTU)
- Hydraulic reservoir
- Accumulator

All components, except for the EDP, are located in the hydraulic equipment bay. The Green system power the following:

- 1 flap motor
- 4 lift spoilers (centre and outer spoilers on the left and right wing)
- Airbrakes
- Landing gear – normal

- Nose gear steering
- Wheel brakes excluding park brake

Green System Standby PTU

The Power transfer unit (PTU) is an alternative power source for the Green system. The PTU is a back-to-back hydraulic motor and pump. It can support all Green system services except for the standby AC/DC generator. The motor is powered by the Yellow system pressure and is connected by a drive shaft to a pump in the Green system. The PTU is controlled from the hydraulics overhead panel by a two-position switch. When the switch is in the ON position it is automatically activated if Green system pressure falls below 2600 psi.

With the switch in the OFF position, the motor is isolated from the Yellow system by a motorised valve. Movement of the valve is indicated by an amber PTU VALVE annunciator on the flight deck hydraulics panel. The PTU may also be used during ground servicing to pressurise the Green hydraulic system, provided the hydraulic reservoir is fully charged with air.

Green System Standby AC/DC Generator

The Green hydraulic system can support the electrical system in the event of low electrical power. A standby AC/DC generator, driven by a hydraulic motor is powered by the Green system and is controlled by a three-position switch on the flight deck overhead electrical panel. The generator can be selected ON or OFF manually but is usually in automatic standby (ARM) mode. The generator is normally isolated from the system pressure by a solenoid operated selector valve.

When the standby AC/DC pump is operating its selector valve is opened, and at the same time Green system services are isolated by their shut-off valve. Green system services are therefore not available while the generator is operating and the Green system LO PRESS annunciator is indicated by a white light on the overhead electrical panel.

Green System Reservoir

The Green system reservoir has the same capacity as the Yellow system and is charged with bleed air from No. 3 engine. Its features are exactly the same as the Yellow system reservoir.

Accumulator

The Green system accumulator is identical to the Yellow system accumulator. It maintains stability in the Green systems during operation of the PTU and also assists the EDP for initial run-up of the standby AC/DC generator.

4.15.2 The BAE SYSTEMS Hawk 200 Hydraulic System

The BAE SYSTEMS Hawk 200 is a single-engine, single-seat multi-role attack aircraft in which the hydraulic power is provided by two independent systems. Both power the flying controls by means of tandem actuators at the ailerons and tailplane. The number 1 system provides power to the rudder, which can also be manually operated.

The number 1 system also provides power for utility services such as flaps, airbrakes, landing gear and wheel brakes. The number 2 system is dedicated to the operation of the flying control surfaces. In the event of engine or hydraulic pump failure, a ram air turbine driven pump automatically extends from the top rear fuselage into the airstream. This powers the flying control system down to landing speed.

A pressurised nitrogen accumulator is provided to operate the flaps and landing gear in an emergency, and wheel brake pressure is maintained by a separate accumulator. The Hawk 200 hydraulic system is shown in Figure 4.17 and the ram air turbine is shown in Figure 4.18.

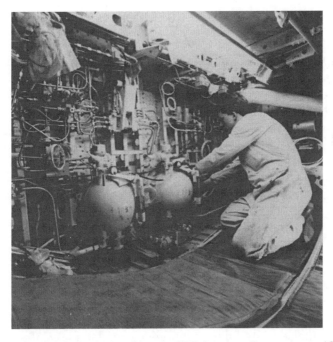

Figure 4.17 The BAE Systems Hawk 200 hydraulic system (Courtesy of BAE Systems)

4.15.3 Tornado Hydraulic System

The Tornado is a twin-engine, two-seat, high-performance aircraft designed for ground attack as the IDS version, or for air defence as the ADV version.

Figure 4.18 The BAE Systems Hawk 200 ram air turbine extended (Courtesy of BAE Systems)

Its hydraulic system is a 4000 psi fully duplicated system shown diagramatically in Figure 4.19. The high operating pressure allows the use of small diameter piping, and the system is low weight despite the duplicated pipe routings required for battle damage tolerance. The two pumps are mounted on the engine gearboxes and incorporate depressurising valves. During engine start the hydraulic system is depressurised to reduce engine power offtake to allow rapid engine starting. A cross-drive is provided between the two RB 199 engines, which allows either engine to power both hydraulic pumps should one engine fail.

The pumps are driven by two independent accessory drive gearboxes or AMADs, one connected by a power offtake shaft to the right-hand engine, and the other similarly connected to the left-hand engine. This allows the hydraulic pumps, together with the fuel pumps and independent drive generators, to be mounted on the airframe and separated from the engine by a firewall. This means that the Tornado hydraulic system is completely contained within the airframe. Not only is this a safety improvement, but it also improves engine change time, since the engine can be removed without the need to disconnect hydraulic pipe couplings.

The engine intake ramp, taileron, wing-sweep, flap and slat actuators are all fed from both systems. Should any part of the utility system become damaged, isolating valves operate to give priority to the primary control actuators.

The undercarriage is powered by the number 2 system and in the event of a failure the gear can be lowered by means of an emergency nitrogen bottle. A hand pump is provided to charge the brake and canopy actuators.

Skin mounted pressure and contents gauges are provided adjacent to the charging points and all filters are hand tightened.

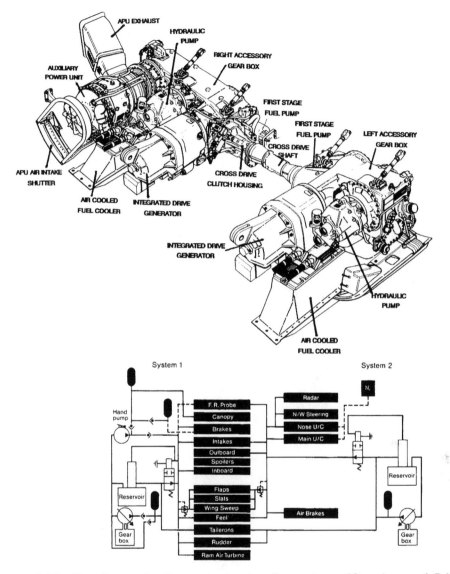

Figure 4.19 The Panavia Tornado hydraulic system (Courtesy of BAE Systems)

4.16 Civil Transport Comparison

The use of 3000 psi hydraulics systems in civil transports is widespread and the Avro RJ systems have been described in depth. However, as a way of examining different philosophies a comparison is made between an Airbus narrow body – the A320 family and a Boeing wide body – the B767. It is usual for three independent hydraulic systems to be employed, since the hydraulic power is needed for flight control system actuation. Hydraulic power is produced by pumps driven by one of the following methods of motive power:

- Engine driven
- Electrically driven
- Air turbine/bleed air driven
- Ram air turbine driven

4.16.1 Airbus A320

The aircraft is equipped with three continuously operating hydraulic systems called Blue, Green and Yellow. Each system has its own hydraulic reservoir as a source of hydraulic fluid.

- The Green system (System 1) is pressurised by an Engine Driven Pump (EDP) located on No. 1 engine which may deliver 37 gallon per minute (US gpm) or 140 L/min
- The Blue system (System 2) is pressurised by an electric motor-driven pump capable of delivering 6.1 gpm or 23 L/min. A Ram Air Turbine (RAT) can provide up to 20.6 gpm or 78 L/min at 2175 psi in emergency conditions
- The Yellow system (System 3) is pressurised by an EDP driven by No. 2 Engine. An electric motor driven pump is provided which is capable of delivering 6.1 gpm or 23 L/min for ground servicing operations. This system also has a handpump to pressurise the system for cargo door operation when the aircraft is on the ground with electrical power unavailable

Each channel has the provision for the supply of ground-based hydraulic pressure during maintenance operations. Each main system has a hydraulic accumulator to maintain system pressure in the event of transients. See Figure 4.20.

Each system includes a leak measurement valve (shown as L in a square on the diagram), and a priority valve (shown as P in a square).

- The leak measurement valve is positioned upstream of the primary flight controls and is used for the measurement of leakage in each flight control system circuit. They are operated from the ground maintenance panel
- In the event of a low hydraulic pressure, the priority valve maintains pressure supply to essential systems by cutting of the supply to heavy load users

The bi-directional Power Transfer Unit (PTU) enables the Green or the Yellow systems to power each other without the transfer of fluid. In flight in the event that only one engine is running, the PTU will automatically operate when

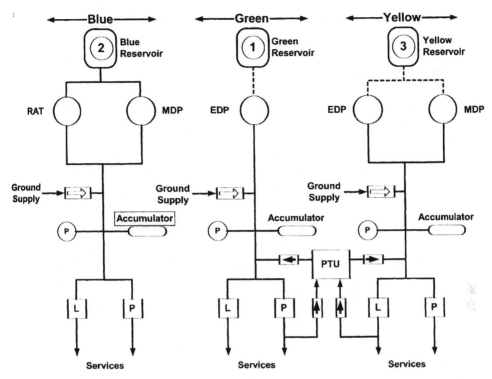

Figure 4.20 Simplified A320 family hydraulic system

the differential pressure between the systems is greater than 500 psi. On the ground, while operating the yellow system using the electric motor driven pump, the PTU will also allow the Green system to be pressurised.

The RAT extends automatically in flight in the event of failure of both engines and the APU. In the event of an engine fire, a fire valve in the suction line between the EDP and the appropriate hydraulic reservoir made be closed, isolating the supply of hydraulic fluid to the engine.

Pressure and status readings are taken at various points around the systems which allows the composition of a hydraulic system display to be shown on the Electronic Crew Alerting and Monitoring (ECAM).

4.16.2 Boeing 767

The B767 also has three full-time independent hydraulic systems to assure the supply of hydraulic pressure to the flight controls and other users. These are the left, right and centre systems serviced by a total of eight hydraulic pumps.

- The left system (Red system) is pressurised by an EDP capable of delivering 37.5 gpm or 142 litres/minute.A secondary or demand electric motor driven pump capable of delivering 7 gpm or 26.5 L/min is turned on automatically in the event that the primary pump cannot maintain pressure
- The right system (Green system) has a similar configuration to the left system
- The centre system (Blue system) uses two electric driven motor pumps, each with the capability of delivering 7 gpm or 26.5 L/min as the primary supply. An Air-Driven Pump (ADP) with a capacity of 37 gpm or 140.2 L/min is used as a secondary or demand pump for the centre system. The centre system also has an emergency RAT rated at 11.3 gpm or 42.8 L/min at 2140 psi

See Figure 4.21 for a simplified diagram of the B767 hydraulic system. Primary flight control actuators, autopilot servo-valves and spoilers receive hydraulic power from each of the three independent hydraulic systems. The stabiliser, yaw dampers, elevator feel units and the brakes are operated from two systems. A Power Transfer Unit (PTU) between the left and right systems provides a third source of power to the horizontal stabiliser.

A motorised valve (M) located between the delivery of ACMP #1 and ACMP #2 may be closed to act an isolation valve between the ACMP #1 and ACMP #2/ADP delivery outputs.

Hydraulic systems status and a synoptic display may be portrayed on the Engine Indication & Crew Alerting System (EICAS) displays situated between the Captain and First Officer on the instrument console. A number of maintenance pages may also be displayed.

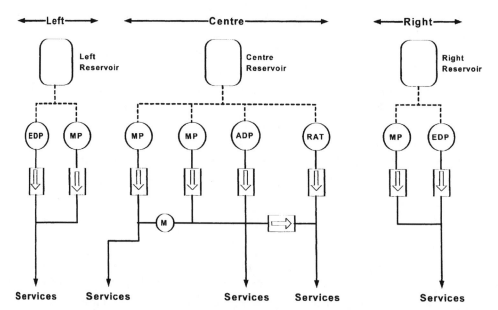

Figure 4.21   Simplified B767 hydraulic system

The supply schedule for the different pumps is given in Table 4.1 below:

Table 4.1 B767 simplified hydraulic schedule

System	Hydraulic power summary		Operating conditions
	Pump Continuous	Pump demand	
Left or right	EDP		Basic system pressure
		ACMP	Supplements EDP
Centre	ACMP #1		Basic system pressure – maintains isolated system pressure
	ACMP #2		Basic system pressure – does not operate when one engine is out or left and right ACMPs on
Centre		ADP	Supplements ACMPs #1 & #2
Centre (Emergency)		RAT	Operates when deployed

The RAT supplies emergency power in flight once the engine speed (N2) has fallen below 50% on both engines and the airspeed is in excess of 80 kts. The RAT may only be restowed on the ground.

While this description outlines the B767 system at a top level, the systems on the B747-400 and B777 also use a combination of engine driven (EDP), air driven (ADP) and electric motor driven pumps and a RAT albeit in different architectures with a different pump configuration. The Boeing philosophy appears to favour fewer accumulators but use more pumps with a more diverse selection of prime pump energy.

Neese (1991) usefully summarises the key hydraulic system characteristics of virtually all wide-body, narrow-body and turboprop/commuter aircraft flying today.

4.17 Landing Gear Systems

The Raytheon/BAE 1000 is representative of many modern aircraft; its landing gear is shown in Figures 4.22 and 4.23. It consists of the undercarriage legs and doors, steering and wheels and brakes and anti-skid system. All of these functions can be operated hydraulically in response to pilot demands at cockpit mounted controls.

4.17.1 Nose Gear

The tricycle landing gear has dual wheels on each leg. The hydraulically operated nose gear retracts forward into a well beneath the forward equipment

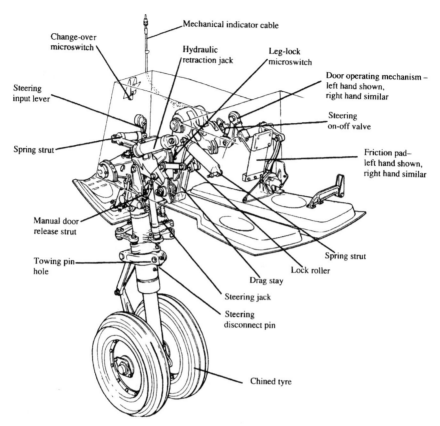

Figure 4.22 The Raytheon 1000 nose landing-gear (Courtesy of Raytheon)

bay. Hinged nose-wheel doors, normally closed, are sequenced to open when lowering or retracting the nose gear. The advantage of the doors being normally closed is twofold. First, the undercarriage bay is protected from spray on take-off and landing, and secondly there is a reduction in drag. A small panel on the leg completes enclosure on retraction and a mechanical indicator on the flight deck shows locking of the gear.

4.17.2 Main Gear

The main gear is also hydraulically operated and retracts inwards into wheel bays. Once retracted the main units are fully enclosed by means of fairings attached to the legs and by hydraulically operated doors. Each unit is operated by a single jack and a mechanical linkage maintains the gear in the locked position without hydraulic assistance. The main wheel doors jacks are controlled by a sequencing mechanism that closes the doors when the gear is fully extended or retracted. Figure 4.24 shows the landing gear sequence for the BAE 146 and also shows the clean lines of the nose wheel bay with the doors shut.

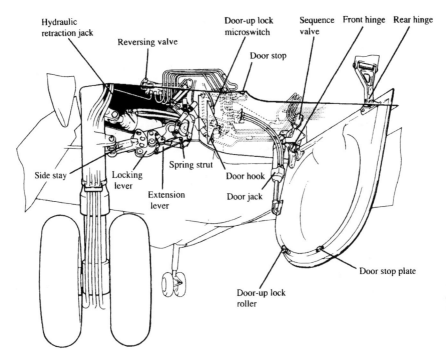

Hydraulic
retraction jack
Reversing valve
Door-up lock
microswitch
Sequence Front hinge Rear hinge
valve
Door stop
Side stay
Locking
lever
Spring strut
Extension
lever
Door hook
Door jack
Door stop plate
Door-up lock
roller

Figure 4.23 The Raytheon 1000 main landing-gear (Courtesy of Raytheon)

4.17.3 Braking Anti-Skid and Steering

Stopping an aircraft safely at high landing speeds on a variety of runway surfaces and temperatures, and under all weather conditions demands an effective braking system. Its design must take into account tyre to ground and brake friction, the brake pressure/volume characteristics, and the response of the aircraft hydraulic system and the aircraft structural and dynamic characteristics. Simple systems are available which provide reasonable performance at appropriate initial and maintenance costs. More complex systems are available to provide minimum stopping distance performance with features such as auto-braking during landing and rejected take-off, additional redundancy and self test. Some of the functional aspects of brakes and steering are illustrated in Figure 4.25.

The normal functions of landing, deceleration and taxying to dispersal or the airport gate require large amounts of energy to be applied to the brakes. Wherever possible, lift dump and reverse thrust will used to assist braking. However it is usual for a large amount of heat to be dissipated in the brake pack. This results from the application of brakes during the initial landing deceleration, the use of brakes during taxying, and the need to hold the aircraft on brakes for periods of time at runway or taxiway intersections.

When the aircraft arrives at the gate the brakes, and the wheel assembly will be very hot. This poses a health and safety risk to ground crew working in the vicinity of the wheels during the turnaround. This is usually dealt with by training.

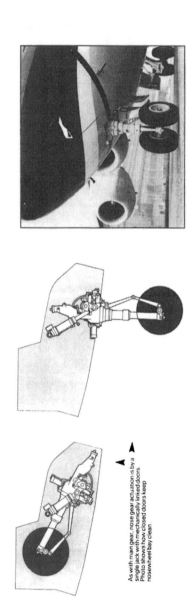

As with main gear, nose gear actuation is by a
single jack with mechanically linked doors.
Photo shows how closed doors keep
nosewheel bay clean.

Landing gear extension sequence (front
view, left leg). A single hydraulic jack effects
extension/retraction of each leg; there is
thus no jack sequencing. When viewed from
rear, locking/bracing geometry and ease of
oleo removal are clear.

Figure 4.24   The 146 landing-gear sequence (Courtesy of BAE Systems)

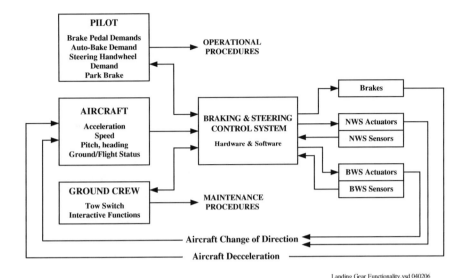

Figure 4.25 Brake control system – functional elements

A more serious operational issue is that the aircraft cannot depart the gate until the brake and wheel assembly temperature cools to a value that will not support ignition of hydraulic fluid. This is to ensure that, during the taxi back to the take-off runway, further brake applications will not raise the temperature of the brake pack to a level that will support ignition if a leak of fluid occurs during retraction. Departure from the gate, therefore, may be determined by brake temperature as indicated by a sensor in the brake pack rather than by time taken to disembark and embark passengers. Some aircraft address this issue by installing brake cooling fans in the wheel assembly to ventilate the brakes. An alternative method is to install fire detection and suppression systems in the wheel bays.

There are events that can raise the temperature of the brakes to the extent that a fire may occur and the tyres can burst. Examples of this are an aborted take-off (maximum rejected take-off) or an immediate go around and heavy landing. In both circumstances the aircraft will be fully laden with passengers and fuel. Thermal plugs will operate to deflate the tyres and fire crews will attend the aircraft to extinguish the fire while the passengers disembark.

One of the simplest and most widely known anti-skid system is the Dunlop Maxaret unit which consists of a hydraulic valve assembly regulated by the dynamics of a spring loaded g sensitive flywheel. Figure 4.26 shows an axle mounted Maxaret together with a modulator.

Rotation of the flywheel is by means of a self-aligning drive from the hub of the wheel, allowing the entire unit to be housed within the axle and protecting the unit from the effects of weather and stones thrown up by the aircraft wheels. Skid conditions are detected by the overrun of the flywheel which opens the Maxaret valve to allow hydraulic pressure to dissipate. A combination of flow sensitive hydraulic units and switches in the oleo leg provide modulation

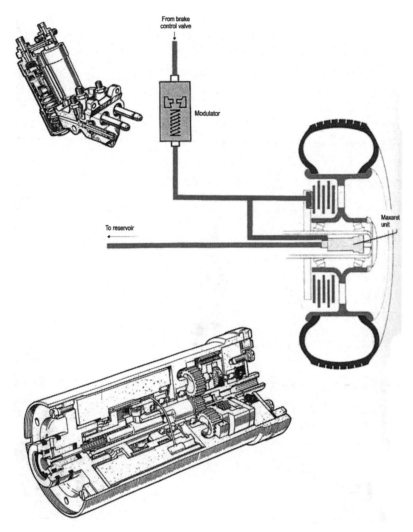

Figure 4.26 The Dunlop Maxaret anti-skid system (Courtesy of Dunlop Aerospace International)

of pressure for optimum braking force and protection against inadvertent application of the brakes prior to touchdown. This ensures that the aircraft does not land with the brakes applied by only allowing the braking system to become active after the oleo switches have sensed that the oleo is compressed. This condition is known as 'weight-on-wheels'. Without this protection the effect of landing with full braking applied could lead to loss of control of the aircraft; at a minimum a set of burst tyres.

4.17.4 Electronic Control

Electronic control of braking and anti-skid systems has been introduced in various forms to provide different features. An electronic anti-skid system with adaptive pressure control is shown in Figure 4.27.

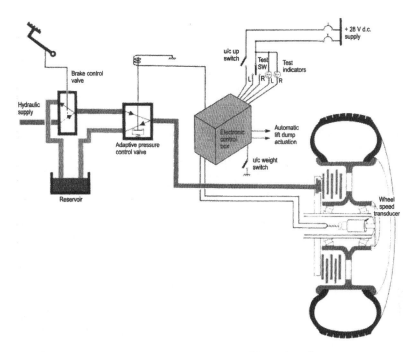

Figure 4.27 Electronic anti-skid system with adaptive pressure control (Courtesy of Dunlop Aerospace International)

In this system the electronic control box contains individual wheel deceleration rate skid detection circuits with cross reference between wheels and changeover circuits to couple the control valve across the aircraft should the loss of a wheel speed signal occur. If a skid develops the system disconnects braking momentarily and the adaptive pressure coordination valve ensures that brake pressure is re-applied at a lower pressure after the skid than the level which allowed the skid to occur. A progressive increase in brake pressure between skids attempts to maintain a high level of pressure and braking efficiency.

The adaptive pressure control valve dumps hydraulic pressure from the brake when its first stage solenoid valve is energised by the commencement of a skid signal. On wheel speed recovery the solenoid is de-energised and the brake pressure re-applied at a reduced pressure level, depending on the time interval of the skid. Brake pressure then rises at a controlled rate in search of the maximum braking level, until the next incipient skid signal occurs.

4.17.5 Automatic Braking

A more comprehensive system is the Dunlop automatic brake control system illustrated in Figure 4.28, which allows an aircraft to be landed and stopped without pilot braking intervention. During automatic braking a two-position three-way solenoid valve is energised following wheel spin-up to feed system pressure via shuttle valves directly to the anti-skid valves where it is modulated

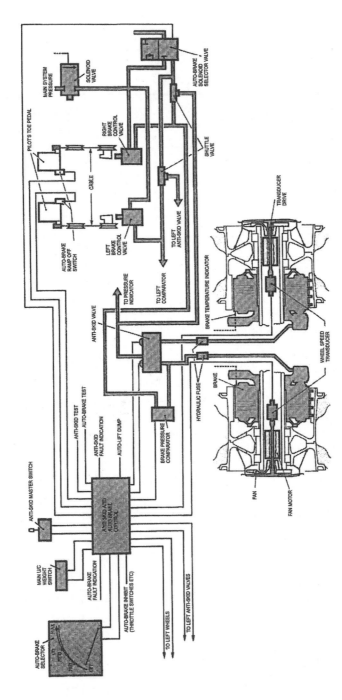

Figure 4.28 An automatic brake control system (Courtesy of Dunlop Aerospace International)

and passed to the brakes. Signals from the auto-braking circuit are responsible for this modulation of pressure at the brake to match a preselected deceleration. However, pilot intervention in the anti-skid control circuit or anti-skid operation will override auto-brake at all times to cater for variations in runway conditions.

In the interest of safety a number of prerequisites must be satisfied before auto-braking is initiated:

- Auto-brake switch must be ON and required deceleration selected
- Anti-skid switch must be ON and operative
- Throttle must be correctly positioned
- Hydraulic pressure must be available
- Brake pedals must not be depressed
- Wheels must be spun up

With all these conditions satisfied auto-braking will be operational and will retard the aircraft at a predetermined rate unless overridden by anti-skid activity. At any time during the landing roll the auto-braking may be overridden by the pilot by advancing the throttle levers for go-around, or by normal application of the brakes.

4.17.6 Multi-Wheel Systems

The systems described thus far apply to most aircraft braking systems. However, large aircraft have multi-wheel bogies and sometimes more than two main gears. The B747-400 has four main oleos, each with a total of four wheels each. The B777 has two main bogies with six wheels each. These systems tend to be more complex and utilise multi-lane dual redundant control. The B777 main gear shown in Figure 4.29 is an example.

For control purposes the wheels are grouped in four lines of three wheels, each corresponding to an independent control channels as shown in the figure. Each of the lines of three wheels – 1, 5, 9; 2, 6, 10 and so on – is controlled by a dual redundant controller located in the Brake System Control Unit (BSCU). Brake demands and wheel speed sensor readings are grouped by each channel and interfaced with the respective channel control. Control channels have individual power supplies to maintain channel segregation and integrity. The BSCU interfaces with the rest of the aircraft by means of left and right A629 aircraft systems data buses. This system is supplied by the Hydro-Aire division, part of Crane Aerospace, and is indicative of the sophistication which modern brake systems offer for larger systems.

The landing gear configuration for the Airbus A380 is shown in Figure 4.30. Goodrich provide two six-wheel under-fuselage landing gear and the two four-wheeled wing-mounted landing gear. The wing-mounted landing gear is slightly forward of the fuselage-mounted gear. The wheels on the main landing gear are fitted with carbon brakes.

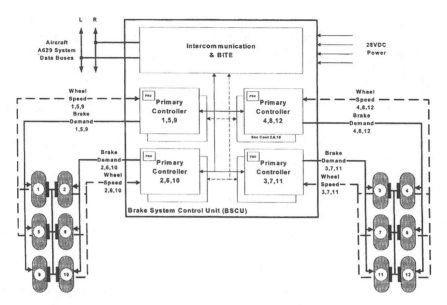

Figure 4.29   Simplified Boeing 777 braking configuration

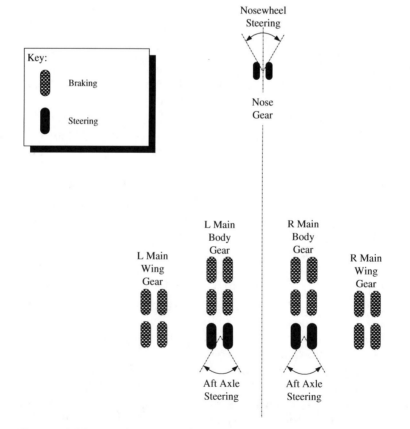

Figure 4.30   The Airbus A380 landing gear configuration

The twin wheel nose landing gear is supplied by Messier-Dowty. The steering control is via the nose gear and via the rear axle of the fuselage landing gear. The gear allows U-turn manoeuvres on a 60 m-wide runway. Manoeuvrability is improved by having a hydraulically steerable aft axle which helps the aircraft attain tight turns without applying unacceptable torsion loads to the main oleo.

The aircraft can manoeuvre on 23 m-wide taxiways and 45 m-wide runways. The French aerospace company Latecoere, based in Toulouse, developed the External and Taxi Aid Camera System (ETACS). The ETACS consists of five video cameras and an onboard computer. The cameras are installed on the top of the tailfin and under the fuselage and the image data is relayed to cockpit displays to assist the crew in ground manoeuvres. The Honeywell terrain guidance and on-ground navigation systems are integrated to the aircraft's flight management system.

The braking system for the A380 is shown in Figure 4.31. This system is provided by Messier-Bugatti. This system is based on self-adapting braking algorithms that were successfully introduced on the A340-500/600. These allow optimised braking by managing the braking function wheel by wheel and landing by landing based on the prevailing conditions of runway, tyres and brakes. Each wheel is thus continuously and independently controlled in real time, taking account of its individual parameters and its particular environment. Both the number of wheels and the rapidity required in the feedback loop for controlling the wheel speed, necessitated introducing three dedicated computers, called RDCs (Remote Data Concentrator), each equipped with specific operating software. They are connected to the IMA by a digital bus.

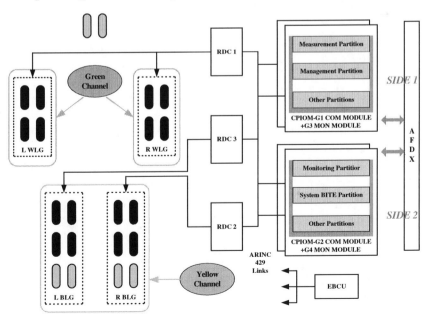

Figure 4.31   A380 brake control system

4.17.7 Brake Parachute

Military aircraft often require assistance to achieve a high-speed landing on short runways. A brake parachute can be used to provide this facility. The system can be armed in flight and commanded by a weight on wheels switch when the main wheels touch down. Figure 4.32 shows an F-117 with brake parachute deployed. The chute is jettisoned on to the runway and must be collected before the next aircraft attempts a landing.

Figure 4.32 F117 deploying brake parachute (Courtesy of US Air Force/ Senior Airman Darnell Cannady)

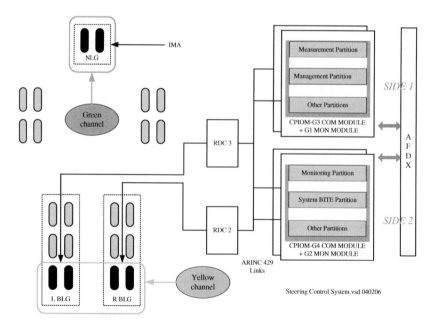

Figure 4.33 A380 steering control system

Steering

Nosewheel steering is normally not engaged for landing – the rudder can be used until forward speed makes it ineffective. At this point steering is engaged manually or automatically. Steering motors respond to demands from the rudder pedals when nose wheel steering is selected. The angular range of the wheels, and the rate of change of steering angle are selected to enable the aircraft to steer on runways and taxi-ways with no risk of the aircraft over-steering or scrubbing the tyres.

An example of the Airbus A380 steering system is shown in Figure 4.33. The A380 steers with the nose-wheels, and also the after wheels of the main gear. This enables the aircraft to complete a 180° turn within 56.5 m, safely within the standard 60 m runway width.

References

[1] SAE Aerospace Information Report (AIR) 5005, 'Aerospace – Commercial Aircraft Hydraulic Systems', March 2000.

Further Reading

Crane, Dale. (1979) Aircraft Wheels, Brakes and Anti-skid Systems. Aviation Maintenance Publications.
Currey, Norman S. (1988). *Landing Gear Design – Principles and Practice*. AIAA Education.
Emerging Technologies in Aircraft Landing Gear Systems. Society of Automotive Engineers, Sept. 1997.
Green, W.L. (1985) *Aircraft Hydraulic Systems: An Introduction to the Analysis of Systems and Components*. John Wiley & Sons Ltd.
Neese, William A. (1991) *Aircraft Hydraulic Systems*. Krieger Publishing.

5

Electrical Systems

5.1 Introduction

Electrical systems have made significant advances over the years as aircraft have become more dependent upon electrically powered services. A typical electrical power system of the 1940s and 1950s was the twin 28 VDC system. This system was used a great deal on twin engined aircraft; each engine powered a 28 VDC generator which could employ load sharing with its contemporary if required. One or two DC batteries were also fitted and an inverter was provided to supply 115 VAC to the flight instruments.

The advent of the V-bombers changed this situation radically due to the much greater power requirements – one the Vickers Valiant – incorporated electrically actuated landing gear. They were fitted with four 115 VAC generators, one being driven by each engine. To provide the advantages of no-break power these generators were paralleled which increased the amount of control and protection circuitry. The V-bombers had to power high loads such as radar and electronic warfare jamming equipment. However, examination of the Nimrod maritime patrol aircraft (derived from the de-Havilland Comet) shows many similarities. As a yardstick of the rated power generated; the Victor (see Figure 5.1) was fitted with four 73 kVA AC generators while the Nimrod was fitted with four 60 kVA generators.

5.1.1 Electrical Power Evolution

Since that time, electrical systems have evolved in the manner shown in Figure 5.2. In the UK, the introduction of powerful new AC electrical systems paved the way for the introduction of electrically powered power flying controls. Four channel AC electrical systems utilised on the Avro Vulcan B2 and Handley Page Victor V-Bombers and the Vickers VC10 transport aircraft

Aircraft Systems: Mechanical, electrical, and avionics subsystems integration. 3rd Edition   Ian Moir and Allan Seabridge
© 2008 John Wiley & Sons, Ltd

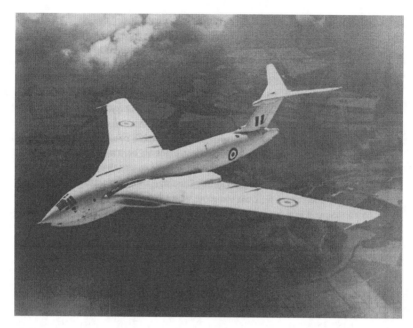

Figure 5.1 Handley Page Victor Bomber, Mk 2 (Courtesy of Handley Page Association)

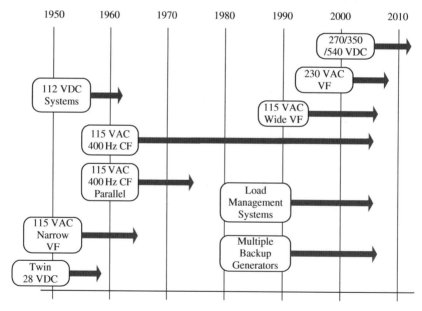

Figure 5.2 Electrical system evolution

utilised flight control actuators powered by the aircraft AC electrical system rather than centralised aircraft hydraulic systems.

Aircraft such as the McDonnell Douglas F-4 Phantom introduced high power AC generation systems to a fighter application. In order to generate constant frequency 115 VAC at 400 Hz a constant speed drive or CSD is required to negate the aircraft engine speed variation over approximately 2:1 speed range (full power speed:flight idle speed). These are complex hydro-mechanical devices which by their very nature are not highly reliable. Therefore the introduction of constant frequency AC generation systems was not without accompanying reliability problems, particularly on fighter aircraft where engine throttle settings are changed very frequently throughout the mission.

The advances in high power solid state switching technology together with enhancements in the necessary control electronics have made variable-speed/constant frequency (VSCF) systems a viable proposition in the last decade. The VSCF system removed the unreliable CSD portion; the variable frequency or frequency wild power from the AC generator being converted to 400 Hz constant frequency 115 VAC power by means of a solid state VSCF converter. VSCF systems are now becoming more commonplace: the F-18 fighter uses such a system and some versions of the Boeing 737-500 did use such a system, not with a lot of success in that particular case. In addition, the Boeing 777 airliner utilises a VSCF system for backup AC power generation.

In US military circles great emphasis is being placed by the US Air Force and the US Navy into the development of 270 VDC systems. In these systems high power generators derive 270 VDC power, some of which is then converted into 115 VAC 400 Hz or 28 VDC required to power specific equipments and loads. This approach has been adopted on the Lockheed Martin F-22 and F-35 aircraft. This is claimed to be more efficient than conventional methods of power generation and the amount of power conversion required is reduced with accompanying weight savings. These developments are allied to the 'more-electric aircraft' concept where it is intended to ascribe more aircraft power system activities to electrical means rather than use hydraulic or high pressure bleed air which is presently the case. The fighter aircraft of tomorrow will therefore need to generate much higher levels of electrical power than at present. Schemes for the use of 270 VDC are envisaging power of 250 to 300 kW and possibly as much as 500 kW per channel; several times the typical level of 50 kVA per channel of today.

At the component level, advances in the development of high power contactors and solid state power switching devices are improving the way in which aircraft primary and secondary power loads are switched and protected. These advances are being married to microelectronic developments to enable the implementation of new concepts for electrical power management system distribution, protection and load switching. The use of electrical power has progressed to the point where the generation, distribution and protection of electrical power to the aircraft electrical services or loads now comprises one of the most complex aircraft systems. This situation was not always so.

The move towards the higher AC voltage is really driven by the amount of power the electrical channel is required to produce. The sensible limit for DC systems has been found to be around 400 amps due to the limitations of feeder size and high power protection switchgear; known as contactors. Therefore for a 28 VDC system delivering 400 amps, the maximum power the channel may deliver is about 12 kW, well below the requirements of most aircraft today. This level of power is sufficient for General Aviation (GA) aircraft and some of the smaller business jets. However, the requirements for aircraft power in business jets, regional aircraft and larger transport aircraft is usually in the range 20 to 90 kVA per channel and higher. The requirement for more power has been matched in the military aircraft arena. More recent aircraft have also readopted VF generation since this is the most reliable method at the generation level, though additional motor controllers may be needed elsewhere in the system to mitigate the effects of frequency variation. Power levels have increased steadily with the Airbus A380 utilising 150 kVA per channel and the Boeing 787 being even more electric with 500 kVA per channel. The A380 and B787 electrical systems are described later in this chapter. The status of More-Electric Aircraft (MEA) and More Electric Engine (MEE) technologies and architectures are described in detail in Chapter 10 – Advanced Systems.

5.2 Aircraft Electrical System

The generic parts of a typical Alternating Current (AC) aircraft electrical system are shown in Figure 5.3 comprising the following:

- Power generation
- Primary power distribution and protection
- Power conversion and energy storage
- Secondary power distribution and protection

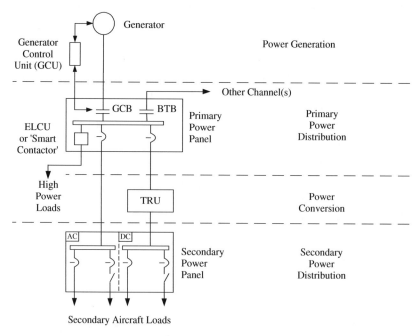

Figure 5.3 Generic aircraft AC electrical system

At this stage it is worth outlining the major differences between AC and DC power generation. Later in the chapter more emphasis is placed upon more recent AC power generation systems.

5.3 Power Generation

5.3.1 DC Power Generation

DC systems use generators to develop a DC voltage to supply aircraft system loads; usually the voltage is 28 VDC but there are 270 VDC systems in being which will be described later in the chapter. The generator is controlled – the technical term is regulated – to supply 28 VDC at all times to the aircraft loads such that any tendencies for the voltage to vary or fluctuate are overcome. DC generators are self-exciting, in that they contain rotating electro-magnets that generate the electrical power. The conversion to DC power is achieved by using a device called a commutator which enables the output voltage, which would appear as a simple sine wave output, to be effectively half-wave rectified and smoothed to present a steady DC voltage with a ripple imposed.

In aircraft applications the generators are typically shunt-wound in which the high resistance field coils are connected in parallel with the armature as shown in Figure 5.4.

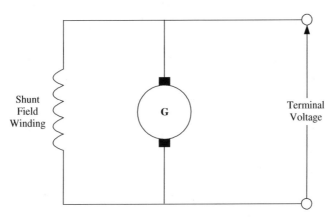

Figure 5.4   Shunt-wound DC generator

The natural load characteristic of the shunt-wound generator is for the voltage to 'droop' with the increasing load current, whereas the desired characteristic is to control the output at a constant voltage – nominally 28 VDC. For this purpose a voltage regulator is used which modifies the field current to ensure that terminal voltage is maintained while the aircraft engine speed and generator loads vary. The principle of operation of the DC voltage regulator is shown in Figure 5.5 and is described later in the chapter.

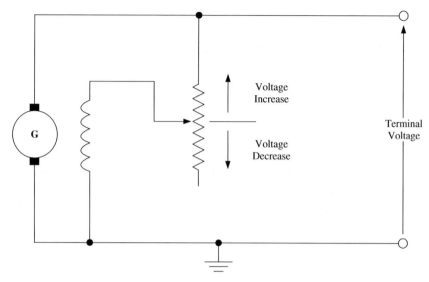

Figure 5.5   DC voltage regulator

5.3.2 AC Power Generation

An AC system uses a generator to generate a sine wave of a given voltage and, in most cases, of a constant frequency. The construction of the alternator is simpler than that of the DC generator in that no commutator is required. Early AC generators used slip rings to pass current to/from the rotor windings however these suffered from abrasion and pitting, especially when passing high currents at altitude. Modern AC generators work on the principle shown in Figure 5.6 which is known as a compound generator.

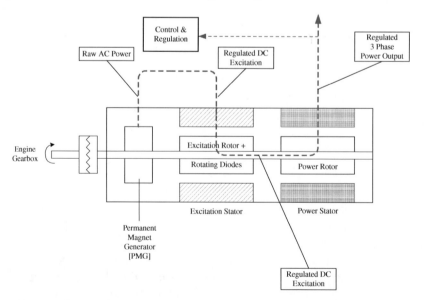

Figure 5.6   Principle of operation of modern AC generator

This AC generator may be regarded as several machines sharing the same shaft. From right to left as viewed on the diagram they comprise:

- A Permanent Magnet Generator (PMG)
- An excitation stator surrounding an excitation rotor containing rotating diodes
- A power rotor encompassed by a power stator

The flow of power through this generator is highlighted by the dashed line. The PMG generates 'raw' (variable frequency, variable voltage) power sensed by the control and regulation section that is part of the generator controller. This modulates the flow of DC current into the excitation stator windings and therefore controls the voltage generated by the excitation rotor. The rotation of the excitation rotor within the field produced by the excitation stator windings is rectified by means of diodes contained within the rotor and supplies a regulated and controlled DC voltage to excite the power rotor windings. The rotating field generated by the power rotor induces an AC voltage in the power stator that may be protected and supplied to the aircraft systems.

Most AC systems used on aircraft use a three-phase system, that is the alternator generates three sine waves, each phase positioned 120° out of phase with the others. These phases are most often connected in a star configuration with one end of each of the phases connected to a neutral point as shown in Figure 5.7. In this layout the phase voltage of a standard aircraft system is 115 VAC, whereas the line voltage measured between lines is 200 VAC. The standard for aircraft frequency controlled systems is 400 cycles/sec or 400 Hz.

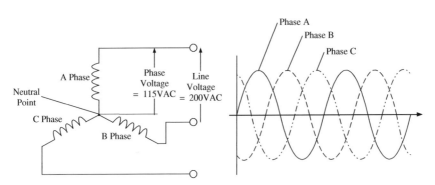

Figure 5.7   Star connected 3 phase AC generator

The descriptions given above outline the two primary methods of power generation used on aircraft for many years. The main advantage of AC power is that it operates at a higher voltage; 115 VAC rather than 28 VDC for the DC system. The use of a higher voltage is not an advantage in itself, in fact higher voltages require better standards of insulation. It is in the transmission of power that the advantage of higher voltage is most apparent. For a given amount of power transmission a higher voltage relates to an equivalent lower current. The lower the current the lower are losses such as voltage drops (proportional to current) and power losses (proportional to current squared). Also as current conductors are generally heavy is can be seen that the reduction in current also saves weight; a very important consideration for aircraft systems.

5.3.3 Power Generation Control

The primary elements of power system control are:

- DC systems

  - Voltage regulation
  - Parallel operation
  - Protection functions

- AC systems

  - Voltage regulation
  - Parallel operation
  - Supervisory functions

DC System Generation Control

Voltage Regulation
DC generation is by means of shunt-wound self-exciting machines as briefly outlined above. The principle of voltage regulation is outlined in Figure 5.5. This shows a variable resistor in series with the field winding such that variation of the resistor alters the resistance of the field winding; hence the field current and output voltage may be varied. In actual fact the regulation is required to be an automatic function that takes account of load and engine speed. The voltage regulation needs to be in accordance with the standard used to specify aircraft power generation systems, namely MIL-STD-704D. This standard specifies the voltage at the point of regulation and the nature of the acceptable voltage drops throughout the aircraft distribution, protection and wiring system. DC systems are limited to around 400 amps or 12 kW per channel maximum for two reasons:

- The size of conductors and switchgear to carry the necessary current becomes prohibitive
- The brush wear on brushed DC generators becomes excessive with resulting maintenance costs if these levels are exceeded

Parallel Operation
In multi-engined aircraft each engine will be driving its own generator and in this situation it is desirable that 'no-break' or uninterrupted power is provided

in cases of engine or generator failure. A number of sensitive aircraft instruments and navigation devices which comprise some of the electrical loads may be disturbed and may need to be restarted or re-initialised following a power interruption. In order to satisfy this requirement generators are paralleled to carry an equal proportion of the electrical load between them. Individual generators are controlled by means of voltage regulators that automatically compensate for variations. In the case of parallel generator operation there is a need to interlink the voltage regulators such that any unequal loading of the generators can be adjusted by means of corresponding alterations in field current. This paralleling feature is more often known as an equalising circuit and therefore provides 'no break' power in the event of a major system failure. A simplified diagram showing the main elements of DC parallel operation is at Figure 5.8.

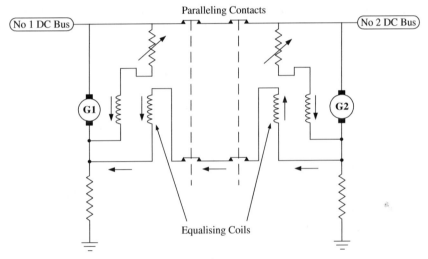

Figure 5.8 DC generator parallel operation

Protection Functions
The primary conditions for which protection needs to be considered in a DC system are as follows:

- *Reverse current.* In a DC system it is evident that the current should flow from the generator to the busbars and distribution systems. In a fault situation it is possible for current to flow in the reverse direction and the primary system components need to be protected from this eventuality. This is usually achieved by means of reverse current circuit breakers or relays. These devices effectively sense reverse current and switch the generator out of circuit, thus preventing any ensuing damage
- *Overvoltage protection.* Faults in the field excitation circuit can cause the generator to over-excite and thereby regulate the supply voltage to an erroneous overvoltage condition. This could then result in the electrical loads

being subject to conditions that could cause permanent damage. Overvoltage protection senses these failure conditions and opens the line contactor taking the generator offline

- *Undervoltage protection.* In a single generator system undervoltage is a similar fault condition as the reverse current situation already described. However, in a multi-generator configuration with paralleling by means of an equalising circuit, the situation is different. Here an undervoltage protection capability is essential as the equalising circuit is always trying to raise the output of a lagging generator; in this situation the undervoltage protection is an integral part of the parallel load sharing function

AC Power Generation Control

Voltage Regulation

As has already been described, AC generators differ from DC machines in that they require a separate source of DC excitation for the field windings although the system described earlier does allow the generator to bootstrap the generation circuits. The subject of AC generator excitation is a complex topic for which the technical solutions vary according to whether the generator is frequency-wild or constant frequency. Some of these solutions comprise sophisticated control loops with error detectors, pre-amplifiers and power amplifiers.

Parallel Operation

In the same way that DC generators are operated in parallel to provide 'no break' power, AC generators may also be controlled in a similar fashion. This technique only applies to constant frequency AC generation as it is impossible to parallel frequency-wild or Variable Frequency (VF) AC generators. In fact many of the aircraft loads such as anti/de-icing heating elements driven by VF generators are relatively frequency insensitive and the need for 'no break' power is not nearly so important. To parallel AC machines the control task is more complex as both real and reactive (imaginary) load vectors have to be synchronised for effective load sharing. No break power transfer is also important during start up /shutdown in the transition from/to ground power, and/or APU generated power, to/from aircraft main generator power, to avoid malfunction or resetting of electrically powered equipment.

The sharing of real load depends upon the relative rotational speeds and hence the relative phasing of the generator voltages. Constant speed or constant frequency AC generation depends upon the tracking accuracy of the constant speed drives of the generators involved. In practice real load sharing is achieved by control laws which measure the degree of load imbalance by using current transformers and error detection circuitry, thereby trimming the constant speed drives such that the torques applied by all generators are equal.

The sharing of reactive load between the generators is a function of the voltage generated by each generator as for the DC parallel operation case. The generator output voltages depend upon the relevant performance of the voltage regulators and field excitation circuitry. To accomplish reactive load sharing requires the use of special transformers called mutual reactors, error detection circuitry and pre-amplifiers/power amplifiers to adjust the field excitation current. Therefore by a using a combination of trimming the speed of the Constant Speed Drives (CSDs) and balancing the field excitation to the generators, real and reactive load components may be shared equally between the generators. Refer to Figure 5.9. This has the effect of providing a powerful single vector AC power supply to the aircraft AC system providing a very 'stiff' supply in periods of high power demand. Perhaps the biggest single advantage of paralleled operation is that all the generators are operating in phase synchronism, therefore in the event of a failure there are no change-over transients.

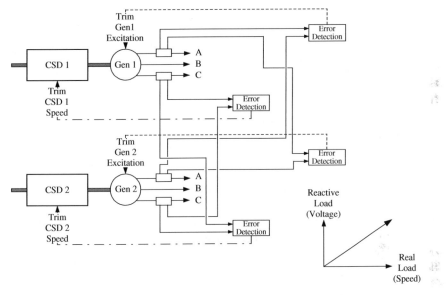

Figure 5.9   AC generator parallel operation

Supervisory and Protection Functions
Typical supervisory or protection functions undertaken by a typical AC generator controller or GCU are listed below:

- Overvoltage
- Undervoltage
- Under/over excitation
- Under/over frequency
- Differential current protection
- Correct phase rotation

The overvoltage, undervoltage and under/over-excitation functions are similar to the corresponding functions described for DC generation control. Under/over frequency protection is effectively executed by the real load sharing function already described above for AC parallel operation. Differential current protection is designed to detect a short-circuit busbar or feeder line fault which could impose a very high current demand on the short-circuited phase. Differential current transformers sense the individual phase currents at differing parts of the system. These are connected so that detection circuitry will sense any gross difference in phase current (say in excess 30 amps per phase) resulting from a phase imbalance and disconnect the generator from the busbar by tripping the Generator Control Breaker (GCB). Phase rotation checks for the correct rotation: R > Y > G of the supply in case any connections have been cross-wired.

Modern AC Electrical Power Generation Types

So far basic DC and AC power generating systems have been described. The DC system is limited by currents greater than 400 amps and the constant frequency AC method using an Integrated Drive Generator (IDG) has been mentioned. In fact there are many more power generation types in use today. A number of recent papers have identified the issues and projected the growth in aircraft electric power requirements in a civil aircraft setting, even without the advent of more-electric systems. However not only are aircraft electrical system power levels increasing but the diversity of primary power generation types is increasing.

The different types of electrical power generation currently being considered are shown in Figure 5.10. The Constant Frequency (CF) 115 VAC, three-phase, 400 Hz generation types are typified by the Integrated Drive Generator (IDG), Variable Speed Constant Frequency (VSCF) Cycloconverter and DC Link options. Variable Frequency (VF) 115 VAC, three-phase power generation – sometimes termed 'frequency wild' – is also a more recent contender,

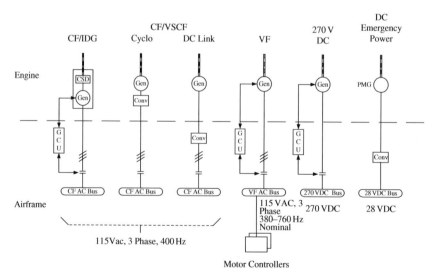

Figure 5.10 Electrical power generation types

and although a relatively inexpensive form of power generation, it has the disadvantage that some motor loads may require motor controllers. Military aircraft in the US are inclining towards 270 VDC systems. Permanent Magnet Generators (PMGs) are used to generate 28 VDC emergency electrical power for high integrity systems.

Figure 5.10 is also interesting in that it shows the disposition between generation system components located on the engine and those within the airframe. Without being drawn into the partisan arguments regarding the pros and cons of the major types of power generation in use or being introduced today it is worth examining the main contenders:

- Constant frequency using an IDG
- Variable frequency
- Variable Speed Constant Frequency (VSCF) options

Constant Frequency/IDG Generation

The main features of CF/IDG power are shown in Figure 5.11. In common with all the other power generation types this has to cater for a 2:1 ratio in engine speed between maximum power and ground idle. The Constant Speed Drive (CSD) in effect acts as an automatic gearbox, maintaining the generator shaft speed at a constant rpm which results in a constant frequency output of

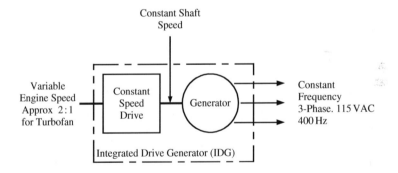

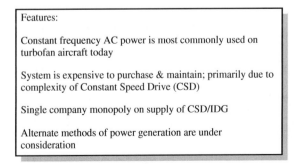

Figure 5.11   Constant frequency / IDG generation

400 Hz, usually within ∼ 10 Hz or less. The drawback of the hydro-mechanical CSD is that it needs to be correctly maintained in terms of oil charge level and oil cleanliness. Also to maintain high reliability frequent overhauls may be necessary.

That said, the IDG is used to power the majority of civil transport aircraft today as shown in Table 5.1.

Variable Frequency Generation

Variable Frequency (VF) power generation as shown in Figure 5.12 is the simplest and most reliable form of power generation. In this technique no attempt is made to nullify the effects of the 2:1 engine speed ratio and the power output, though regulated to 115 VAC, suffers a frequency variation typically from 380 to 720 Hz. This wide band VF power has an effect on frequency sensitive aircraft loads, the most obvious being the effect on AC electric motors that are used in many aircraft systems. There can therefore be a penalty to be paid in the performance of other aircraft systems such as fuel, ECS and hydraulics. In many cases variations in motor/pump performance may be accommodated but in the worst cases a motor controller may be needed to restore an easier control situation. Major airframe manufacturers such as Airbus and Boeing place the burden upon equipment suppliers to ensure that major electrical components perform to specification throughout the anticipated frequency range and the aircraft power quality – such as power factor – is not adversely affected.

VF is being widely adopted in the business jet community as their power requirements take them above the 28 VDC/12 kW limit of twin 28 VDC

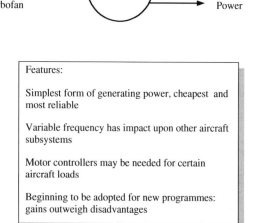

Figure 5.12 Variable frequency power generation

systems. Aircraft such as Global Express had VF designed in from the beginning. Other recent VF power users are the Airbus A380 and Boeing 787.

VSCF Generation

Figure 5.13 shows the concept of the VSCF converter. In this technique the variable frequency power produced by the generator is electronically converted

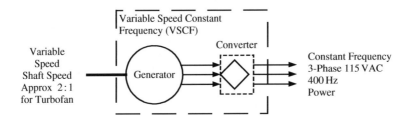

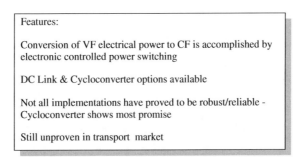

Figure 5.13 VSCF power generation

by solid state power switching devices to constant frequency 400 Hz, 115 VAC power. Two options exist:

- *DC link*: In the DC link the raw power is converted to an intermediate DC power stage – the DC link – before being electronically converted to three-phase AC power. DC link technology has been used on the B737, MD-90 and B777 but has yet to rival the reliability of CF or VF power generation
- *Cycloconverter*: The cycloconverter uses a different principle. Six phases are generated at relatively high frequencies in excess of 3000 Hz and the solid state devices switch between these multiple phases in a predetermined and carefully controlled manner. The effect is to electronically commutate the input and provide three phases of constant frequency 400 Hz power. Though this appears to be a complex technique it is in fact quite elegant and cycloconverter systems have been successfully used on military aircraft in the US: F-18, U-2 and the F-117 stealth fighter. As yet no civil applications have been used. The cycloconverter concept is revisited later in the chapter

As suggested earlier in Figure 5.10, each of these techniques may locate the power conversion section on the engine or in the airframe. Bonneau (1998) examines the implications of moving the VSCF converter from the engine to the airframe in a civil aircraft context [1].

Table 5.1 Recent civil & military aircraft power system developments

Generation type	Civil application		Military application
IDG/CF	B777	2 × 120 kVA	Eurofighter Typhoon
[115 VAC / 400 Hz]	A340	4 × 90 kVA	
	B737NG	2 × 90 kVA	
	MD-12	4 × 120 kVA	
	B747-X	4 × 120 kVA	
	B717	2 × 40 kVA	
	B767-400	2 × 120 kVA	
VSCF (Cycloconverter)			F-18C/D 2 × 40/45 kVA
[115 VAC / 400 Hz]			F-18E/F 2 × 60/65 kVA
VSCF (DC Link)	B777	2 × 20 kVA	
[115 VAC / 400 Hz]	(Backup)		
	MD-90	2 × 75 kVA	
VF	Global Ex	4 × 40 kVA	Boeing JSF 2 × 50 kVA
[115 VAC / 380–760 Hz	Horizon	2 × 20/25	[X-32A/B/C]
typical]	kVA		
	A380	4 × 150 kVA	
VF	B787	4 × 250 kVA	
230 VAC			
270 VDC			F-22 Raptor 2 × 70 kVA
			Lockheed-Martin F-35 –
			Under Review

Table 5.1 lists the power generation types of developed and proposed for civil and military (fighter) aircraft platforms throughout the 1990s. Not only are the electrical power levels increasing in this generation of aircraft but the diversity of electrical power generation methods introduce new aircraft system issues which need to be addressed. For example, the B777 standby VSCF and the MD-90 VCSF converters, being located in the airframe, increase the ECS requirements since waste heat is dissipated in the airframe whereas the previous IDG solution rejected heat into the engine oil system. Similarly the adoption of Variable Frequency (VF) can complicate motor load and power conversion requirements. The adoption of 270 VDC systems by the US military has necessitated the development of a family of 270 VDC protection devices since conventional circuit breakers cannot be used at such high voltages.

Switched Reluctance Machines

Most primary electrical AC power generators today are based upon the compound generator concept described earlier in the chapter. These have proved to be reliable with a generator MTBF $\sim 25\,000$ hours though the reliability of the generator channel has to bear in mind the other elements in the generation package. For example, in conventional constant frequency (CF) applications the generator is combined with a Constant Speed Drive (CSD) unit which has reliability and maintenance limitations. Alternatively the generator may be combined with Variable Speed Constant Frequency (VSCF) electronics or may act as a variable frequency (VF) machine with power electronics or motor controllers downstream to control specific loads.

The compound generator is complex, being effectively three generators on the same shaft and multiple sets of windings and rotating diode packs mean that there are limits to carrying the technology further. Compound machines effectively have an upper speed limit as there comes a point where it is not practicable to package the rotating elements within speed or weight constraints. Furthermore, as the use of starter/generators becomes a realistic option for larger systems and more-electric system generation demands increase so there is a need to consider other options.

A likely solution to these issues is the Switched Reluctance (SR) machine, the principle of which is illustrated in Figure 5.14. SR machines have been subject to research and development activities for a number of years and a number of demonstration programmes have proved the concept on both sides of the Atlantic. The SR machine has a solid rotor so has none of the

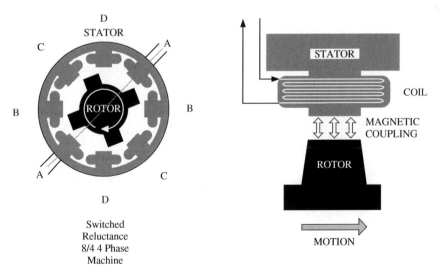

Figure 5.14   Switched reluctance machine – principle of operation

encumbrances of the compound generator; the stator is robust and has the only windings within the machine. The machine is therefore easy to manufacture and is robust. In the example shown there are four pairs of poles on the stator and four on the rotor. As the rotor pole is driven past the stator pole there is magnetic coupling between the two and if a winding is placed round the stator pole then this may be used to generate electrical power. This winding will be associated with another winding diametrically opposite – A to A as shown in the figure and AC power will be generated. This power will be multi-phase and the configuration depicted is called an 8/4 4-phase machine.

Power electronics is needed to condition this power and turn it into electrical power of a suitable quality for use in aircraft applications. It is relatively easy to convert the power into 270 VDC or ± 270 VDC or even 540 VDC. These are the typical voltages being explored in some of the advanced more-electric aircraft and engine technology demonstration programmes presently under way. The simplicity and robustness of the SR machine allows it to be considered for use within the engine as opposed to being located on an aircraft accessory gearbox.

The fact that the SR machine is relatively simple in operation and that power electronics is available allows the machine's flexibility to be used to the full. Figure 5.15 illustrates how a SR machine may be configured to act both as a starter and a generator.

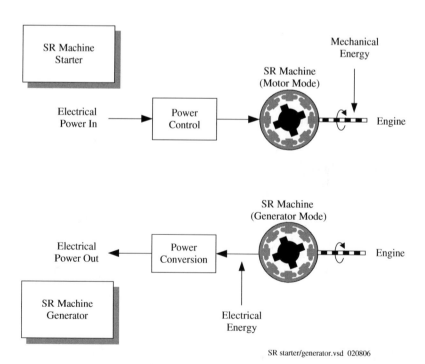

Figure 5.15 Switched reluctance machine – modes of operation

SR Machine – Starter

The top half of the diagram shows power being applied to a power controller embracing power switching electronics. If the power is sequentially fed to the various stator windings then the induced magnetic field will cause the SR machine to motor. This mechanical energy may be harnessed to the engine shaft during the engine start cycle and cause the engine to spool up to a speed where the combustion ignites and the engine becomes self-sustaining.

SR Machine – Generator

In the SR machine generation mode the converse is true. Mechanical energy from the engine is extracted and by switching and conditioning the winding outputs the power conversion electronics can supply high quality electrical power for the aircraft electrical system. The same switching power electronics used for SR start can be reconfigured to be used for power generation. A 270VDC SR starter generator was demonstrated on the JIST program and is now incorporated into the F-35 Lightning II electrical system.

5.4 Primary Power Distribution

The primary power distribution system consolidates the aircraft electrical power inputs. In the case of a typical civil airliner the aircraft may accept power from the following sources:

- Main aircraft generator; by means of a Generator Control Breaker (GCB) under the control of the GCU
- Alternate aircraft generator – in the event of generator failure – by means of a Bus Tie Breaker under the control of a Bus Power Control Unit (BPCU)
- APU generator; by means of an APU GCB under the control of the BPCU
- Ground power; by means of an External Power Contactor (EPC) under the control of the BPCU
- Backup converter, by means of a Converter Control Breaker (CCB) under the control of the VSCF Converter (B777 only)
- RAT generator when deployed by the emergency electrical system

The power switching used in these cases is a power contactor or breaker. These are special high power switches that usually switch power in excess of 20 amps per phase. As well as the power switching contacts auxiliary contacts are included to provide contactor status – 'Open' or 'Closed – to other aircraft systems.

Higher power aircraft loads are increasingly switched from the primary aircraft bus bars by using Electronic Load Control Units (ELCUs) or 'smart contactors' for load protection. Like contactors these are used where normal rated currents are greater than 20 amperes per phase, i.e. for loads of around

7 kVA or greater. Figures 5.16a shows the comparison of a line contactor such as a GCB with an ELCU or 'smart contactor' in Figure 5.16b. The latter has in-built current sensing coils that enable the current flowing in each of the three phases to be measured. Associated electronics allow the device trip characteristics to be more closely matched to those of the load. Typical protection characteristics

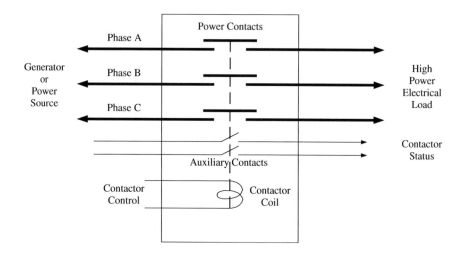

Contactor.vsd   13/1/99

Figure 5.16a   Power contactor

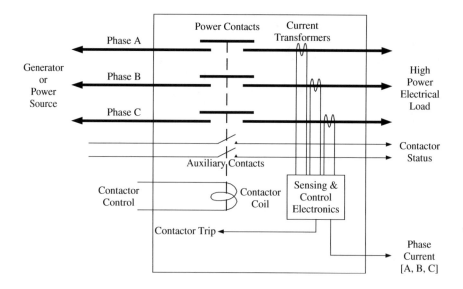

ELCU.vsd 13/1/99

Figure 5.16b   ELCU or 'Smart Contactor'

embodied within the electronics are I^2t, modified I^2t and differential current protection. Boyce reference [2] explains more about 'smart' contactors.

5.5 Power Conversion and Energy Storage

This chapter so far has addressed the primary generation of electrical power and primary power distribution and protection. There are, however, many occasions within an aircraft electrical system where it is required to convert power from one form to another. Typical examples of power conversion are:

• Conversion from DC to AC power – this conversion uses units called inverters to convert 28 VDC to 115 VAC single phase or three-phase power
• Conversion from 115 VAC to 28 VDC power – this is a much used conversion using units called Transformer Rectifier Units (TRUs)
• Conversion from one AC voltage level to another; a typical conversion would be from 115 VAC to 26 VAC
• Battery charging – as previously outlined it is necessary to maintain the state of charge of the aircraft battery by converting 115 VAC to a 28 VDC battery charge voltage
• In more recent military platforms such as F-22 and F-35 utilising 270 VDC; conversion to 115 VAC, 3 phase, 400 Hz AC and 28 VDC is required to power legacy equipments originally designed to operate using these voltages

5.5.1 Inverters

Inverters convert 28 VDC power into 115 VAC single phase electrical power. This is usually required in a civil application to supply captain's or first officer's instruments following an AC failure. Alternatively, under certain specific flight conditions, such as autoland, the inverter may be required to provide an alternative source of power to the flight instruments in the event of a power failure occurring during the critical autoland phase. Some years ago the inverter would have been a rotary machine with a DC motor harnessed in tandem with an AC generator. More recently the power conversion is likely to be accomplished by means of a static inverter where the use of high power, rapid switching, Silicon Controlled Rectifiers (SCRs) will synthesise the AC waveform from the DC input. Inverters are therefore a minor though essential part of many aircraft electrical systems.

5.5.2 Transformer Rectifier Units (TRUs)

TRUs are probably the most frequently used method of power conversion on modern aircraft electrical systems. Most aircraft have a significant 115 VAC three-phase AC power generation capability inherent within the electrical system and it is usual to convert a significant portion of this to 28 VDC by the use of TRUs. TRUs comprise star primary and dual star/delta secondary transformer windings together with three-phase full wave rectification and smoothing to provide the desired 115 VAC/28 VDC conversion. A typical TRU will convert a large amount of power, for example the Boeing 767 uses two TRUs each of which supply a rated load of 120 amps (continuous) with a five-minute rating of 180 amps. TRUs dissipate a lot of heat and are therefore forced air cooled. The Boeing 767 unit is packaged in a 6 MCU ARINC 600 case and weights around 24 lb. Figure 5.17 shows a typical TRU.

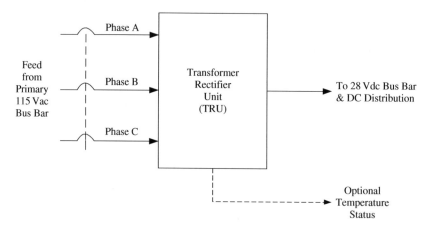

Figure 5.17   Transformer Rectifier Unit (TRU)

TRUs are usually simple, unregulated units; that is the voltage is not controlled to maintain 28 VDC as load is increased and accordingly the load characteristic tends to 'droop'. In some specialist military applications this feature is not desirable and regulated TRUs are used. TRUs are usually operated in isolation; however, when regulated they may also be configured to operate in parallel in a similar way to the parallel operation of DC generators. Johnson *et al.* relates to the development of a regulated TRU [3].

5.5.3 Auto-Transformers

In certain parts of an electrical system simple auto-transformers may be used to provide a simple voltage step-up or step-down conversion. An example of this is the 115 V/26 VAC transformation used to provide 26 VAC aircraft lighting supplies direct from main 115 VAC busbars in the easiest way.

5.5.4 Battery Chargers

Battery chargers share many of the attributes of TRUs and are in fact dedicated units whose function is purely that of charging the aircraft battery. In some systems the charger may also act as a standby TRU providing a boosted source of DC power to the battery in certain system modes of operation. Usually, the task of the battery charger is to provide a controlled charge to the battery without overheating and for this reason battery temperature is usually closely monitored.

5.5.5 Batteries

The majority of this section has described power generation systems, both DC and AC. However it neglects an omnipresent element – the battery. This effectively provides an electrical storage medium independent of the primary generation sources. Its main purposes are:

- To assist in damping transient loads in the DC system
- To provide power in system startup modes when no other power source is available
- To provide a short-term high-integrity source during emergency conditions while alternative/backup sources of power are being brought on line

The capacity of the aircraft battery is limited and is measured in terms of ampere-hours. This parameter effectively describes a current/time capability or storage capacity. Thus a 40 ampere-hour battery when fully charged would have the theoretical capacity of feeding a 1 ampere load for 40 hours or a 40 ampere load for 1 hour. In fact the capacity of the battery depends upon the charge sustained at the beginning of the discharge and this is a notoriously difficult parameter to quantify. Most modern aircraft systems utilise battery chargers to maintain the battery charge at moderately high levels during normal system operation thereby assuring a reasonable state of charge should solo battery usage be required.

The battery most commonly used is the nickel-cadmium (Ni-Cd) type which depends upon the reaction between nickel oxides for the anode, cadmium for the cathode and operating in a potassium hydroxide electrolyte. Lead-acid batteries are not favoured in modern applications due to corrosive effects. To preserve battery health it is usual to monitor its temperature which gives a useful indication of over-charging and if thermal runaway is likely to occur.

5.6 Secondary Power Distribution

5.6.1 Power Switching

In order to reconfigure or to change the state of a system it is necessary to switch power at various levels within the system. At the high power levels that prevail at the primary power part of the system, power switching is accomplished by high power electromagnetic devices called contactors. These devices can switch hundreds of amps and are used to switch generator power on to the primary busbars in both DC and AC systems. The devices may be arranged so that they magnetically latch, that is they are magnetically held in a preferred state or position until a signal is applied to change the state. In other situations a signal may be continuously applied to the contactor to hold the contacts closed and removal of the signal causes the contacts to open. Primary power contactors and ELCUs have been described earlier in the chapter.

For switching currents below 20 amps or so relays are generally used. These operate in a similar fashion to contactors but are lighter, simpler and less expensive. Relays may be used at certain places in the primary electrical system. However, relays are more likely to be employed for switching of medium and high power secondary aircraft loads or services.

For lower currents still where the indication of device status is required, simple switches can be employed. These switches may be manually operated by the crew or they may be operated by other physical means as part of the aircraft operation. Such switches are travel limit switches, pressure switches, temperature switches and so on.

5.6.2 Load Protection

Circuit Breakers

Circuit breakers perform the function of protecting a circuit in the event of an electrical overload. Circuit breakers serve the same purpose as fuses or current limiters. A circuit breaker comprises a set of contacts which are closed during normal circuit operation. The device has a mechanical trip mechanism which is activated by means of a bi-metallic element. When an overload current flows, the bi-metallic element causes the trip mechanism to activate, thereby opening the contacts and removing power from the circuit. A push button on the front of the unit protrudes showing that the device has tripped. Pushing in the push button resets the breaker but if the fault condition still exists the breaker will trip again. Physically pulling the button outwards can also allow the circuit breaker to break the circuit, perhaps for equipment isolation or aircraft maintenance reasons. Circuit breakers are rated at different current values for use in differing current carrying circuits. This enables the trip characteristic to be matched to each circuit. The trip characteristic also has to be selected to coordinate with the feeder trip device upstream. Circuit breakers are literally used by the hundred in aircraft distribution systems; it is not unusual to find 500–600 devices throughout a typical aircraft system. Figure 5.18 shows a circuit breaker and a typical trip characteristic.

Solid State Power Controllers

The availability of high power solid state switching devices has been steadily increasing for a number of years, both in terms of variety and rating. More recent developments have led to the availability of solid state power switching devices which provide a protection capability as well as switching power. These devices known as Solid State Power Controllers or SSPCs effectively

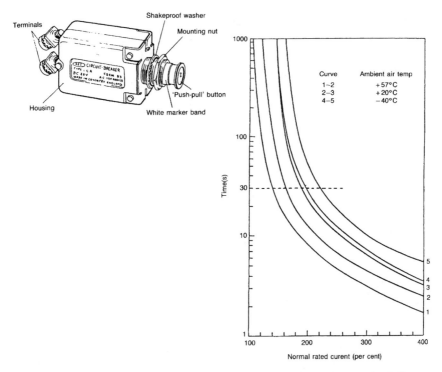

Terminals

Shakeproof washer

Mounting nut

Housing

'Push-pull' button

White marker band

Curve	Ambient air temp
1–2	+57°C
2–3	+20°C
4–5	−40°C

Time(s)

Normal rated curent (per cent)

Figure 5.18 Typical circuit breaker and trip characteristic

combine the function of a relay or switch and a circuit breaker. There are disadvantages with the devices available at present; they are readily available up to a rating of 22.5 amps for use with DC loads; however, the switching of AC loads may only be carried out at lower ratings and with a generally unacceptable power dissipation. Another disadvantage of SSPCs is that they are expensive and costwise may not be comparable with the relay/circuit breaker combination they replace. They are, however, predicted to be more reliable than conventional means of switching and protecting small and medium-sized electrical loads and are likely to become far more prevalent in use in some of the aircraft electrical systems presently under development. SSPCs are also advantageous when utilised in high duty cycle applications where a relay may wear out.

Present devices are rated at 5, 7.5, 12.5 and 22.5 amps and are available to switch 28 VDC and 270 VDC. Layton summarises the development and capabilities of SSPCs and power management units embodying SSPCs to date [5].

Circuit Breaker v SSPC Protection

It is generally accepted that the use of Solid State Power Controllers (SSPCs) provides an enormous improvement over conventional circuit breaker or circuit breaker plus relay combinations. Apart from cost, a key consideration in using SSPCs is also the fact that they offer improved trip accuracy compared to conventional MIL STD (MS) circuit breakers (see Figure 5.19). The thing to note for any electrical protection device is that there are certain (fault) conditions for which a device must trip: shown above the trip characteristics in the diagram. Conversely there are other conditions for which a device must not trip, otherwise nuisance trips will occur when there is no enduring fault: shown below the trip characteristics.

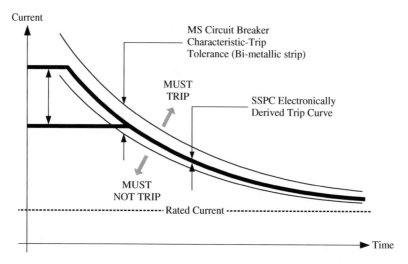

Figure 5.19   Comparison of circuit breaker and SSPC protection

Military Specification (MS) Circuit Breakers

Because of the fact that it is based upon the operation of a simple bi-metallic strip device the circuit breaker is relatively cheap. The disadvantage is that production tolerances will lead inevitably to some dispersion between the device trip point as shown on the diagram. In certain applications this may lead the designer towards difficult compromises, especially in a more complex system where trip coordination between various protection devices has to be considered. In the event that the circuit breaker status needs to be remotely provided, an additional monitor is required.

SSPCs

In an SSPC the trip curve is determined electronically and is therefore more accurate; accordingly the trip tolerance may be achieved within tighter boundaries as shown in the figure. Devices that embody electronic trip characterisation have other advantages. Different trip strategies other than I²R may be implemented. For certain loads a modified I²R characteristic may be employed or for other loads with a high inrush current on start-up it may be necessary to increase the trip current threshold for short durations as shown. SSPCs have a further advantage that they more readily provide status information and later advanced versions may be serial data bus addressable.

5.7 Typical Aircraft DC System

A generic distribution system is shown in Figure 5.20. In this case a twin 28 VDC system is shown which might be typical for a twin-engine commuter aircraft requiring less than ~ 12 kW per channel.

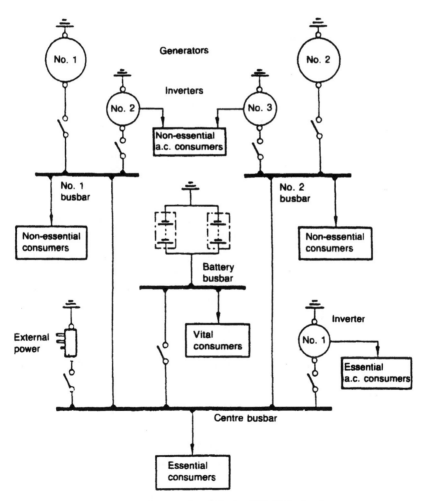

Figure 5.20 Typical Twin 28VDC System

The main elements of this electrical system are:

- Two 28 VDC generators operating in parallel to supply No. 1 and No. 2 main DC busbars. These busbars feed the non-essential DC services
- Two inverters operate, one off each of the DC busbars to provide 115 VAC 400 Hz to non-essential AC services
- Both No. 1 and No. 2 busbars feed power to a centre or essential busbar which provides DC power for the aircraft essential DC services. An inverter powered off this busbar feeds essential 115 VAC loads. A 28 VDC external power source may also feed this busbar when the aircraft is on the ground without the engines running
- The aircraft battery feeds the battery busbar from which are fed vital services. The battery may also be connected to the DC essential busbar is required

To enable a system such as this is to be afforded suitable protection requires several levels of power switching and protection:

- Primary power generation protection of the type described earlier and which includes reverse current and under/over voltage protection under the control of the voltage regulator. This controls the generator feed contactors which switch the generator output on to the No. 1/No. 2 DC busbars
- The protection of feeds from the main buses, i.e. the protection of the feeds to the essential busbar. This may be provided by a circuit breaker or a 'smart' contactor may be used to provide the protection. (Note: The operation of 'smart' contactors will be described later in the chapter.)
- The use of circuit breakers to protect individual loads or groups of loads fed from the supply or feeder busbars

The cardinal principle is that fault conditions should be contained with the minimum of disruption to the electrical system. Furthermore, faults that cause a load circuit breaker to trip should not cause the next level of protection also to trip which would be a cascade failure. Thus the trip characteristics of all protection devices should be coordinated to ensure that this does not occur.

5.8 Typical Civil Transport Electrical System

A typical civil transport electrical power system is shown in Figure. 5.21. This is a simplified representation of the Boeing 767 aircraft electrical power system that is described in detail in Wall [4].

The primary AC system comprises identical left and right channels. Each channel has an Integrated Drive Generator (IDG) driven from the accessory gearbox of the respective engine. Each AC generator is a three-phase 115 V 400 Hz machine producing 90 kVA and is controlled by its own Generator Control Unit (GCU). The GCU controls the operation of the GCB closing the GCB when all operating parameters are satisfactory and opening the GCB when fault conditions prevail. Two Bus Tie Breakers (BTBs) may be closed to tie both buses together in the event that either generating source is lost. The BTBs can also operate in conjunction with the External Power Contactor (EPC) or the Auxiliary Power Breaker (APB) to supply both main AC buses power or the 90 kVA APU generator may also feed the ground handling and ground servicing buses by means of changeover contactors. The control of the BTBs, EPB, and the ground handling/servicing contactors is carried out by a unit called a Bus Power Control Unit (BPCU). The APU may also be used as a primary power source in flight on certain aircraft in the event that either left or right IDG is lost.

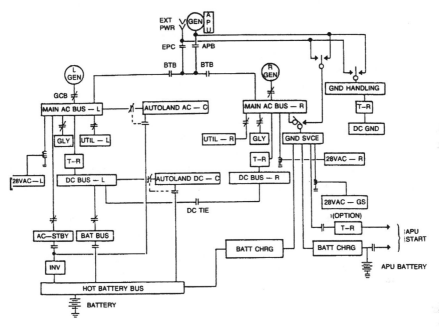

Figure 5.21 Simplified Boeing 767 electrical power system (Courtesy of Boeing)

Each of the main AC buses feeds a number of sub-buses or power conversion equipment. TRUs convert 115 VAC to 28 VDC to feed the left and right DC buses respectively. In the event that either main AC bus or TRU should fail, a DC bus-tie contactor closes to tie the left and right DC buses together. The main AC buses also feed the aircraft galleys (a major electrical load) by means of 'smart' contactors. The utility buses are also fed via contactors from each of the main AC buses. In the event of a major electrical system failure the galley loads and non-essential utility bus loads may be shed under the supervision of the BPCU. Both main AC buses feed 26 VDC buses via auto-transformers. Other specific feeds from the left main AC bus are: a switched feed to the autoland AC bus (interlocked with a switched feed from the standby inverter); and a switched feed to the AC standby bus. Dedicated feeds from the right main AC bus are: via the air/ground changeover contactor to the ground services bus feeding the APU TRU and battery charger; and via the main battery charger to the hot battery bus. The left DC bus also supplies a switched feed to the autoland DC bus (interlocked with a switched feed from the hot battery bus). The hot battery bus also has the capability of feeding the autoland AC bus via the standby inverter.

To the uninitiated this may appear to be overly complex; however, the reason for this architecture is to provide three independent lanes of AC and DC conversion for use during autoland conditions. These are:

- Left main AC bus (disconnected from the autoland AC bus) via the left TRU to the left DC bus (which in this situation will be disconnected from the autoland DC bus
- Right main AC bus via the right TRU, to the right DC bus
- Right main AC bus via the ground services bus and main battery charger to the hot battery bus and thence to the autoland DC bus (now disconnected from the left DC bus). Also from the hot battery bus via the standby inverter to the autoland AC bus (now disconnected from the left main AC bus)

This provides the three independent lanes of electrical power required. It might be argued that two lanes are initially derived from the right main AC bus and therefore the segregation requirements are not fully satisfied. In fact, as the hot battery is fed from the main aircraft battery, this represents an independent source of stored electricity, provided that an acceptable level of charge is maintained. This latter condition is satisfied as the battery charger is fed at all times the aircraft is electrically powered from the ground services bus from either an air or ground source. The battery capacity is such that all standby loads may be powered for 30 minutes following primary power loss.

5.9 Electrical Loads

Once the aircraft electrical power has been generated and distributed then it is available to the aircraft services. These electrical services cover a range of functions spread geographically throughout the aircraft depending upon their task. While the number of electrical services is legion they may be broadly subdivided into the following categories:

- Motors and actuation
- Lighting services
- Heating services
- Subsystem controllers and avionics systems

5.9.1 Motors and Actuation

Motors are obviously used where motive force is needed to drive a valve or an actuator from one position to another depending upon the requirements of the appropriate aircraft system. Typical uses for motors are:

- Linear actuation: electrical position actuators for engine control; trim actuators for flight control systems
- Rotary actuation: electrical position actuators for flap/slat operation
- Control valve operation: electrical operation of fuel control valves; hydraulic control valves; air control valves; control valves for ancillary systems
- Starter motors: provision of starting for engine, APU and other systems that require assistance to reach self-sustaining operation
- Pumps: provision of motive force for fuel pumps, hydraulic pumps; pumping for auxiliary systems
- Gyroscope motors: provision of power to run gyroscopes for flight instruments and autopilots; in modern avionics systems gyroscopic sensors are increasingly likely to be solid state and therefore will not require an AC supply
- Fan motors: provision of power to run cooling fans for the provision of air to passengers or equipment

Many of the applications for which electric motors are used are not continuously rated; that is, the motor can only be expected to run for a small proportion of the time. Others, such as the gyroscope and cooling fan motors, may be run continuously throughout the period of operation of the aircraft and the sizing/rating of the motor has to be chosen accordingly. The following categorises the characteristics of the DC and AC motor types commonly used for aircraft applications.

5.9.2 DC Motors

A DC motor is the inverse of the DC generator described earlier in this chapter. It comprises armature field windings and commutator/brushgear and is similarly self-excited. The main elements of importance in relation to motors are the speed and torque characteristics, i.e. the variations of speed and torque with load respectively. Motors are categorised by their field winding configuration (as for generators) and typical examples are series-wound, shunt-wound and compound-wound (a combination of series and shunt-wound). Each of these types of motor offers differing performance characteristics that may be matched to the application for which they are intended.

A specialised form of series motor is the split-field motor where two sets of series windings of opposite polarity are each used in series with the armature but parallel with each other.

Either one set of field windings or the other may receive power at any one time and therefore the motor may run bi-directionally, depending upon which winding is energised. When used in conjunction with suitable switches or relays this type of motor is particularly useful for powering loads such as fuel system valves where there may be a requirement to change the position of various valves several times during flight. Limit switches at the end of the actuator travel prevent the motor/actuator from over-running once the desired position has been reached. Split-field motors are commonly used for linear and rotary position actuators when used in conjunction with the necessary position feedback control.

DC motors are most likely to be used for linear and rotary actuation, fuel valve actuation and starter functions. DC brushless motors with associated control electronics are becoming extensively employed.

5.9.3 AC Motors

AC motors used for aircraft applications are most commonly of the 'induction motor' type. An induction motor operates upon the principle that a rotating magnetic field is set up by the AC field current supplied to two or more stator windings (usually three-phase). A simple rotor, sometimes called a 'squirrel cage', will rotate under the effects of this rotating magnetic field without the need for brushgear or slip rings; the motor is therefore simple in construction and reliable. The speed of rotation of an induction motor depends upon the frequency of the applied voltage and the number of pairs of poles used. The advantage of the induction motor for airborne uses is that there is always a source of constant frequency AC power available and for constant rated applications it offers a very cost-effective solution. Single-phase induction motors also exist; however, these require a second set of phase windings to be switched in during the start phase, as single-phase windings can merely sustain and not start synchronous running.

AC motors are most likely to be used for continuous operation, i.e. those applications where motors are continuously operating during flight, such as fuel booster pumps, flight instrument gyroscopes and air conditioning cooling fans.

5.9.4 Lighting

Lighting systems represent an important element of the aircraft electrical services. A large proportion of modern aircraft operating time occurs during night or low-visibility conditions. The availability of adequate lighting is essential to the safe operation of the aircraft. Lighting systems may be categorised as follows:

External Lighting Systems

- Navigation lights
- Strobe lights and High Intensity Strobe Lights (HISL)
- Landing/taxi lights
- Formation lights
- Inspection lights (wing/empennage/engine anti-ice)
- Emergency evacuation lights
- Logo lights
- Searchlights (for search and rescue or police aircraft)

Internal Lighting Systems

- Cockpit/flight deck lighting (general, spot, flood and equipment panel)
- Passenger information lighting
- Passenger cabin general and personal lighting
- Emergency/evacuation lighting
- Bay lighting (cargo or equipment bays for servicing)

Lighting may be powered by 28 VDC or by 26 VAC provided by auto-transformer from the main AC buses and is mainly achieved by means of conventional filament bulbs. These filaments vary from around 600 watts for landing lights to a few watts for minor internal illumination uses. Some aircraft instrument panels or signs may use electroluminescent lighting which is a phosphor layer sandwiched between two electrodes, the phosphor glows when supplied with AC power.

5.9.5 Heating

The use of electrical power for heating purposes on aircraft can be extensive. The highest power usage relates to electrically powered anti-icing or de-icing systems which can consume many tens of kVAs. This power does not have to be frequency stable and can be frequency-wild and therefore much easier and cheaper to generate. Anti/de-icing elements are frequently used on the tailplane and fin leading edges, intake cowls, propellers and spinners. The precise mix of electrical and hot air (using bleed air from the engines) anti/de-icing methods varies from aircraft to aircraft. Electrical anti/de-icing systems are high current consumers and require controllers to time, cycle and switch the heating current between heater elements to ensure optimum use of the heating capability and to avoid local overheating.

Windscreen heating is another important electrical heating service. In this system the heating element and the controlling thermostat are embedded in the windscreen itself. A dedicated controller maintains the temperature of the element at a predetermined value which ensures that the windscreen is demisted at all times.

5.9.6 Subsystem Controllers and Avionics Systems

As aircraft have become increasingly complex so has the sophistication of the aircraft subsystems increased. Many have dedicated controllers for specific system control functions. For many years the aircraft avionics systems, embracing display, communication and navigation functions, have been packaged into Line Replaceable Units (LRUs) which permit rapid removal should a fault occur. Many of the aircraft subsystem controllers are now packaged into similar LRUs due to increased complexity and functionality and for the same reasons of rapid replacement following a failure. These LRUs may require DC or AC power depending upon their function and modes of operation. Many may utilise dedicated internal power supply units to convert the aircraft power to levels better suited to the electronics that require ±15 VDC and +5 VDC. Therefore these LRUs represent fairly straightforward and, for the most part, fairly low power loads. However there are many of them and a significant proportion may be critical to the safe operation of the aircraft. Therefore two important factors arise: first, the need to provide independence of function by distribution of critical LRUs across several aircraft busbars, powered by both DC and AC supplies. Secondly, the need to provide adequate sources of emergency power such that, should a dire emergency occur, the aircraft has sufficient power to supply critical services to support a safe return and landing.

5.9.7 Ground Power

For much of the period of aircraft operation on the ground a supply of power is needed. Ground power may be generated by means of a motor-generator set where a prime motor drives a dedicated generator supplying electrical power to the aircraft power receptacle.

The usual standard for ground power is 115 VAC three-phase 400 Hz, which is the same as the power supplied by the aircraft AC generators. In some cases, and this is more the case at major airports, an electrical conversion set adjacent to the aircraft gate supplies 115 VAC three-phase power that has been derived and converted from the national electricity grid. The description given earlier in this chapter of the Boeing 767 system explained how ground power could be applied to the aircraft by closing the EPC.

The aircraft system is protected from substandard ground power supplies by means of a ground power monitor. This ensures that certain essential parameters are met before enabling the EPC to close. In this way the ground power monitor performs a similar function to a main generator GCU. Typical parameters which are checked are undervoltage, overvoltage, frequency and correct phase rotation.

5.10 Emergency Power Generation

In certain emergency conditions the typical aircraft power generation system already described may not meet all the airworthiness authority requirements and additional sources of power generation may need to be used to power the aircraft systems. The aircraft battery offers a short-term power storage capability, typically up to 30 minutes. However, for longer periods of operation the battery is insufficient. The operation of twin-engined passenger aircraft on Extended Twin OPerationS (ETOPS) flights now means that the aircraft has to be able to operate on one engine while up to 180 minutes from an alternative or diversion airfield. This has led to modification of some of the primary aircraft systems, including the electrical system, to ensure that sufficient integrity remains to accomplish the 180 minute diversion while still operating with acceptable safety margins. The three standard methods of providing backup power on civil transport aircraft are:

- Ram Air Turbine (RAT)
- Backup power converters
- Permanent Magnet Generators (PMGs)

5.10.1 Ram Air Turbine

The Ram Air Turbine or RAT is deployed when most of the conventional power generation system has failed or is unavailable for some reason. The RAT is an air-driven turbine, normally stowed in the aircraft ventral or nose section that is extended either automatically or manually when the emergency commences. The passage of air over the turbine is used to power a small emergency generator of limited capacity, usually enough to power the crew's essential flight instruments and a few other critical services – see Figure 5.22. Typical RAT generator sizing may vary from 5 to 15 kVA depending upon the aircraft. The RAT also powers a small hydraulic power generator for similar hydraulic system emergency power provision. Once deployed then the RAT remains extended for the duration of the flight and cannot be restowed without maintenance action on the ground. The RAT is intended to furnish the crew with sufficient power to fly the aircraft while attempting to restore the primary generators or carry out a diversion to the nearest airfield. It is not intended to provide significant amounts of power for a lengthy period of operation.

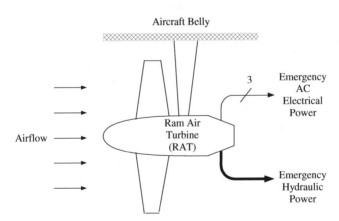

Figure 5.22   Ram Air Turbine (RAT)

5.10.2 Backup Power Converters

The requirements for ETOPS have led to the need for an additional method of backup power supply, short of deploying the RAT that should occur in only the direst emergency. The use of backup converters satisfies this requirement and is used on the B777. Backup generators are driven by the same engine accessory gearbox but are quite independent of the main IDGs. See Figure 5.23.

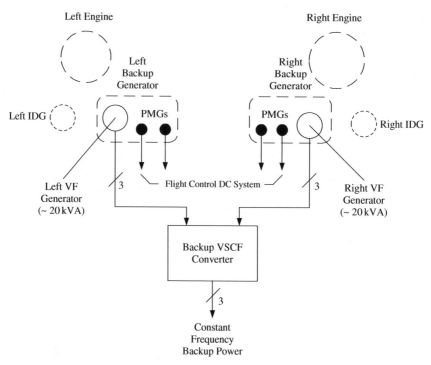

Figure 5.23 Simplified backup VSCF converter system

The backup generators are VF and therefore experience significant frequency variation as engine speed varies. The VF supply is fed into a backup converter which, using the DC link technique, first converts the AC power to DC by means of rectification. The converter then synthesises three-phase 115 VAC 400 Hz power by means of sophisticated solid state power switching techniques. The outcome is an alternative means of AC power generation which may power some of the aircraft AC busbars; typically the 115 VAC transfer buses in the case of the Boeing 777. In this way substantial portions of the aircraft electrical system may remain powered even though some of the more sizeable loads such as the galleys and other non-essential loads may need to be shed by the Electrical Load Management System (ELMS).

Tenning presents the entire Boeing 777 electrical system [6].

5.10.3 Permanent Magnet Generators (PMGs)

The use of PMGs to provide emergency power has become prominent over the last decade or so. PMGs may be single phase or multi-phase devices. Figure 5.24 shows a three-phase PMG with converter, obviously within reason the more phases used the easier the task to convert the power to regulated 28 VDC.

As can be seen, the backup converter hosts PMGs which may supply several hundred watts of independent generated power to the flight control DC system where the necessary conversion to 28 VDC is undertaken. It was already

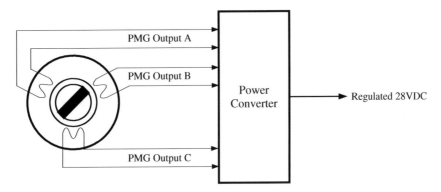

Figure 5.24 Three phase PMG with converter

explained earlier in the chapter that AC generators include a PMG to bootstrap the excitation system. Also PMGs – also called Permanent Magnet Alternators (PMAs) – are used to provide dual independent on-engine supplies to each lane of the FADEC. As an indication of future trends it can therefore be seen that on an aircraft such as the B777 there are a total of 13 PMGs/PMAs across the aircraft critical control systems – flight control, engine control and electrical systems. See Figure 5.25.

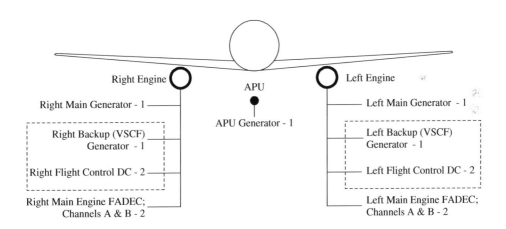

Total PMGs used on B777: 13

Figure 5.25 Boeing 777 PMG/PMA Complement

Tenning [6] gives an early description of the use of a PMG and Rinaldi [7], describes some of the work being undertaken in looking at higher levels of PMG power generation.

Some military aircraft use Emergency Power Units (EPUs) for the supply of emergency power.

5.11 Recent Systems Developments

In recent years a number of technology advances have taken place in the generation, switching and protection of electrical power. These new developments are beginning to have an impact upon the classic electrical systems that have existed for many years, probably for the first time since WWII. This has resulted in the availability of new devices that in turn have given credibility to new system concepts, or at least provide the means for advanced systems concepts that could not previously be implemented. These techniques and concepts embrace the following:

- Electrical Load Management System (ELMS)
- Variable Speed Constant Frequency (VSCF) – Cycloconverter
- 270 VDC systems
- More-Electric Aircraft (MEA)

5.11.1 Electrical Load Management System (ELMS)

The Boeing 777 Electrical Load Management System (ELMS) developed and manufactured by GE Aviation set new standards for the Industry in terms of electrical load management. The general layout of the ELMS is shown in Figure 5.26. The system represents the first integrated electrical power distribution and load management system for a civil aircraft.

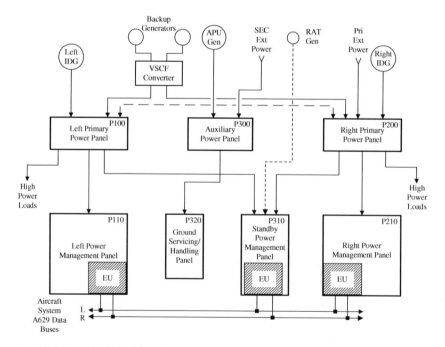

Figure 5.26 B777 Electrical Load Management System (ELMS) (Courtesy of Smiths Group – now GE Aviation)

The system comprises seven power panels, three of which are associated with primary power distribution:

- P100 – Left Primary Power Panel distributes and protects the left primary loads
- P200 – Right Primary Power Panel distributes and protects the right primary loads
- P300 – Auxiliary Power Panel distributes and protects the auxiliary primary loads

The secondary power distribution is undertaken by four secondary power panels:

- P110 – Left Power Management Panel distributes and protects power, and controls loads associated with the left channel
- P210 – Right Power Management Panel distributes and protects power, and controls loads associated with the right channel
- P310 – Standby Power Management Panel distributes and protects power, and controls loads associated with the standby channel
- P320 – Ground Servicing/Handling Panel distributes and protects power associated with ground handling

Load management and utilities systems control is exercised by mean of Electronic Units (EUs) mounted within the P110, P210 and P310 power management panels. Each of these EUs interfaces with the left and right aircraft systems ARINC 629 digital data buses and contain a dual redundant architecture for reasons of dispatch availability. The EUs contain a modular suite of Line Replaceable Modules (LRMs) that can readily be replaced when the door is open. A total of six module types is utilised to build a system comprising an overall complement of 44 modules across the three EUs. This highly modular construction with multiple use of common modules reduced development risk and resulted in highly accelerated module maturity at a very early stage of airline service. LRMs typically have mature in-service Mean Time Between Failures (MTBF) $\sim 200\,000$ hours as reported by Haller, Weale and Loveday (1998) [9]. See Figure 5.27 for a diagrammatic portrayal of the modular concept.

The load management and utilities control features provided by ELMS are far in advance of any equivalent system in airline service today. Approximately 17–19 Electrical Load Control Units (ELCUs) – depending upon aircraft configuration – supply and control loads directly from the aircraft main AC buses. These loads can be controlled by the intelligence embedded within the ELMS EUs. A major advance is the sophisticated load shed/load optimisation function which closely controls the availability of functions should a major electrical power source fail or become unavailable. The system is able to reconfigure the loads to give the optimum distribution of the available power. In the event that electrical power is restored, the system is able to re-instate loads according to a number of different schedules. The system is therefore able to make the optimum use of power at all times rather than merely shed loads in an emergency.

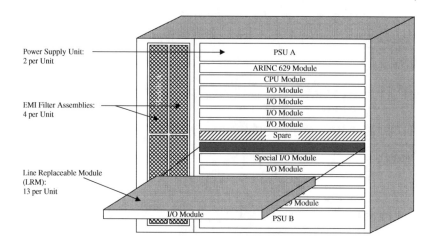

Power Supply Unit:
2 per Unit

EMI Filter Assemblies:
4 per Unit

Line Replaceable Module
(LRM):
13 per Unit

PSU A
ARINC 629 Module
CPU Module
I/O Module
I/O Module
I/O Module
I/O Module
Spare
Special I/O Module
I/O Module
I/O Module
9 Module
I/O Module
PSU B

Figure 5.27   B777 ELMS EU concept (Courtesy of Smiths Group – now GE Aviation)

The benefits conferred by ELMS have proved to be significant with substantial reduction in volume, wiring and connectors, weight, relays and circuit breakers. Due to the inbuilt intelligence, use of digital data buses, maintainability features and extensive system Built-In Test (BIT), the system build and on-aircraft test time turned out to $\sim 30\%$ of that experienced by contemporary systems.

A large number of utilities management functions are embedded in the system making it a true load management rather than merely an electrical power distribution system. Key functions are the load optimisation function already described, fuel jettison, automatic RAT deployment and many others. Figure 5.28 presents an overview of some of the more important functions.

5.11.2 Variable Speed Constant Frequency (VSCF)

The principle of VSCF has already been outlined in the backup converter description earlier in the chapter. There are considerable benefits to be accrued by dispensing with the conventional AC power generation techniques using IDGs to produce large quantities of frequency stable 400 Hz 115 VAC power. The constant speed element of the IDG is generally fairly unreliable compared to the remainder of the generation system. The techniques are now available through the use of VSCF to produce significant quantities of primary

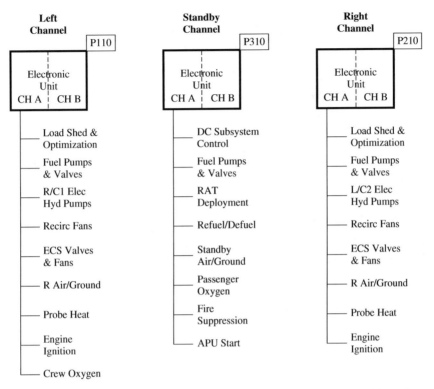

Figure 5.28 B777 ELMS subsystem functional overview (Courtesy of Smiths Group – now GE Aviation)

AC by means of frequency-wild power generation accompanied by suitable power conversion. In particular, the VSCF Cycloconverter version developed by Leland Electrosystems, a part of GE Aviation is a mature technology. Over 4000 cycloconverter systems are in service with the US Military:F-18C/D, F-117A, TR-1 and U-2 and the later versions are fitted to the F-18E/F.

Theory of VSCF Cycloconverter System Operation

The VSCF system consists of a brushless generator and a solid state frequency converter. The converter assembly also has a filter capacitor assembly and control and protection circuit. A simplified block diagram for the VSCF system is shown in Figure 5.29. The generator is driven by the accessory gearbox and produces AC output voltage at variable frequency proportional to the gearbox speed. The converter converts the variable frequency into constant 400 Hz, three-phase power by using an SCR-based cycloconverter. The filter assembly filters out high frequency ripple in the output voltage. The GCU regulates the output voltage and provides protection to the system.

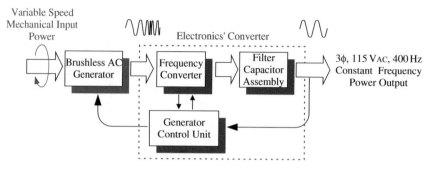

Figure 5.29 Simplified VSCF cycloconverter system diagram

Generator Operation

The function of the generator is to convert mechanical power from the aircraft turbine engine to electrical power suitable for electronic conversion. The electronic converter processes the generator output electrical power into high-quality 400 Hz electrical power. See Figure 5.30.

The brushless, self-excited generator comprises three AC machines:

- Permanent magnet generator
- Exciter generator
- Main generator

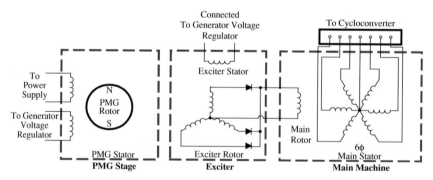

Figure 5.30 VSCF generator electrical schematic (Courtesy of Leland Electrosystems / GE Aviation)

The Permanent Magnet Generator (PMG) provides electrical power for all control circuitry and the exciter field as soon as the rotor is rotating at minimum speed. The PMG also provides raw electrical power for the Main Line Contactors (MLC). The integral PMG makes the generator self-contained; thus, it does not require any external power for excitation. The PMG is a synchronous machine with flux excitation provided by the permanent magnets contained inside the rotor assembly. The PMG stator contains two separate and electrically isolated windings in a laminated, slotted, magnetic steel core. AC voltages are induced in the stator windings as the flux provided by the PM rotor sweeps past the stator. The PM rotor is driven directly by the gearbox output shaft.

The output of one of the single-phase windings of the PMG stator is fed into the generator voltage regulator. The generator voltage regulator rectifies and modulates the PMG output. This output provides proper current for the exciter field winding, allowing generation of AC voltage on the exciter rotor. The output of the second single-phase winding is used for the converter power supply.

The exciter is a brushless synchronous machine with a DC excited stator and a three-phase wound rotor. The exciter stator winding receives controlled DC current from the rectified PMG output through the generator voltage regulator. This in turn develops the AC power in the three-phase rotor windings as they rotate past the exciter generator stator winding, inducing an AC voltage in the three-phase windings of the exciter's rotor. The magnitude of this rectified AC voltage is proportional to the speed of the shaft and to the DC excitation current on the exciter's stator winding. The rotor output is rectified with three silicon rectifiers mounted inside the rotor shaft. The exciter and rectifiers are used to eliminate brushes anywhere in the generator. The rectified exciter output supplies field current for the main generator.

The main generator is a wound rotor, synchronous machine with a 16-pole rotor and a six-phase stator. The connections between the exciter rotor windings, three rectifier diodes and the main rotor field winding are all on the rotor. The six-phase stator output winding is star connected. All six phase leads and the neutral connection are brought out to the terminal block. The wound rotor, when excited with DC current supplied by the exciter, establishes magnetic flux in the air gap between the rotor and the stator. This magnetic flux, when driven by the gearbox's shaft, induces alternating voltage into the six-phase windings of the stator. The magnitude of this AC stator voltage is proportional to the speed of the rotor and the DC current supplied by the exciter rotor. The magnitude of the rotor DC current in turn depends upon the excitation current provided by the generator voltage regulator to the exciter's stator. Therefore, the magnitude of the exciter's stator current determines the magnitude of the main generator stator's AC voltage output. The frequency of the main generator's output is dependent upon the shaft speed. With 16 poles, the frequency of the main generator varies from 1660 Hz to 3500 Hz as the input speed is varied from 12 450 to 26 250 rpm. The main generator output supplies a variable frequency, six-phase AC power to the cycloconverter for further processing.

The neutral ends of each of the six stator windings are connected to the neutral through Current Transformers (CTs). The CTs sense the current in each winding and compare it with the current in each phase in the converter. If any current differential is detected in the zone between the generator neutral and the converter, the system de-energises quickly by means of the High Frequency Differential Protection (HFDP) circuit, preventing damage to any of the generator windings.

All connections between the generator and frequency converter are internal to the VSCF package so the converter cannot be subjected to abnormal phase rotation unless the generator rotation is reversed. The Generator Over-Current (GOC) protection will de-energise the system in the event of reversed generator rotation.

The electrical schematic for the generator is shown in Figure 5.31.

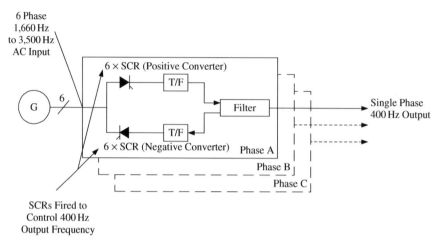

Figure 5.31   VSCF cycloconverter principle

This section describes the Cycloconverter design and operation as configured for a 30/40 kVA rating. This review concentrates on the most critical aspects of a variable speed constant frequency (VSCF) system, i.e. the power flow section and switch module control circuits.

The frequency conversion system consists of three frequency converters, one for each phase (Figure 5.31). The generator delivers six-phase, variable frequency power to each converter. Each frequency converter consists of a cycloconverter (12 silicon controlled rectifiers) and its associated control circuits: modulators, mixer, firing wave generator, reference wave generator, feedback control circuit, and low-pass filter. The SCRs are controlled by the modulators. They compare the cosine firing wave with the processed reference wave to generate appropriately timed SCR gating signals. The low-pass output filter attenuates the ripple frequency components.

Negative feedback is used to improve the linearity of the cycloconverter and to reduce the output impedance. Thus, the cycloconverter is a high power amplifier producing an output wave that is a replica of the reference sine wave. The actual feedback loop has multiple feedback paths to improve the waveform, reduce the DC content, and lower the output impedance. The mixer amplifier adds the feedback signals in the correct proportions.

The 400 Hz output voltage is regulated with individual phase voltage regulators that adjust the 400 Hz reference wave amplitudes. Consequently, the voltage unbalance in the line-to-neutral output voltages is negligible even with large unbalanced loads.

The unfiltered output of the two rectifier banks – solid jagged lines in Figure 5.32 – shows the conduction period where the rectifiers are connected to the generator lines. The heavy, smooth lines are the filtered output of the cyclo-converter. Both rectifier banks are programmed to operate over the entire 360° of the output wave, and each bank can supply either voltage polarity. The positive half of the output voltage wave is formed by operating either the positive bank in the rectifying mode or the negative bank in the inverting mode.

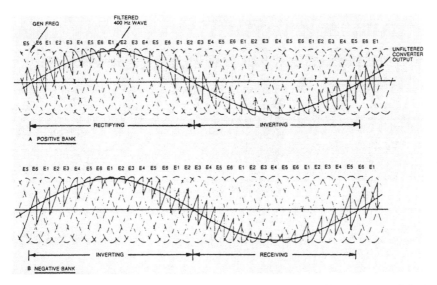

Figure 5.32 VSCF 400 Hz waveform formulation (Courtesy of Leland Electrosystems / GE Aviation)

Figure 5.33 Leland VSCF cycloconverter assemblies (Courtesy of Leland Electrosystems / GE Aviation)

The negative half of the output wave is formed in reverse fashion. The rectifying and inverting modes define the direction of power flow; towards the load in the rectifying mode and toward the source in the inverting mode.

Some of the physical attributes of the 60/65 kVA machine are shown in Figures 5.33 and 5.34. This particular version also embodies PMGs capable of supplying three independent channels of 28 VDC regulated power to feed flight control and other essential loads. A simplified version of the F-18E/F electrical system is shown in Figure 5.35.

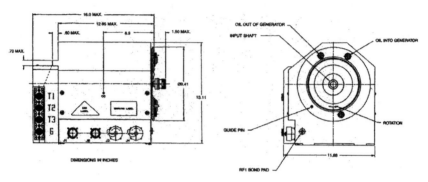

Figure 5.34 Leland VSCF cycloconverter – dimensions (Courtesy of Leland Electrosystems / GE Aviation)

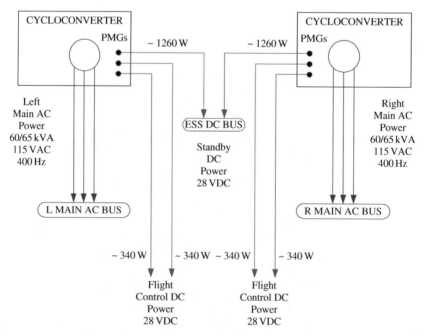

Figure 5.35 Simplified F-18E/F electrical power system

5.11.3 270 VDC Systems

An initiative which has been underway for a number of years in the US military development agencies is the 270 VDC system. The US Navy has championed this concept and the technology has developed to the point that some of the next generation of US combat aircraft will have this system imposed as a tri-Service requirement. The aircraft involved are the US Air Force Advanced Tactical Fighter (ATF) (now the Lockheed F-22 Raptor), the former US Navy Advanced Tactical Aircraft (ATA) or A-12, and the US Army Light Helicopter (LHX or LH) (now known as RAH-66 Comanche). More recent projects noted in Table 5.1 included the Joint Strike Fighter (JSF) offerings from Lockheed Martin (X-35A/B/C) and the Boeing (X-32A/B/C), although the latter was reportedly a predominantly VF 115 VAC system with some power conversion for 270 VDC loads. The selected version of JSF – the Lockheed Martin F-35 Lightning II uses 270 VDC for the primary electrical system.

The use of 270 VDC is an extrapolation of the rationale for moving from 28 VDC to 115 VAC: reduction in the size of current carrying conductors thereby minimising weight, voltage drop and power dissipation. There are, however, a number of disadvantages associated with the use of 270 VDC. 270 VDC components are by no means commonplace; certainly were not so at the beginning of development and even now are not inexpensive. Also, a significant number of aircraft services will still require 28 VDC or 115 VAC supplies and the use of higher voltages places greater reliance on insulation techniques to avoid voltage breakdown. The US military addressed these technical issues through a wide range of funded technology development and demonstrator programmes. Some of these are also directed at the greater use of electrical power on the combat aircraft, possibly to supplant conventional secondary power and hydraulic power systems or at least to augment them to a substantial degree. The term for these developments is the More-Electric Aircraft (MEA), implying a much greater if not total use of electrical power for aircraft systems.

The high DC voltage poses a risk in military aircraft of increased possibility of fire resulting from battle damage in carbon-fibre composite aircraft. Care must be taken to reduce the risk of arcing at high altitudes or in humid salt laden air conditions such as tropical or maritime environments. There is also a potential lethal hazard to ground crew during servicing operations. All these must be taken into account in design.

One of the problems in moving to 270 VDC is that there is still a need for the conventional 115 VAC and 28 VDC voltages for some equipment as mentioned above. The 270 VDC aircraft therefore becomes a somewhat hydrid system as shown in Figure 5.36 that may lose some of the original 270 VDC advantages.

5.11.4 More-Electric Aircraft (MEA)

For at least the last twenty years a number of studies have been under way in the US that have examined the all-electric aircraft. As stated earlier, aircraft developed in the UK in the late 1940s/early 1950s, such as the V-Bombers,

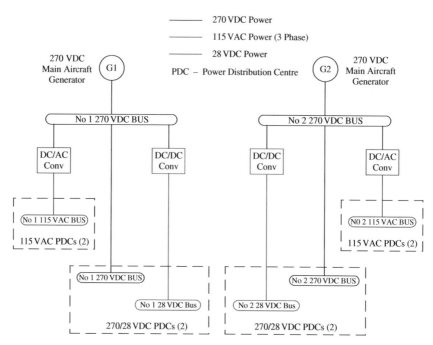

Figure 5.36 Simplified F-22 electrical system

utilised electric power to a greater extent than present day aircraft. In the 1980s, a number of studies promoted by NASA, the US Navy, US Air Force development agencies, and undertaken by Lockheed and Boeing, addressed the concept in detail. The topic is covered in this book under Chapter 10 – Advanced Systems, since the implications of the MEA are more embracing than merely organising the aircraft electrical system in a different manner. The concept addresses more energy-efficient ways of converting and utilising aircraft power in the broadest sense and therefore has a far-reaching effect upon overall aircraft performance [8].

More-electric technology has progressed tremendously over the past decade and More-Electric Aircraft (MEA) and More-Electric Engines (MEE) developments are described in full together with some of the associated applications in Chapter 10 – Advanced Systems.

5.12 Recent Electrical System Developments

Three major aircraft programmes under way illustrate in different ways the architectures and concepts that have evolved since the turn of the millennium.These projects are:

- Airbus A380
- Airbus 400M
- Boeing 787

Each of these systems is described.

5.12.1 Airbus A380 Electrical System Overview

The A380 was the first large civil aircraft in recent times to re-adopt variable frequency (VF), or 'frequency wild' as it was formerly called, since some of the turboprop airliners of the 1950s and early 1960s.

A380 Power Generation System Overview

AC power generation
The key characteristics of the A380 electrical power generation systems are as follows:

- 4 × 150 kVA VF Generators (370–770 Hz). VF generators are reliable but do not offer a no-break power capability
- 2 × 120 kvA CF APU Generators (nominal 400 Hz)
- 4 × External Power Connections (400 Hz) for ground power
- 1 × 70 kVA Ram Air Turbine for emergency use

See Figure 5.37.

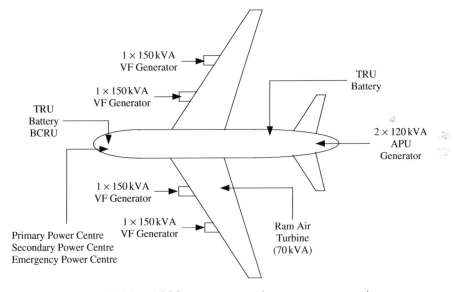

Figure 5.37   A380 – power system components

The 150 kVA per primary power channel represented an increase over previous civil aircraft. Hitherto the most powerful had been the Boeing 777 with 120 kVA (CF) plus 20 kVA (VSCF Backup) representing 140 kVA per channel.

The AC power system architecture is shown in Figure 5.38. Each of the main 150 kVA AC generators is driven by the associated engine. The two APU generators are driven by the respective Auxiliary Power Unit (APU). Each main generator supplies power to the appropriate AC bus under the control of the GCU. Each main AC bus can also accept a ground power input for servicing and support activities on the ground. Because the aircraft generators are variable frequency (VF) and the frequency of the AC power depends upon the speed of the appropriate engine, the primary AC buses cannot be paralled.

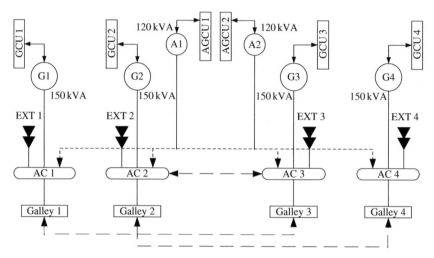

Figure 5.38   A380 AC electrical power architecture

The aircraft galleys which form a large proportion of the aircraft load are split between each of the four AC buses as shown.

DC System

The key characteristics of the A380 DC power conversion and energy storage system are outlined below:

- 3 × 300A Battery Charge Regulator Units BCRU; these are regulated TRUs
- 1 × 300A TRU
- 3 × 50 Ah Batteries
- 1 × Static Inverter

The DC system provides a no-break power capability thereby permitting key aircraft systems to operate without power interruption during changes in system configuration. Most control computers or IMA cabinets are DC powered and the use of DC paralleling techniques facilitates the provision of no-break power for these crucial elements. See Figure 5.39.

The figure shows how the AC buses 1 to 4 (AC1 to AC4) feed the main DC system power conversion units. The Ram Air Turbine (RAT) feeds the AC Ess bus as do main AC buses AC1 and AC4. The AC Ess bus in turn feeds an AC Emer bus which can also be powered from the DC Ess bus through a static inverter. AC1/AC4, AC2 and AC3 respectively feed the DC Ess, DC1 and DC 2 buses that are regulated to 28Vdc since the BCRUs are effectively regulated

TRUs. Each of these buses has an associated 50Ah battery whose charge is maintained by the charging function of the BCRU.

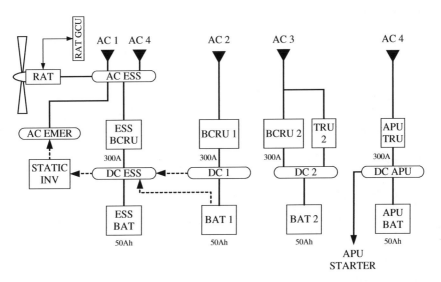

Figure 5.39   A380 DC electrical power architecture

For APU Starting the following dedicated subsystem is provided:

- 1 × 300 A APU TRU
- 1 × 50 Ah TRU Battery

Electrical System Control

The control of the electrical system is vested in a combination of dedicated units and some of the IMA CPIOMs as described below:

Control resident in dedicated control units:

- 4 × Main generator GCUs
- 2 X APU generator GCUs
- 1 X RAT generator GCU

Control resident in IMA CPIOMs:

- Electrical load management function – controlling load shed
- Secondary load monitoring function – monitoring the status of secondary power distribution devices

System Segregation

In broad terms the total aircraft electrical system is segregated as shown in Figure 5.40 into four main channels:

- E1 Channel powered by AC generators 1 and 2
- E2 Channel powered by AC generators 3 and 4
- E3 Channel powered by the RAT and the static inverter
- APU Channel associated with APU start

The E1, E2, and E3, channels each have an associated main generator, BCRU and 50Ah battery associated with them to give effectively three independent channels of power:

- E1 Channel: AC2 + BCRU1 + Battery 1
- E2 Channel: AC3 + BCRU2 + Battery 2
- E3 Channel: AC2(AC4) + Ess BCRU + Ess Battery

The effect of this electrical channel segregation may be better understood by cross-referring to the A380 FCS description in Chapter 1 – Flight Controls.

Power Distribution System

The power switching and protection devices that form the aircraft power distribution system are vested in the following electrical panels:

- 1 × Integrated Primary Electrical Power Distribution Centre (PEPDC)
- 2 × Secondary Electrical Power Distribution Centres (SEPDCs) for aircraft loads; these panels are used to distribute power to smaller electrical loads consuming < 15A per phase or less than 5 kVA
- 6 × Secondary Electrical Power Distribution Boxes (SEPDBs) distributing power to domestic loads; domestic loads are those associated with the cabin and passenger comfort as opposed to aircraft systems and are described below; these units are geographically dispersed within the aircraft to be close to their respective loads, thereby minimising feeder weight
- Solid State Power Controllers (SSPCs) are used in preference to thermal circuit breakers for secondary power distribution

  Domestic loads include: cabin lighting $\sim 15\,$kVA; galleys $\sim 120 - 240\,$kVA, intermittent load depending upon the meal service); galley cooling $\sim 90\,$kVA, permanent load; In Flight Entertainment (IFE) $\sim 50 - 60\,$kVA or about 100 W/seat permanent load

- The power distribution functions embedded in the appropriate CPIOM IMA modules are:

  - Electrical load management function assuring optimum loading of the aircraft buses according to the electrical power resource available and shedding load as appropriate
  - Circuit breaker monitoring function where circuit breakers are used

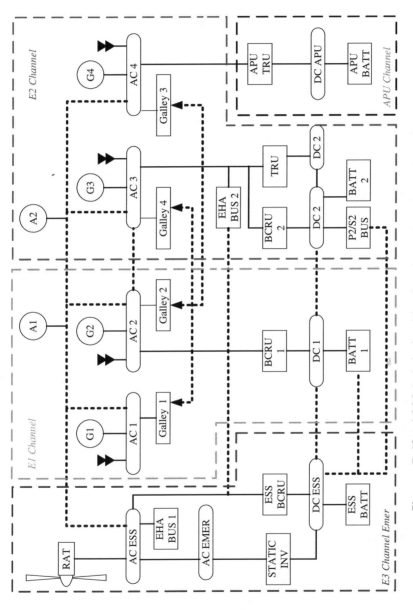

Figure 5.40 A380 total electrical system showing segregation

5.12.2 A400M

The A400M is a European joint project to develop a military transport to replace a number of platforms. The A400M borrows much of the electrical power technology from the A380 and also uses the common avionics IMA/CPIOM architecture.

The key points of the A400M AC architecture are:

- $4 \times 75\,kVA$ VF generators operating over 390–620 Hz frequency range
- $1 \times 90\,kVA$ APU generator operating at a nominal 400 Hz
- $1 \times 43\,kVA$ RAT
- $1 \times 90\,kVA$ ground power connection

The DC system has almost identical features to the A380 system:

- $3 \times$ Battery Charger Rectifier Units (BCRUs) rated at 400 A
- 1×300 A TRU which also supports APU starting
- 3×40 Ah NiCd batteries

The higher rating of the BCRUs (400 A versus 300 A) results from the higher DC loads on the military platform. DC paralleling techniques provide DC no-break power as for the A380.

5.12.3 B787 Electrical Overview

The Boeing 787 now in the late stages of prototype build has many novel more-electric aircraft features. The aircraft is a large step towards the all-electric airplane – one in which all systems are run by electricity. Bleed air from the engines has essentially been eliminated and while hydraulic actuators are still used, the majority of their power comes from electricity.

In breaking with five decades of practice, Boeing claims that electric compressors are better suited for the cabin than engine bleed and have many savings.

Boeing 787 Electrical Power System

The B787 electrical power system is portrayed at a top-level in Figure 5.41. A key feature is the adoption of three-phase 230 VAC electric power compared with the conventional three-phase 115 VAC arrangement usually used. The increase in voltage by a factor of 2:1 decreases feeder losses in the electrical distribution system and allows significant wiring reduction. The use of higher 230 VAC phase voltage, or 400 VAC line-to-line, does require considerable care during design to avoid the possible effects of partial discharge, otherwise known as 'corona'.

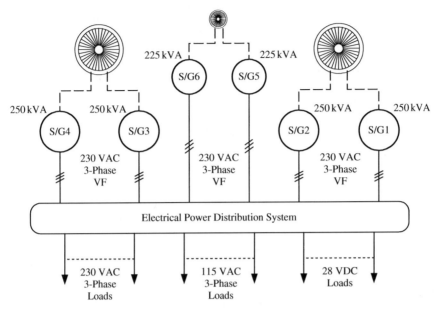

Figure 5.41 Boeing 787 top-level electrical system

The salient features of the B787 electrical power system are:

- 2 × 250 kVA starter/generators per engine, resulting in 500 kVA of generated power per channel. The generators are variable frequency (VF) reflecting recent industry traits in moving away from constant frequency (CF) 400 Hz power
- 2 × 225 kVA APU starter/generators, each starter/generator driven the APU. Each main generator feeds its own 230 VAC main bus before being fed into the power distribution system. As well as powering 230 VAC loads, electrical power is converted into 115 VAC and 28 VDC power to feed many of the legacy subsystems that require these more conventional supplies

A summary of the B787 electrical loads is given in Figure 5.42. As bleed air is no longer used within the airframe there are no air feeds to the environmental control system, cabin pressurisation system, wing anti-icing system as well as other air-powered subsystems. The only bleed air taken from the engine is low-pressure fan air used to anti-ice the engine cowl. Tapping bleed air off the engine compressor is extremely wasteful, especially as engine pressure ratios and bypass ratios increase on modern engines such as the General Electric GeNex and Rolls-Royce Trent 1000. An additional saving is removal of the overhead of providing large ducts throughout the airframe to transport the air; typically 8 inch diameter ducts are required between engine and airframe and 7 inch ducts between APU and airframe and in the Air Driven Pump (ADP) feed. In some parts of the airframe overheat detection systems are required to warn the flight crew of hot gas leaks.

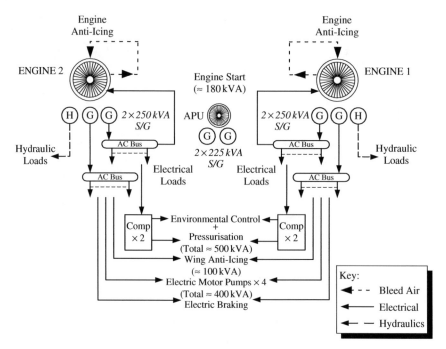

Figure 5.42 Boeing 787 electrical loads

The main more-electric loads in the B787 system are:

- *Environmental Control System (ECS) & Pressurisation.* The removal of bleed air means that air for the ECS and pressurisation systems needs to be pressurised by electrical means; on the B787 four large electrically driven compressors are required drawing in the region of 500 kVA
- *Wing Anti-Icing.* Non-availability of bleed air means that wing anti-icing has to be provided by electrical heating mats embedded in the wing leading edge. Wing anti-icing requires in the order of 100 kVA of electrical power
- *Electric Motor Pumps.* Some of the aircraft hydraulic Engine Driven Pumps (EDPs) are replaced by electrically driven pumps. The four new electrical motor pumps require ~ 100 kVA each giving a total load requirement of 400 kVA

A further outcome of the adoption of the 'bleedless engine' is that the aircraft engines cannot be started by the conventional means: high pressure air. The engines use the in-built starter/generators for this purpose and require ~ 180 kVA to start the engine

The introduction of such high-powered electrical machines has a significant impact upon the aircraft electrical distribution system. The electrical power distribution system is shown in Figure 5.43.

Primary power electrical power distribution is undertaken by four main distribution panels, two in the forward electrical equipment bay and two others

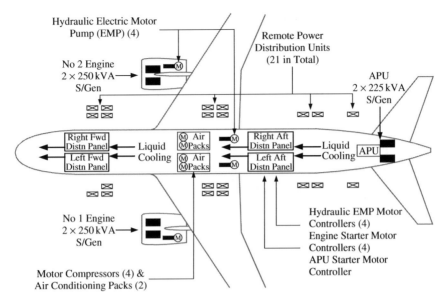

Figure 5.43 Boeing 787 electrical power distribution system

in the aft electrical equipment bay. The aft power distribution panels also contain the motor controllers for the four Electrical Motor Pumps (EMPs); two of the associated pumps are located in the engine pylons and two in the aircraft centre section. Also located within the aft distribution panels are the engine starter motor controllers (4) and APU starter motor controller (1). The high levels of power involved and associated power dissipation generate a lot of heat and the primary power distribution panels are liquid cooled.

The electrically powered air conditioning packs are located in the aircraft centre section.

Secondary power distribution is achieved by using Remote Power Distribution Units (RPDUs) located at convenient places around the aircraft. In all there are a total of 21 RPDUs located in the positions indicated in Figure 5.43.

5.13 Electrical Systems Displays

The normal method of displaying electrical power system parameters to the flight crew has been via dedicated control and display panels. On a fighter or twin-engined commuter aircraft the associated panel is likely to be fairly small. On a large transport aircraft the electrical systems control and display would have been achieved by a large systems panel forming a large portion of the flight engineer's panel showing the status of all the major generation and power conversion equipment. With the advent of two crew flight deck operations, of which the Boeing 757, 767, 747-400 and Airbus A320 and indeed most modern aircraft are typical examples, the electrical system selection panel was moved into the flight crew overhead panel. EICAS or ECAM systems now permit the display of a significant amount of information by the use of:

- Synoptic displays
- Status pages
- Maintenance pages

These displays show in graphic form the system operating configuration together with the status of major system components, key system operating parameters and any degraded or failure conditions which apply. The maximum use of colour will greatly aids the flight crew in assimilating the information displayed. The overall effect is vastly to improve the flight crew/system interface giving the pilots a better understanding of the system operation while reducing the crew workload.

References

[1] Bonneau, V. (1998) 'Dual-Use of VSCF Cycloconverter', FITEC'98, London.
[2] Boyce, J.W. 'An Introduction to Smart Relays'. Paper presented at the SAE AE-4 Symposium.
[3] Johnson, W., Casimir, B., Hanson, R., Fitzpatrick, J., Pusey, G., 'Development of 200 Ampere Regulated Transformer Rectifier', SAE, Mesa
[4] Wall, M.B., 'Electrical Power System of the Boeing 767 Airplane'.
[5] Layton, S.G., 'Solid State Power Control', ERA Avionics Conference, London Heathrow.
[6] Tenning, C., 'B777 Electrical System', RAeS Conference, London.
[7] Rinaldi, M.R., 'A Highly Reliable DC Power Source for Avionics Subsystems', SAE Conference.
[8] Mitcham, A.J. (1999) 'An Integrated LP Shaft Generator for the More-Electric Aircraft', IEE Colloquium, London.
[9] Haller, J.P., Weale, D.V., Loveday, R.G. (1998) 'Integrated Utilities Control for Civil Aircraft', FITEC'98, London.

6

Pneumatic Systems

6.1 Introduction

The modern turbofan engine is effectively a very effective gas generator and this has led to the use of engine bleed air for a number of aircraft systems, either for reasons of heating, provision of motive power or as a source of air for cabin conditioning and pressurisation systems. Bleed air is extracted from the engine compressor and after cooling and pressure reduction/regulation it is used for a variety of functions.

In the engine, high pressure bleed air is used as the motive power – sometimes called 'muscle power' – for many of the valves associated with the bleed air extraction function. Medium-pressure bleed air is used to start the engine in many cases, either using air from a ground power unit, APU or cross-bled from another engine on the aircraft which is already running. Bleed air is also used to provide anti-ice protection by heating the engine intake cowling and it is also used as the motive power for the engine thrust reversers.

On the aircraft, bleed air tapped from the engine is used to provide air to pressurise the cabin and provide the source of air to the cabin conditioning environmental control system. A proportion of bleed air is fed into air conditioning packs which cools the air dumping excess heat overboard; this cool air is mixed with the remaining warm air by the cabin temperature control system such that the passengers are kept in a comfortable environment. Bleed air is also used to provide main wing anti-ice protection.

Bleed air is also used for a number of ancillary functions around the aircraft: pressurising hydraulic reservoirs, providing hot air for rain dispersal from the aircraft windscreen, pressurising the water and waste system and so on. In some aircraft Air Driven Pumps (ADPs) are used as additional means of providing aircraft hydraulic power.

Pitot static systems are also addressed in the pneumatic chapter, as although this is a sensing system associated with measuring and providing essential air data parameters for safe aircraft flight, it nonetheless operates on pneumatic

Aircraft Systems: Mechanical, electrical, and avionics subsystems integration. 3rd Edition   Ian Moir and Allan Seabridge
© 2008 John Wiley & Sons, Ltd

principles. Pitot systems have been used since the earliest days of flight using pneumatic, capsule based mechanical flight instruments. The advent of avionics technology led first to centralised Air Data Computers (ADCs) and eventually on to the more integrated solutions of today such as Air Data & Inertial Reference System (ADIRS).

Pneumatic power is the use of medium pressure air to perform certain functions within the aircraft. While the use of pneumatic power has been ever present since aircraft became more complex, the evolution of the modern turbojet engine has lent itself to the use of pneumatic power, particularly on the civil airliner.

The easy availability of high pressure air from the modern engine is key to the use of pneumatic power as a means of transferring energy or providing motive power on the aircraft. The turbojet engine is in effect a gas generator where the primary aim is to provide thrust to keep the aircraft in the air. As part of the turbojet combustion cycle, air is compressed in two or three stage compressor sections before fuel is injected in an atomised form and then ignited to perform the combustion process. The resulting expanding hot gases are passed over turbine blades at the rear of the engine to rotate the turbines and provide shaft power to drive the LP fan and compressor sections. When the engine reaches self-sustaining speed the turbine is producing sufficient shaft power to equal the LP fan/compressor requirements and the engine achieves a stable condition – on the ground this equates to the ground idle condition. The availability of high pressure, high temperature air bled from the compressor section of the engine lends itself readily to the ability to provide pneumatic power for actuation, air conditioning or heating functions for other aircraft subsystems.

Other areas of the aircraft use pneumatic principles for sensing the atmosphere surrounding the aircraft for instrumentation purposes. The sensing of air data is crucial to ensuring the safe passage of the aircraft in flight.

6.2 Use of Bleed Air

The use of the aircraft engines as a source of high-pressure, high-temperature air can be understood by examining the characteristics of the turbojet, or turbofan engine as it should more correctly be described. Modern engines 'bypass' a significant portion of the mass flow past the engine and increasingly a small portion of the mass flow passes through the engine core or gas generation section. The ratio of bypass air to engine core air is called the bypass ratio and this can easily exceed 10:1 for the very latest civil engines; much higher than the 4 or 5:1 ratio for the previous generation.

The characteristics of a modern turbofan engine are shown in Figure 6.1. This figure shows the pressure (in psi) and the temperature (in degrees centigrade) at various points throughout the engine for three engine conditions:ground idle, take-off power and in the cruise condition.

It can be seen that in the least stressful condition – ground idle – the engine is in a state of equilibrium but that even at this low level the compressor

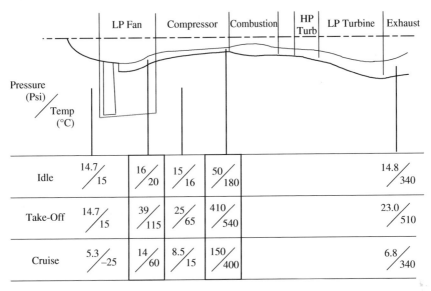

	LP Fan	Compressor	Combustion	HP Turb	LP Turbine	Exhaust
Idle	14.7/15	16/20	15/16	50/180		14.8/340
Take-Off	14.7/15	39/115	25/65	410/540		23.0/510
Cruise	5.3/−25	14/60	8.5/15	150/400		6.8/340

Figure 6.1   Characteristics of a modern turbofan engine

air pressure is 50 psi and the temperature 180 °C. At take-off conditions the compressor air soars to 410 psi/540 °C. In the cruise condition the compressor air is at 150 psi/400 °C. The engine is therefore a source of high pressure and high temperature air that can be 'bled' from the engine to perform various functions around the aircraft. The fact that there are such considerable variations in air pressure and temperature for various engine conditions places an imposing control task upon the pneumatic system. Also the variations in engine characteristics between similarly rated engines of different manufactures poses additional design constraints. Some aircraft such as the Boeing 777 offer three engine choices, Pratt & Whitney, General Electric and Rolls-Royce, and each of these engines has to be separately matched to the aircraft systems, the loads of which may differ as a result of operator specified configurations.

As well as the main aircraft engines the Auxiliary Power Unit (APU) is also a source of high pressure bleed air. The APU is in itself a small turbojet engine, designed more from the viewpoint of an energy and power generator than a thrust provider which is the case for the main engines. The APU is primarily designed to provide electrical and pneumatic power by a shaft driven generator and compressor. The APU is therefore able to provide an independent source of electrical power and compressed air while the aircraft is on the ground, although it can be used as a backup provider of power while airborne. Some aircraft designs are actively considering the use of in-flight operable APUs to assist in in-flight engine re-lighting and to relieve the engines of offtake load in certain areas of the flight envelope.

It is also usual for the aircraft to be designed to accept high pressure air from a ground power cart, for aircraft engine starting.

These three sources of pneumatic power provide the muscle or means by which the pneumatic system is able to satisfy the aircraft demands. In a simplified form the pneumatic system may be represented by the interrelationships shown in Figure 6.2 below.

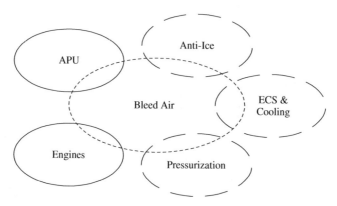

Figure 6.2   Relationship of HP air with major aircraft systems

This simplified drawing – the ground air power source is omitted – shows how the aircraft High Pressure (HP) air sources provide bleed air which forms the primary source for the three major aircraft air related systems:

- *Ice protection*: the provision of hot air to provide anti icing of engine nacelles and the wing, tailplane or fin leading edges; or to dislodge ice that has formed on the surfaces
- *ECS and cooling*: the provision of the main air source for environmental temperature control and cooling
- *Pressurisation*: the provision of a means by which the aircraft may be pressurised, giving the crew and passengers a more comfortable operating environment

A simplified representation of this relationship is shown in Figure 6.3. This example shows a twin-engine configuration typical of many business jets and regional jet transport aircraft.

Bleed air from the engines is passed through a Pressure-Reducing Shut-Off Valve (PRSOV) which serves the function of controlling and, when required, shutting off the engine bleed air supply. Air downstream of the PRSOV may be used in a number of ways:

- By means of a cross flow Shut-Off Valve (SOV) the system may supply air to the opposite side of the aircraft during engine start or if the opposite engine is inoperative for any reason
- A SOV from the APU may be used to isolate the APU air supply

- SOVs provide isolation as appropriate to the left and right air conditioning packs and pressurisation systems
- Additional SOVs provide the means by which the supply to left and right wing anti-icing systems may be shut off in the event that these functions are not required

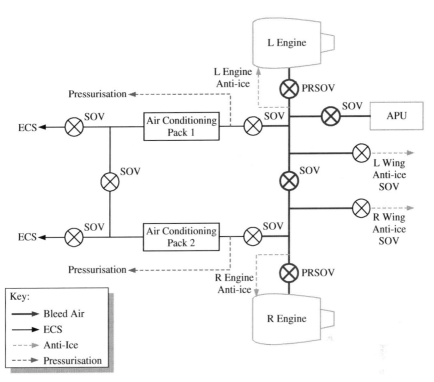

Figure 6.3 Simplified bleed air system and associated aircraft systems

This is a simplified model of the use of engine bleed air in pneumatic systems. A more comprehensive list of those aircraft systems with which bleed air is associated are listed as follows with the accompanying civil ATA chapter classification:

- Air conditioning (ATA Chapter 21)
- Cargo compartment heating (ATA Chapter 21)
- Wing and engine anti-icing (ATA Chapter 30)
- Engine start (ATA Chapter 80)
- Thrust reverser (ATA Chapter 78)
- Hydraulic reservoir pressurisation (ATA Chapter 29)
- Rain repellent nozzles – aircraft windscreen (ATA Chapter 30)
- Water tank pressurisation and toilet waste (ATA Chapter 38)
- Air driven hydraulic pump (ADP) (ATA Chapter 29)

Several examples will be examined within this pneumatic systems chapter. However, before describing the pneumatically activated systems it is necessary to examine the extraction of bleed air from the engine in more detail.

6.3 Engine Bleed Air Control

Figure 6.4 gives a more detailed portrayal of the left-hand side of the aircraft bleed air system, the right side being an identical mirror image of the left-hand side.

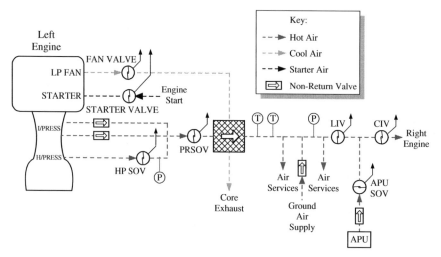

Figure 6.4 Typical aircraft bleed air system – left hand side

Air is taken from an intermediate stage or high pressure stage of the engine compressor depending upon the engine power setting. At lower power settings, air is extracted from the high pressure section of the compressor while at higher power settings the air is extracted from the intermediate compressor stage. This ameliorates to some degree the large variations in engine compressor air pressure and temperature for differing throttle settings as already shown in Figure 6.1. A pneumatically controlled High Pressure Shut-Off Valve (HP SOV) regulates the pressure of air in the engine manifold system to around 100 psi and also controls the supply of bleed air from the engine.

The Pressure-Reducing Shut-Off Valve (PRSOV) regulates the supply of the outlet air to around 40 psi before entry into the pre-cooler. Flow of cooling air through the pre-cooler is regulated by the fan valve which controls the temperature of the LP fan air and therefore of the bleed air entering the aircraft system. Appropriately located pressure and temperature sensors allow the engine bleed air temperature and pressure to be monitored and controlled within specified limits.

A typical PRSOV is shown in Figure 6.5a; an example of a Harrier II valve which is solenoid controlled and pneumatically operated and which controls temperature, flow and pressure is shown in Figure 6.5b.

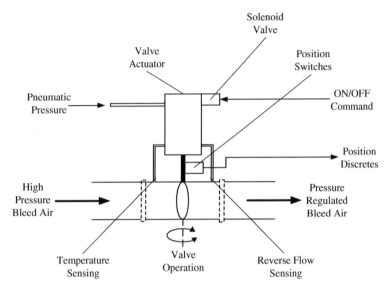

Figure 6.5a Typical Pressure-Reducing Shut-Off Valve (PRSOV)

Figure 6.5b Harrier II pneumatic valve (Courtesy of Honeywell Normalair-Garret Ltd)

The PRSOV performs the following functions:

- On/off control of the engine bleed system
- Pressure regulation of the engine supply air by means of a butterfly valve actuated by pneumatic pressure
- Engine bleed air temperature protection and reverse flow protection
- Ability to be selected during maintenance operations in order to test reverse thrust operation

The PRSOV is pneumatically operated and electrically controlled. Operation of the solenoid valve from the appropriate controller enables the valve to control the downstream pressure pneumatically to ~40 psi within predetermined limits. The valve position is signalled by means of discrete signals to the bleed air controller and pressure switches provide over and under-pressure warnings. The various pressure, flow and discrete signals enable the bleed air controller Built-In Test (BIT) to confirm the correct operation of the PRSOV and fan control valve combination. This ensures that medium pressure air (~40 psi) of the correct pressure and temperature is delivered to the pre-cooler and thence downstream to the pneumatic and air distribution system.

Downstream of the PRSOV and pre-cooler, the air is available for the user subsystems, a number of which are described below.

A number of isolation valves or SOVs are located in the bleed air distribution system. These valves are usually electrically initiated, pneumatically operated solenoid valves taking 28 VDC electrical power for ON/OFF commands and indication. A typical isolation valve is shown in Figure 6.6. The valve shaft runs almost vertically across the duct as shown in the diagram and the valve mechanism and solenoid valve is located on the top of the valve.

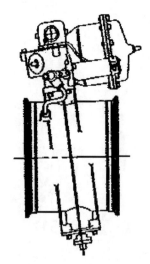

Figure 6.6   Bleed air system isolation valve

6.4 Bleed Air System Indications

It is common philosophy in civil aircraft bleed air systems, in common with other major aircraft subsystems, to display system synoptic and status data to the flight crew on the Electronic Flight Instrument System (EFIS) displays. In the case of Boeing aircraft the synoptics are shown on the Engine Indication and Crew Alerting System (EICAS) display whereas for Airbus aircraft the Electronic Crew Alerting and Monitoring (ECAM) displays are used. Both philosophies display system data on the colour displays located on the central display console where they may be easily viewed by both Captain and First Officer. A typical bleed air system synoptic is shown in Figure 6.7.

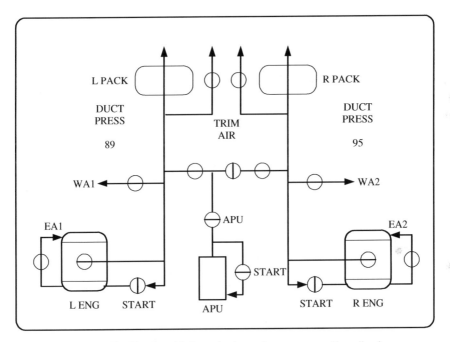

Figure 6.7 Typical bleed air system synoptic display

The synoptic display as shown portrays sufficient information in a pictorial form to graphically show the flight crew the operating status of the system. In the example, both main engines are supplying bleed air normally but the APU is isolated. The cross-flow valve is shut, as are both engine start valves. The wing and engine anti-ice valves are open, allowing hot bleed air to be fed to the engines and wing leading edge to prevent any ice accretion.

6.5 Bleed Air System Users

The largest subsystem user of bleed air is the air system. Bleed air is used as the primary source of air into the cabin and fulfils the following functions:

- Cabin environmental control – cooling and heating
- Cabin pressurisation
- Cargo bay heating
- Fuel system pressurisation in closed vent fuel system used in some military aircraft

Chapter 7 – Environmental Control – addresses the air systems. However there are other subsystems where the use of engine bleed air is key. These subsystems are:

- Wing and engine anti-ice protection
- Engine start
- Thrust reverser actuation
- Hydraulic system

6.5.1 Wing and Engine Anti-Ice

The protection of the aircraft from the effects of aircraft icing represents one of the greatest and flight critical challenges which confront the aircraft. Wing leading edges and engine intake cowlings need to be kept free of ice accumulation at all times. In the case of the wings, the gathering of ice can degrade the aerodynamic performance of the wing, leading to an increased stalling speed with the accompanying hazard of possible loss of aircraft control. Ice that accumulates on the engine intake and then breaks free entering the engine can cause substantial engine damage with similar catastrophic results. Considerable effort is also made to ensure that the aircraft windscreens are kept clear of ice by the use of window heating so that the flight crew has an unimpeded view ahead. Finally, the aircraft air data sensors are heated to ensure that they do not ice up and result in a total loss of air data information that could cause a hazardous situation or the aircraft to crash. The prevention of ice build-up on the windscreen and air data system probes is achieved by means of electric heating elements. In the case of the wing and engine anti-icing the heating is provided by hot engine bleed air which prevents ice forming while the system is activated.

The principles of wing anti-ice control are shown in Figure 6.8. The flow of hot air to the outer wing leading edges is controlled by the Wing Anti-Ice Valve. The air flow is modulated by the electrically enabled anti-icing controller; this allows air to pass down the leading edge heating duct. This duct can take the form of a pipe with holes appropriately sized to allow a flow of air onto the inner surface of the leading edge – sometimes known as a 'piccolo tube'. The pressure of air in the ducting is controlled to about 20–25 psi. Telescopic ducting is utilised where the ducting moves from fixed wing to movable slat structure and flexible couplings are used between adjacent slat sections. These devices accommodate the movement of the slat sections relative to the main wing structure as the slats are activated. The air is bled out into the leading edge slat section to heat the structure before being dumped overboard.

A pressure switch and an overheat switch protect the ducting downstream of the wing anti-ice valve from over-pressure and over-temperature conditions.

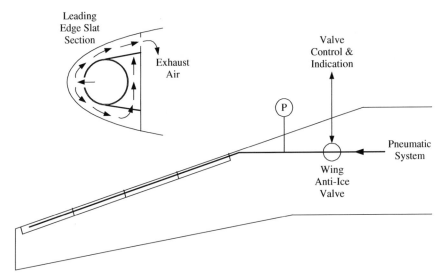

Figure 6.8   Wing anti-ice control

Engine anti-icing is similarly achieved. An Engine Anti-Ice (EAI) valve on the engine fan casing controls the supply of bleed air to the fan cowl in order to protect against the formation of ice. As in the case of the wing anti-ice function, activation of the engine anti-icing system is confirmed to the flight crew by means of the closure of a pressure switch that provides an indication to the display system.

The presence of hot air ducting throughout the airframe in the engine nacelles and wing leading edges poses an additional problem; that is to safeguard against the possibility of hot air duct leaks causing an overheat hazard. Accordingly, overheat detection loops are provided in sensitive areas to provide the crew with a warning in the event of a hot gas leak occurring. An overheat detection system will have elements adjacent to the air conditioning packs, wing leading edge and engine nacelle areas to warn the crew of an overheat hazard – a typical system is shown in Figure 6.9.

The operation of fire detection elements is described in Chapter 8 – Emergency Systems. In a civil airliner the hazardous areas are split into zones as shown in the figure. Each zone is served by two detection loops – Loop A and Loop B. Modern technology is capable not just of locating an overheat situation but locating the point of detection down stream to within about one foot, thereby giving more information as to where the leak has actually occurred. Civil systems employ a dual system to aid dispatch. It is possible to dispatch the aircraft with one loop inoperative for a specific operating period provided that assurance that remaining loop is operating correctly. This feature

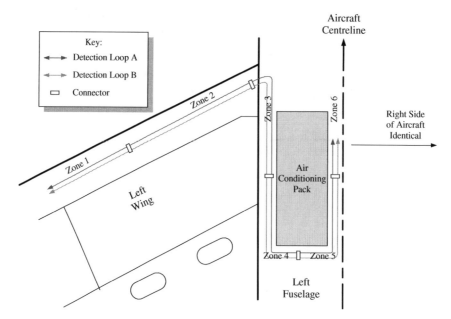

Figure 6.9 Typical overheat warning system

would allow the aircraft to recover to main base in order to have corrective maintenance action carried out.

A number of low speed commercial aircraft employ a method of de-icing based on a flexible rubber leading edge 'boot' that is inflated by air pressure to dislodge ice built up on the surface. The system is operated manually or in response to an ice detector input. The British Aerospace Advanced Turbo Prop (ATP) wing, tailplane and fin leading edges are protected by pneumatic rubber boots actuated by low-pressure engine compressor air. A cycling system is used to reduce the amount of air required. The ice is removed by successive inflation and deflation cycles of the boots. The crew is able to select light or heavy ice removal modes.

6.5.2 Engine Start

The availability of high pressure air throughout the bleed air system lends itself readily to the provision of motive power to crank the engine during the engine start cycle. As can be seen from earlier figures, a start valve is incorporated which can be activated to supply bleed air to the engine starter. On the ground the engines may be started in a number of ways:

- By use of a ground air supply cart
- By using air from the APU – probably the preferred means
- By using air from another engine which is already running

The supply of air activates a pneumatic starter motor located on the engine accessory gearbox. The engine start cycle selection enables a supply of fuel to the engine and provision of electrical power to the ignition circuits. The pneumatic starter cranks the engine to ~15–20% of full speed by which time engine ignition is established and the engine will pick up and stabilise at the ground-idle rpm.

6.5.3 Thrust Reversers

Engine thrust reversers are commonly used to deflect engine thrust forward during the landing roll-out to slow the aircraft and preserve the brakes. Thrust reversers are commonly used in conjunction with a lift dump function, whereby all the spoilers are simultaneously fully deployed, slowing the aircraft by providing additional aerodynamic drag while also dispensing lift. Thrust reversers deploy two buckets, one on each side of the engine, which are pneumatically operated by means of air turbine motor actuators to deflect the fan flow forward, thereby achieving the necessary braking effect when the aircraft has a 'weight-on-wheels' condition. The air turbine motor has an advantage in that it is robust enough to operate in the harsh temperature and acoustic noise environment associated with engine exhaust, where hydraulic or electrical motors would not be sufficiently reliable.

Interlock mechanisms are provided which prevent inadvertent operation of the thrust reversers in flight. The Tornado thrust reversers are selected by rocking the throttle levers outboard in flight. On touchdown a signal is sent by the engine control systems to an air turbine motor connected to a Bowden cable and a screw jack mechanism to deploy the buckets. Most modern civil aircraft with large fans divert fan discharge air (which is the source of most of the thrust). The more benign conditions here allow the use of hydraulic actuators. These systems are described elsewhere in the hydraulics chapter.

6.5.4 Hydraulic Systems

Pneumatic pressure is commonly used to pressurise the aircraft hydraulic reservoirs. Some Boeing aircraft – usually the wide bodies – also use pneumatic power or air-driven hydraulic pumps to augment the normal Engine Driven Pumps (EDPs) and AC Motor Driven Pumps (ACMPs) for certain phases of flight. Figure 6.10 shows a typical centre hydraulic power channel as implemented by the Boeing philosophy – this is shown in a hydraulic system context in Chapter 4 – Hydraulic Systems.

The hydraulic reservoir is pressurised using regulated bleed air from the pneumatic/bleed air system. Supply hydraulic fluid may be pressurised by the two alternate pumps:

- By means of the ACMP powered by three-phase 115 VAC electrical power
- By means of the Air Driven Pump (ADP) using pneumatic power as the source

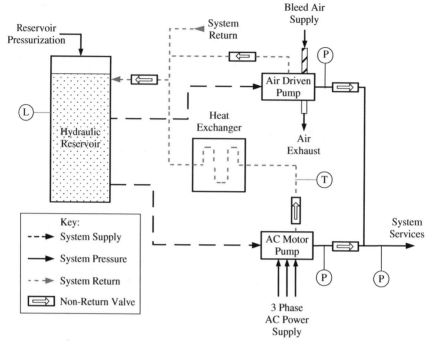

Figure 6.10 Simplified pneumatic system – hydraulic system interaction

Either pump in this hydraulic channel is able to deliver hydraulic pressure to the system services downstream; it is, however, more usual for the ACMP to be used as the primary source of power with the ADP providing supplementary or demand power for specific high demand phases of flight. The ACMP may be activated by supplying a command to a high power electrical contactor, or Electrical Load Management Unit (ELCU), as described in Chapter 5 – Electrical Systems. The pneumatic pressure driving the ADP is controlled by means of a 28 VDC powered solenoid controlled Modulating Shut-Off Valve (MSOV) upstream of the ADP. Hydraulic fluid temperature and pressure is monitored at various points in the system and the system information displayed on system synoptic or status pages as appropriate.

6.6 Pitot Static Systems

By contrast with the bleed air system already described which provides energy or power for a number of diverse aircraft systems, the pitot static system is an instrumentation system used to sense air data parameters of the air through which the aircraft is flying. Without the reliable provision of air data the aircraft is unable safely continue flight. The pitot static system is therefore a high integrity system with high levels of redundancy.

There are two key parameters which the pitot static system senses:

- Total pressure P_t is the sum of local static pressure and the pressure caused by the forward flight of the aircraft. The pressure related to the forward motion of the aircraft by the following formula:

  Pressure $= \frac{1}{2} \rho V^2$ Where ρ is the air density of the surrounding air and V is the velocity

- Static pressure or Ps is the local pressure surrounding the aircraft and varies with altitude

Therefore total pressure, $P_t = Ps + \frac{1}{2} \rho V^2$

The forward speed of the aircraft is calculated by taking the difference between P_t and P_s

An aircraft will have three or more independent pitot and static sensors

Figure 6.11 shows the principle of operation of pitot and sensors.

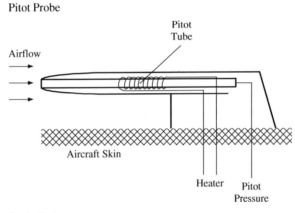

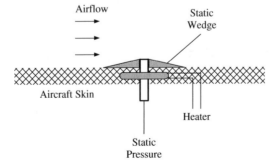

Figure 6.11   Pitot and static sensors

The pitot probe shown in the top diagram is situated such that it faces in the direction of the airflow, thereby being able to sense the variation in aircraft speed using the formula quoted above. The sensing portion of the pitot probe stands proud from the aircraft skin to minimise the effect of the boundary layer. Pitot pressure is required at all stages throughout flight and a heater element is incorporated to prevent the formation of ice that could block the sensor or create an erroneous reading. The pitot heating element is active throughout the entire flight.

The static probe shown in the lower diagram is located perpendicular to the airflow and so is able to sense the static pressure surrounding the aircraft. Like the pitot probe the static probe is provided with a heater element that continuously heats the sensor and prevents the formation of ice.

On some aircraft the pitot and static sensing functions are combined to give a pitot-static probe capable of measuring both dynamic and static pressures. A typical installation on a civil transport aircraft is depicted in Figure 6.12.

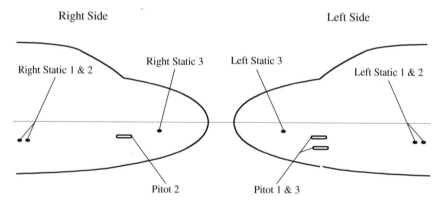

Figure 6.12 Typical pitot and static probe installation

This shows a configuration where three pitot probes are used; pitot 2 on the right side and pitot 1 and pitot 3 on the left side of the aircraft nose. Three static probes are located on the left and right sides of the aircraft. Pitot and static probes are carefully towards the nose of the aircraft such that the sensitive air data measurements are unaffected by other probes or radio antenna. Residual instrumentation errors due to probe location or installation are calibrated during the aircraft development phase and the necessary corrections applied further downstream in the system.

Fine bore tubing carries the sensed air data pressure – pitot and static – to the aircraft instruments or the air data suite. Due to the sensitivity of the sensed data, water drain traps are provided so that extraneous moisture such as condensation may be extracted from the pitot-static lines. Also, following the replacement of any part of the pipework or the destination instrument, leak checks have to be carried out to ensure pipework integrity.

The ways in which the air data is used to portray meaningful data to the crew by means of the aircraft instruments is shown in Figure 6.13.

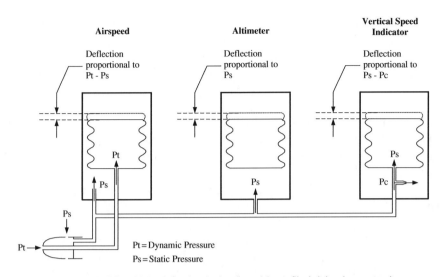

Figure 6.13 Use of air data to drive flight instruments

Three major parameters be calculated from the pitot-static pressure information sensed by the pitot and static probes or by a combined pitot-static probe as shown in the diagram:

- Airspeed may be calculated from the deflection in the left hand instrument where P_t and P_s are differentially sensed. Airspeed is proportionate to $P_t - P_s$ and therefore the mechanical deflection may be sensed and airspeed deduced. This may be converted into a meaningful display to the flight crew value in a mechanical instrument by the mechanical gearing between capsule and instrument dial
- Altitude may be calculated by the deflection of the static capsule in the centre instrument. Again in a mechanical instrument the instrument linkage provides the mechanical scaling to transform the data into a meaningful display
- Vertical speed may be deduced in the right hand instrument where the capsule deflection is proportional to the rate of change of static pressure with reference to a case pressure, P_e. Therefore the vertical speed is zero when the carefully sized bleed orifice between capsule inlet and case allows these pressures to equalise

The examples given above are typical for aircraft instruments used up to about 40 years ago. There are three methods of converting air data into useful aircraft related parameters etc. that the aircraft systems may use:

- On older aircraft conventional mechanical flight instruments may be used, these tend to be relatively unreliable, expensive to repair, and are limited in the information they can provide to an integrated system. Mechanical instruments are also widely used to provide standby or backup instrumentation

- On some integrated systems the pitot-static sensed pressures are fed into centralised Air Data Computers (ADCs). This allows centralisation of the air data calculations into dedicated units with computational power located in electrical bay racks. The ADCs can provide more accurate air data calculations more directly aligned to the requirements of a modern integrated avionics system. When combined with digital computation techniques within the ADC and the use of modern data buses such as MIL-STD-1553B, ARINC 429 and ARINC 629 to communicate with other aircraft systems, higher degrees of accuracy can be achieved and the overall aircraft system performance improved
- More modern civil aircraft developed in the late 1980s and beyond use Air Data Modules (ADMs) located at appropriate places in the aircraft to sense the pitot and static information as appropriate. This has the advantage that pitot-static lines can be kept to a minimum length reducing installation costs and the subsequent maintenance burden. By carefully selecting appropriate architecture greater redundancy and improved fault tolerance may be designed at an early stage, improving the aircraft dispatch availability

An example of a modern air data system using ADMs is shown in Figure 6.14. This architecture equates to the probe configuration installation shown in Figure 6.12, namely, three pitot probes and a total of six static probes, three each on the left and right hand side of the aircraft.

Figure 6.14 shows how these probes are connected to ADMs and the degree of redundancy that can be achieved:

- Each pitot probe is connected to an individual ADM so there is triple redundancy of pitot pressure sensing. Pitot probe 3 also connects to the mechanical standby Airspeed Indicator (ASI) that operates as shown in Figure 6.13
- The four static probes represented by static probes 1 and 2, left and right are connected to individual ADMs effectively giving quadruple redundancy of static pressure. Static probes left and right are physically interconnected and linked to a further ADM while also providing the static pressure sensing for the mechanical standby ASI and standby altimeter – see Figure 6.13
- Each of the eight ADMs shown in this architecture can be identical, since each is merely sensing an air data pressure parameter – pitot or static. The use of pin-programming techniques in the aircraft wiring means that an ADM may be installed in any location and will automatically adopt the personality required for that location
- The ADMs interconnect to the aircraft display and navigation systems by means of ARINC 429 data buses as shown in Figure 6.14

6.6.1 Innovative Methods of Pitot-Static Measurement

Conventional pitot-static sensing methods have been described. More recently the use of pitot-static sensing plates have been adopted; particularly on stealth

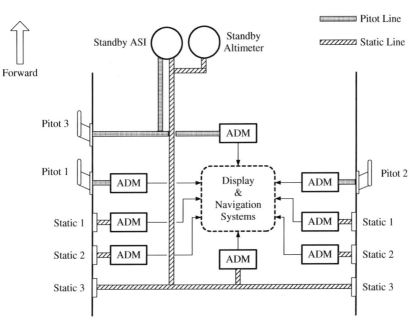

Figure 6.14   Air data system using ADMs

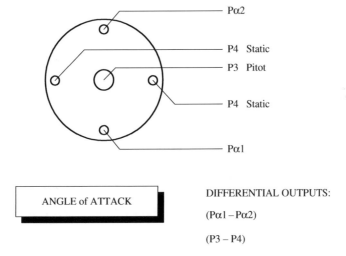

Figure 6.15   Angle of attack measurement

aircraft where the use of conventional pitot-static probes can severely compro-mise the aircraft low observeable radar signature. These pressure plates are able to derive data relating to:

- Pitot pressure
- Static pressure
- Angle of attack (α)
- Angle of sideslip (β)

These sensors are utilised on aircraft such as the B-2 Spirit stealth bomber and have reportedly recently been fitted to the F-22 Raptor. See Figure 6.15 for angle of attack and Figure 6.16 for angle of sideslip measurement configurations.

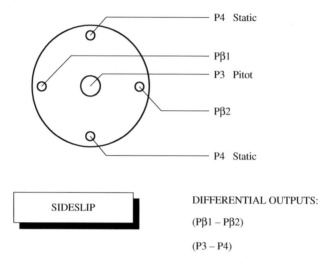

Figure 6.16   Angle of sideslip measurement

Data sheets produced by Goodrich relating to the various pitot-static sensing probes and vanes and the theory and computation behind air data sensing may be found in the following references [1 to 6].

References

[1] Pitot and Pitot-Static Probes; 4080 Lit 04/02 Marketing Publication; Rosemount Aerospace 2002.
[2] Angle of Attack Systems; 4070 Lit 03/02 Marketing Publication; Rosemount Aerospace 2002.
[3] Total Air Temperature Sensors; 4018 Lit 03/02 Marketing Publication; Rosemount Aerospace 2002.
[4] Multifunction Smart Probes; 4083 Lit 03/02 Marketing Publication; Rosemount Aerospace 2002.
[5] Multifunction Probes; 4015 Lit 08/04 Marketing Publication; Rosemount Aerospace 2002.
[6] Air Data Handbook; 4081 Lit 08/02 Marketing Publication; Rosemount Aerospace 2002.

7

Environmental Control Systems

7.1 Introduction

Throughout the operation of an aircraft, whether on the ground or in the air, the crew and passengers must be kept in comfortable conditions.They must be neither too hot nor too cold, they must have air to breathe and they must be kept in comfortable atmospheric pressure conditions. This is by no means easy, given the rapid changes in climatic conditions and internal temperatures seen by aircraft in flight from one destination to another.

A military aircraft may have only a small crew, but the aircraft may be designed to perform in climatic extremes ranging from Arctic to full desert sunlight. A commercial aircraft may carry over 300 fare-paying passengers. In neither case can the human cargo be subjected to extremes of discomfort – passengers will go to another airline and the military crew will not perform at their most effective.

The environmental control system must cope with widely differing temperature conditions, must extract moisture and provide air with optimum humidity, and must ensure that the air in the aircraft always contains a sufficient concentration of oxygen and that it is safe to breathe.

Modern systems do this and more, for the term 'environmental control' also includes the provision of suitable conditions for the avionic, fuel and hydraulic systems by allowing heat loads to be transferred from one medium to another.

In addition to these essentially comfort related tasks, environmental control systems provide de-misting, anti-icing, anti-g and rain dispersal services.

Aircraft Systems: Mechanical, electrical, and avionics subsystems integration. 3rd Edition   Ian Moir and Allan Seabridge
© 2008 John Wiley & Sons, Ltd

7.2 The Need for a Controlled Environment

In the early days of flight, pilots and passengers were prepared to brave the elements for the thrill of flying. However, as aircraft performance has improved and the operational role of both civil and military aircraft has developed, requirements for Environmental Control Systems (ECS) have arisen. They provide a favourable environment for the instruments and equipment to operate accurately and efficiently, to enable the pilot and crew to work comfortably, and to provide safe and comfortable conditions for the fare-paying passengers.

In the past, large heating systems were necessary at low speeds to make up for the losses to the cold air outside the aircraft. With many of today's military aircraft operating at supersonic speeds, the emphasis is more towards the provision of cooling systems, although heating is still required, for example on cold night flights and for rapid warm-up of an aircraft which has been soaked in freezing conditions on the ground for long periods. The retirement of Concorde has eliminated this as an issue for commercial aircraft. Providing sufficient heat for the aircraft air conditioning system is never a problem, since hot air can be bled from the engines to provide the source of conditioning air. The design requirement is to reduce the temperature of the air sufficiently to give adequate conditioning on a hot day. The worst case is that of cooling the pilot and avionics equipment in a high performance military aircraft [1]. The following heat sources give rise to the cooling problem:

7.2.1 Kinetic Heating

Kinetic heating occurs when the aircraft skin heats up due to friction between itself and air molecules. The skin, in turn, heats up the interior of the aircraft such as the cockpit and equipment bays. Skin temperatures can reach up to 100 °C or more in low-level flight at transonic speeds, and even higher temperatures can be reached in supersonic flight at medium and high altitudes. Figure 7.1. shows a typical flight envelope for a high performance military aircraft.

Note that in some flight cases, for example subsonic cruise at altitude on a cold day, kinetic heat loads can actually be negative. This is when heating is required.

Aircraft leading edges feel the full effect of kinetic heating due to friction and reach what are known as ram temperatures. All other surfaces away from the leading edges are subject to slightly lower temperatures termed recovery temperature. For design purposes, the following equations can be used to calculate ram and recovery temperatures:

$T_{rec} = T_{amb} \ (1 + 0.18 \ M^2)$
$T_{ram} = T_{amb} \ (1 + 0.2 \ M^2)$
$T_{rec} = \text{Recovery air temperature } °K$

T_{ram} = Ram air temperature °K
T_{amb} = Ambient air temperature °K
M = Mach number

Unconditioned equipment bays may reach recovery temperatures during flight.

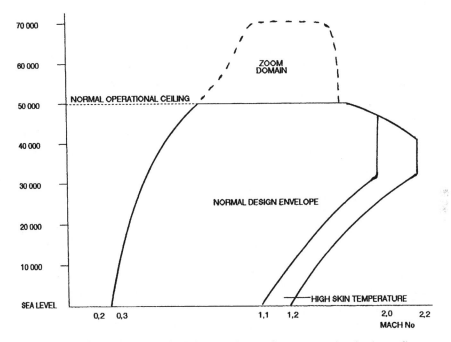

Figure 7.1 Typical flight envelope for a combat aircraft

7.2.2 Solar Heating

Solar radiation affects a military aircraft cockpit directly through the wind-screen and canopy.

Equipment bays and civil aircraft cabins are only affected indirectly. A fighter aircraft is the worst case, since it usually has a large transparent canopy to give the pilot good all round vision, and can fly typically up to twice the maximum altitude of a civil aircraft. At such altitudes solar radiation intensity is much higher. The combined effect of internal heating and direct solar radiation has an effect on the pilot, especially when wearing survival gear and anti-g trousers and vest which requires considerable cooling air in the cockpit.

Solar heating significantly affects both cabin and equipment bays on ground standby, since surfaces exposed to direct solar radiation will typically rise 20 °C above the ambient temperature, depending on the thermal capacity of the surface material. This is of special concern in desert areas of the world where the sun is hot and continuous throughout the day.

7.2.3 Avionics Heat Loads

While advances in technology have led to reductions in heat dissipation in individual electronic components, the increased use of avionics equipment and the development of high density digital electronics have increased the heat load per unit volume of avionics equipment. This has resulted in an overall increase in heat load.

The avionic equipment is generally powered continuously from power up to power down and, hence, dissipates heat continuously. The equipment, usually in standard form equipment boxes, is installed in designated avionic equipment bays in small aircraft, or in equipment cabinets in larger aircraft. Air is ducted to these areas for the specific purpose of cooling equipment and is then recirculated or dumped overboard.

The system must be designed to protect the components of the equipment throughout the aircraft flight envelope, and in whatever climatic conditions the aircraft must operate.

7.2.4 Airframe System Heat Loads

Heat is produced by the environmental control system itself, as well as hydraulic systems, electrical generators, engines and fuel systems components. This takes the form of heat produced as radiation from energy consuming components in the systems such as pumps or motors, or from heat rejected in cooling fluids such as oil. To maintain operating efficiency and to prevent chemical breakdown of fluids with resulting degradation in their performance it is essential to cool these fluids.

7.2.5 The Need for Cabin Conditioning

Design considerations for providing air conditioning in the cockpit of a high performance fighter are far more demanding than those for a subsonic civil airliner cruising between airports.

The cockpit is affected by the sources of heat described above, but a high-performance fighter is particularly affected by high skin temperatures and the effects of solar radiation through the large transparency. However, in designing a cabin conditioning system for the fighter, consideration must also be taken of what the pilot is wearing. If, for example, he is flying on a mission over the sea, he could be wearing a thick rubber immersion suit which grips firmly at the throat and wrists. In addition, the canopy and windscreen will have hot air blown over the inside surfaces to prevent misting which would affect the temperature of the cabin. Another important factor is pilot workload or high stress conditions such as may be caused by a failure, or by exposure to combat. All these factors make it very difficult to cool the pilot efficiently so that his body temperature is kept at a level that he can tolerate without appreciable loss of his functional efficiency.

Commercial aircraft conditioning is provided to maintain a comfortable environment for passengers and cabin crew throughout the flight, including the time required for boarding and taxying to the runway. The system is designed so that air enters the cabin from overhead ducts and is extracted at floor level. The intention is to reduce the risk of air flowing from front to back of the cabin in order to reduce the risk of cross-contamination between passengers. Filtration is required to remove viral and bacterial contamination to further improve the condition of the air. The air volume in the cabin needs to be changed at frequent intervals, usually every two or three minutes.

7.2.6 The Need for Avionics Conditioning

Most aircraft equipment which generates heat will operate quite satisfactorily at a much higher ambient air temperature than can be tolerated by a human. The maximum temperatures at which semi-conductor components can safely operate is above 100 °C, although prolonged operation at this level will seriously affect reliability.

Air conditioning systems are typically designed to provide a maximum conditioned bay temperature of 70 °C, which is considered low enough to avoid significantly affecting the reliability of components. The minimum design equipment operating temperature for worldwide use tends to be about −30 °C. Equipment must also be designed to remain undamaged over a wider temperature range, typically from −40 °C to +90 °C for worldwide use. These figures define the maximum temperature range to which the equipment may be subjected depending on the storage conditions, or in the event that the aircraft is allowed to remain outside for long durations in extreme hot or cold conditions.

7.3 The International Standard Atmosphere (ISA)

An international standard atmosphere has been defined for design purposes. Tables of figures can be found in textbooks which show how values of temperature, pressure and air density vary with altitude. At sea level it is defined as follows:

- Air pressure = 101.3 kpa absolute
- Air temperature = 15 °C
- Air density = 1.225 kg/m

In addition, maximum and minimum ambient air temperatures have been derived from temperatures which have been recorded over a number of years throughout the world. These figures have been used to define a standard to which aircraft can be designed for worldwide operation. Examples are illustrated in Figures 7.2, 7.3 and 7.4, which are to be considered for design purposes only, and should not be considered as realistic atmospheres which could occur at any time.

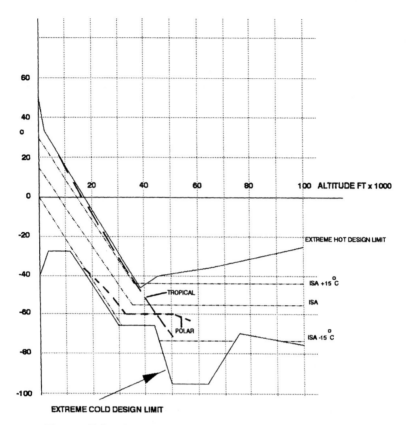

Figure 7.2   Ambient temperature versus altitude

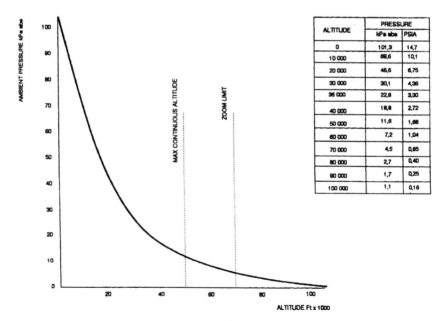

ALTITUDE	PRESSURE	
	kPa abs	PSIA
0	101,3	14,7
10 000	69,6	10,1
20 000	46,6	6,75
30 000	30,1	4,36
36 000	22,8	3,30
40 000	18,8	2,72
50 000	11,6	1,68
60 000	7,2	1,04
70 000	4,5	0,65
80 000	2,7	0,40
90 000	1,7	0,25
100 000	1,1	0,16

Figure 7.3   Ambient pressure variation with altitude

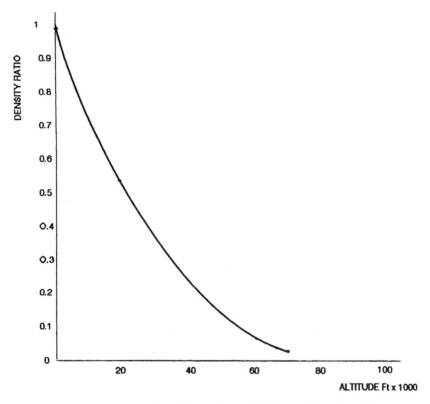

Figure 7.4 Air density ratio variation with altitude

Figure 7.5 shows a distribution of maximum temperatures below and above ISA which are typically encountered throughout the world. These figures are used as a guide for designers of systems which are required to operate in particular areas. Knowledge of all the contributory sources of heat is used to

Figure 7.5 Typical mid-day world temperatures

design the conditioning systems for crew, passenger and equipment. It is also used to ensure that equipment and components are designed to withstand the extremes of temperature likely to be encountered in the aircraft during operations which include flight and also storage and parking in direct desert sunlight or Arctic cold conditions. Equipment specifications will contain realistic maximum and minimum temperatures and tests will be designed to qualify equipment. Further explanation is contained in Chapter 13.

7.4 Environmental Control System Design

This section describes methods of environmental control in common use and, in addition, outlines some recent advances and applications in environmental control system design.

The cooling problem brought about by the heat sources described above must be solved to successfully cool the aircraft systems and passengers in flight. For ground operations some form of ground cooling system is also required.

Heat must be transferred from these sources to a heat sink and rejected from the aircraft. Heat sinks easily available are the outside air and the internal fuel. The outside air is used either directly as ram air, or indirectly as air bled from the engines. Since the available heat sinks are usually at a higher temperature than that required for cooling the systems and passengers, then some form of heat pump is usually necessary.

7.4.1 Ram Air Cooling

Ram air cooling is the process of rejecting aircraft heat load to the air flowing round the aircraft. This can be achieved by scooping air from the aircraft boundary layer or close to it. The air is forced through a scoop which faces into the external air flow, through a heat exchanger matrix and then rejected overboard by the forward motion of the aircraft. The heat exchanger works just like the radiator of a car.

This system has the disadvantage that it increases the aircraft drag because the resistance of the scoop, pipework and the heat exchanger matrix slows down the ram airflow.

The use of ram air as a cooling medium has its limitations, since ram air temperature increases with airspeed and soon exceeds the temperature required for cabin and equipment conditioning. For example, at Mach 0.8 at sea level on a 40 °C day, the ram air temperature is about 80 °C. Ram air is also a source of heating itself as described above (Kinetic heating). In addition, at high altitude the air density becomes very low, reducing the ram air mass flow and hence its cooling capacity. In fact, when conditioning is required for systems which require cooling on the ground, then ram air cooling alone is unsuitable.

However, this situation can be improved by the use of a cooling fan, such as used on a civil aircraft, or a jet pump, mainly used on military aircraft, to enhance ram air flow during taxi-ing or low speed flight. The jet pump enhances ram air cooling in the heat exchanger by providing moving jets of primary fluid bled from the engines to entrain a secondary fluid, the ram air, and move it downstream as shown in Figure 7.6.

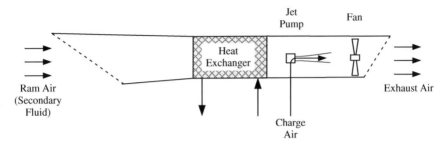

Figure 7.6   Use of fans and jet pumps to increase ram air flow

7.4.2 Fuel Cooling

Fuel cooling systems have limited applications on aircraft for the transfer of heat from a heat source into the aircraft fuel. This is mainly due to the fact that fuel flow is variable and is greatly reduced when the engines are throttled back. However, fuel is much better than air as a cooling medium because it has a higher heat capacity and a higher heat transfer coefficient. Fuel is typically used to cool engine oil, hydraulic oil and gearbox oil.

Figure 7.7 shows a typical fuel and oil cooling system. When the fuel flow is low, the fuel temperature will rise significantly, so recirculation lines are used to pipe the hot fuel back into the fuel tank. Ram air cooled fuel coolers often need to be introduced into the recirculation flow lines to prevent a rapid increase in fuel temperatures in the tank when fuel level is low. This can only be brought into effect in low-speed flight when ram temperatures are low enough. This prevents a rapid rise in the tank fuel temperature during the final taxi after landing, when the tanks are most likely to be almost empty. In a sense this is self-regulating since in high-speed flight the fuel flow is high and hence recirculation is not required. During taxi back ejectors are sometimes required to induce sufficient airflow through the air cooler.

7.4.3 Engine Bleed

The main source of conditioning air for both civil and military aircraft is engine bleed from the high pressure compressor. This provides a source whenever the engines are running. The conditioning air is also used to provide cabin pressurisation.

There are two types of bleed air system: open loop and closed loop. Open loop environmental control systems continually bleed large amounts of air

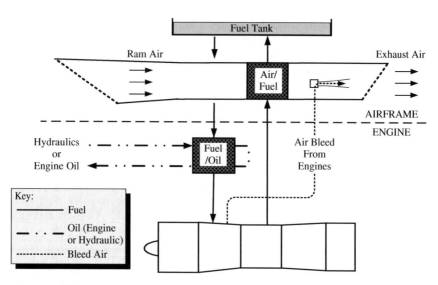

Figure 7.7   Use of fuel as a coolant for hydraulic or engine oil

from the engines, refrigerate it and then use it to cool the passengers and crew, as well as equipment, before dumping the air overboard. Closed loop systems, as shown in Figure 7.8, collect the air once it has been used for cabin conditioning, refrigerate it and recycle it to be used again. In this way bleed air is used only to provide pressurisation, a low venting air supply and sufficient flow to compensate for leaks in the closed loop system. This means that such a system uses considerably less engine bleed air than an open loop system and therefore has a correspondingly reduced effect on engine performance. It

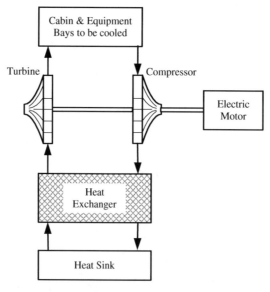

Figure 7.8   Closed loop cooling system

follows that with a closed loop system, a military aircraft has more available thrust at its disposal, or that a civil aircraft is able to operate more efficiently, particularly on long flights.

Since only a small amount of air is bled off from the engines, the need for ram air cooling of the bleed air is reduced. However, to recycle conditioning air it is necessary to seal and pressurise the equipment bays. The cooling air is distributed between equipment using cooling trays with fans to draw equipment exhaust air into the recirculation loop.

Closed loop systems have to date only been used in a few aircraft applications. Not only are there the practical difficulties of collecting and reusing the conditioning air, but closed loop systems also tend to be heavier and more expensive than equivalent open loop systems. As a result the latter, using air cycle refrigeration to cool engine bleed air are most commonly used in aircraft applications. However, some recirculation of cabin air has been introduced on civil aircraft to reduce the ECS cooling penalty. The cabin air is drawn into the recirculation line by a jet pump or fan, and then mixed with refrigerated engine bleed air before being supplied to the cabin inlet at the required temperature. The utilisation of such a recirculation flow can double the efficiency of the system in some cases.

The above method of reducing bleed flow has limited application on high-performance military aircraft because of problems such as the lack of recirculation air available at high altitudes from unpressurised bays and restricted space for ducting.

Therefore, bleed flow reduction on most military aircraft is achieved by modulation of system flow in accordance with demand as described in the following passage.

7.4.4 Bleed Flow and Temperature Control

Typically air at a workable pressure of about 650 kpa absolute (6.5 atmospheres) and a temperature of about 100 °C is needed to provide sufficient system flow and a temperature high enough for such services as rapid demisting and anti-icing. However, the air tapped from the engine high pressure compressor is often at higher pressures and temperatures than required. For example, in a high performance fighter aircraft the air can be at pressures as high as 3700 kpa absolute (37 atmospheres) and temperatures can be over 500 °C, high, enough to make pipes manufactured from conventional materials glow red hot. Tapping air at lower pressures and temperatures from a lower compressor stage would be detrimental to engine performance. On many civil aircraft, different bleed tappings can be selected according to engine speed.

The charge air pressure needs to be reduced as soon as possible to the required working pressure for safety reasons and to reduce the complexity of components since there are problems with sealing valves at such high pressures.

A pressure reducing valve can be used to reduce the pressure of the engine bleed air. This valve controls its downstream pressure to a constant value, no matter what the upstream pressure. The maintenance of this downstream pressure controls the amount of flow from the engines through the environmental control system.

This is acceptable for an aircraft with very few speed variations, such as a civil airliner. However, the faster an aircraft flies the more conditioning air is required, since the greater is the effect of kinetic heating.

In a supersonic aircraft, if the pressure reducing valve was designed to provide sufficient cooling air at high speeds, there would be an excess of flow at low speed. This is wasteful and degrades engine performance unnecessarily. On the Eurofighter Typhoon the environmental control system contains a variable pressure reducing valve which automatically controls its downstream pressure and, therefore, the amount of engine bleed, depending on aircraft speed. This means that the effect of engine bleed on engine performance can be kept to a minimum at all times.

Once the air pressure has been reduced to reasonable working values, the air temperature needs to be reduced to about 100 °C for such services as de-icing and demisting. Heat exchangers are used to reject unwanted heat to a cooling medium, generally ram air as shown in Figure 7.9a. Figure 7.9b shows some typical zones of a passenger cabin in which temperature can be set individually.

In some flight conditions, particularly on cold days, there is so much relatively cool air that the heat exchanger outlet temperature is much less than the 100 °C required for de-icing or de-misting. In such cases the correct proportion

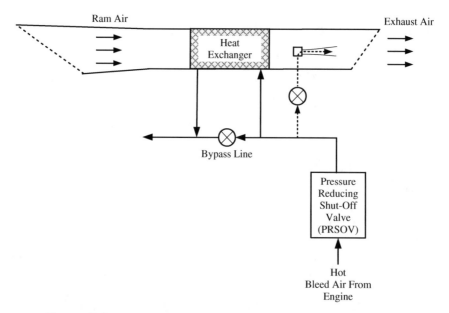

Figure 7.9a Mixing hot air with heat exchanger outlet

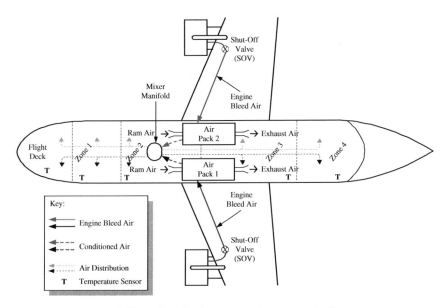

Figure 7.9b   Cabin temperature control system

of hot air from upstream of the heat exchanger is mixed with heat exchanger outlet flow to maintain at least 100 °C mixed air outlet temperatures.

7.5 Cooling Systems

There are two main types of refrigeration systems in use:

- Air cycle refrigeration systems
- Vapour cycle refrigeration systems

7.5.1 Air Cycle Refrigeration Systems

The basic principle is that energy (heat) is removed by a heat exchanger from compressed air which then performs work by passing through a turbine which drives the compressor, and hence energy is transferred resulting in a reduction in temperature and pressure. The resultant air is then at a temperature (and to a small extent pressure) below that at which it entered the compressor.

Air cycle refrigeration systems are used to cool engine bleed air down to temperatures required for cabin and equipment conditioning. Since engine bleed air is generally available, air cycle refrigeration is used because it is the simplest solution to the cooling problem, fulfilling both cooling and cabin pressurisation requirements in an integrated system. However, although lighter and more compact than vapour cycle, air cycle systems have their limitations. Very large air flows are required in high heat load applications which require large diameter ducts with the corresponding problems of installation in the limited space on board an aircraft. Large engine bleed flows are detrimental to engine performance and large aircraft drag penalties are incurred due to the need for ram air cooling.

7.5.2 Turbofan System

This will typically be used in a low-speed civil aircraft where ram temperatures will never be very high. A typical turbofan system is illustrated in Figure 7.10.

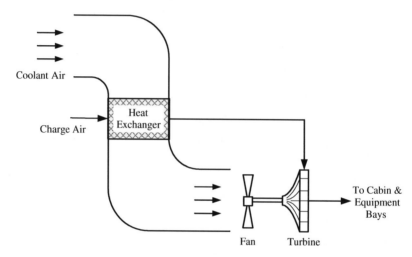

Figure 7.10   Turbofan cooling system

7.5.3 Bootstrap System

Conventional bootstrap refrigeration is generally used to provide adequate cooling for high ram temperature conditions, for example a high performance fighter aircraft.

The basic system consists of a cold air unit and a heat exchanger as shown in Figure 7.11. The turbine of the cold air unit drives a compressor. Both are mounted on a common shaft. This rotating assembly tends to be supported on ball bearings, but the latest technology uses air bearings. This provides a lighter solution which requires less maintenance, for example no oil is required.

Three-rotor cold air units or air cycle machines can be found on most recently designed large aircraft, incorporating a heat exchanger coolant fan on the same shaft as the compressor and turbine. Military aircraft tend to use the smaller and simpler two-rotor cold air unit using jet pumps to draw coolant air through the heat exchanger when the aircraft is on the ground and in low speed flight. Figure 7.12 shows the environmental control system of the British Aerospace Advanced Turbo-Prop (ATP) aircraft as a typical example.

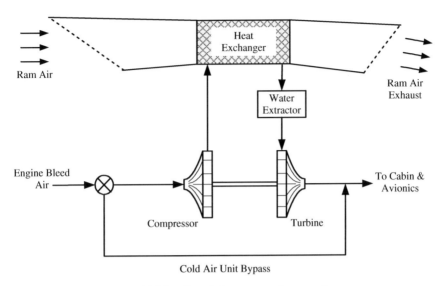

Figure 7.11 Bootstrap cooling system

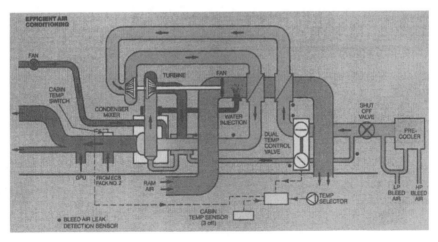

Figure 7.12 Example ECS

The compressor is used to increase the air pressure with a corresponding increase in temperature. The temperature is then reduced in the ram air cooled heat exchanger. This reduction in temperature may lead to water being condensed out of the air, especially when the aircraft is operating in a humid climate. Figure 7.12 shows a water extractor at the turbine inlet which will remove most of the free water, helping to prevent freezing of the turbine blades and water being sprayed into the cabin and equipment bays. As the air expands across the turbine the temperature can drop below 0 °C in certain flight conditions. Figure 7.12 also shows a cold air unit by-pass line which is used to vary turbine outlet temperature to the required value for cabin and equipment cooling. The volume of air flowing round the bypass is varied by a temperature control valve until the air mixture at the turbine outlet is at the required temperature.

Examples of some of the machines presently in use on the Boeing 737 and Boeing 757/Boeing 767 are shown in Figure 7.13.

Figure 7.13 Examples of air cycle machines and air conditioning packs (Courtesy of Honeywell)

7.5.4 Reversed Bootstrap

The reversed bootstrap system is so named because the charge air passes through the turbine of the cold air unit before the compressor. Following initial ram air cooling from a primary heat exchanger the air is cooled further in a regenerative heat exchanger and is then expanded across the turbine with a corresponding decrease in temperature. This air can then be used to cool an air or liquid closed-loop system, for radar transmitter cooling for example. The air then passes through the coolant side of the regenerative heat exchanger before being compressed by the compressor and dumped overboard (Figure 7.14).

7.5.5 Ram Powered Reverse Bootstrap

In some cases equipments may be remotely located where it is not practicable to duct an air supply from the main ECS. In such cases a separate cooling package must be employed. This situation is becoming particularly common on military aircraft with equipment mounted in fin tip or under-wing pods, where it is not possible to find a suitable path to install ducting or pipes. A ram-powered reverse bootstrap air-cycle system can be used to meet such 'standalone' cooling requirements.

The method increases the capability of a ram air cooled system by expanding the ram air through a turbine, so reducing air temperature as shown in Figure 7.15. Therefore cooling can be provided up to much higher air speeds than a purely ram air cooled system. However, cooling is still a problem on

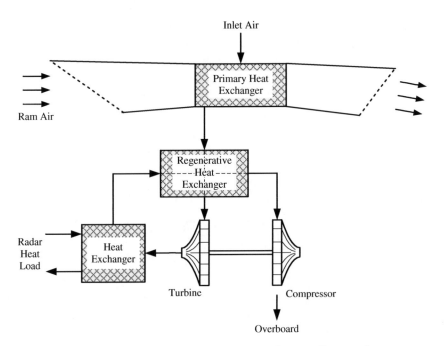

Figure 7.14 Reverse bootstrap refrigeration system

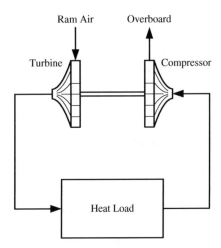

Figure 7.15 Ram powered reverse bootstrap system

the ground and in low speed flight. Therefore this system is typically used as a 'standalone' cooling system for equipment which is operated only during flight.

7.5.6 Vapour Cycle Systems

The vapour cycle system is a closed loop system where the heat load is absorbed by the evaporation of a liquid refrigerant such as Freon® in an evaporator (NB the trade name Freon® is a registered trademark belonging to E.I. du Pont de Nemours & Company (DuPont)). The refrigerant then passes through a compressor with a corresponding increase in pressure and temperature, before being cooled in a condenser where the heat is rejected to a heat sink. The refrigerant flows back to the evaporator via an expansion valve. This system is illustrated in Figure 7.16.

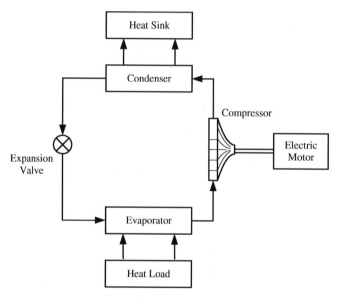

Figure 7.16 Vapour cycle cooling system

Although vapour cycle systems are very efficient, with a coefficient of performance typically five times that of a comparable closed loop air cycle system, applications are limited due to problems such as their limited temperature range and heavy weight compared to air cycle systems. The maximum operating temperatures of many refrigerants are too low, typically between 65 °C and 70 °C, significantly less than the temperatures which are required for worldwide operation.

It should be noted that chlorofluorocarbons (CFC) endanger the ozone layer and are the subject of much debate calling for a limitation in their use.

7.5.7 Liquid Cooled Systems

Liquids such as Coolanol® are now more commonly being used to transport the heat away from avionics equipment. (NB COOLANOL® is a registered trademark of Exxon for Silicate Ester dielectric heat transfer fluids.)

Liquid can easily replace air as a transport medium flowing through the cold wall heat exchanger.

A typical liquid loop consists of an air/liquid heat exchanger which is used to dump the heat load being carried by the liquid into the air conditioning system, a pump and a reservoir as illustrated in Figure 7.17.

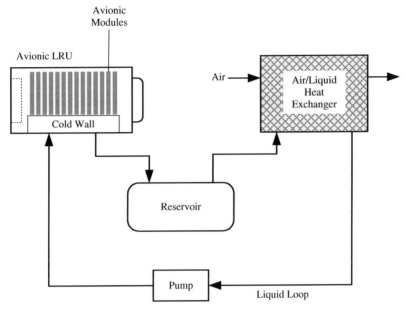

Figure 7.17 Example of a liquid cooling system

The advantages are that it is a more efficient method of cooling the heat source, and the weight and volume of equipment tends to be less than the air conditioning equipment which would otherwise be required. The disadvantages are that it is expensive, and the liquid Coolanol® is toxic. Self sealing couplings must be provided to prevent spillage wherever a break in the piping is required for maintenance purposes. The Boeing AH-64C/D Longbow Apache attack helicopter uses such a vapour cooling system to cool the extended forward avionics bays.

7.5.8 Expendable Heat Sinks

An expendable coolant, typically water, can be carried to provide a heat sink by exploiting the phenomenon of latent heat of vaporisation. A simple system is shown in Figure 7.18.

The liquid refrigerant is stored in a reservoir which supplies an evaporator where the heat load is cooled. The refrigerant is then discharged overboard. This type of system can only be used to cool small heat loads (or large loads for a short time), otherwise the amount of liquid refrigerant that must be carried on board the aircraft would be too large.

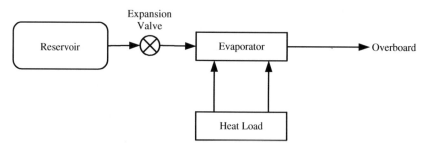

Figure 7.18   Simple expendable heat sink system

7.6 Humidity Control

Passenger comfort is achieved not only by overcoming the problems of cooling and cabin pressurisation, but also by controlling humidity in the passenger cabin. This is only a problem on the ground and at low altitudes, since the amount of moisture in the air decreases with increasing altitude. There is a particular difficulty in hot humid climates. For example, in Northern Europe the typical air moisture content can be 10 grams of water per kg of air, but in some parts of the Far East moisture contents of more than 30 grams per kg can be encountered. In a hot, humid climate the cabin inlet air supply temperature needs to be cold to keep the passengers and aircrew comfortable. Without good humidity control this can result in a wet mist being supplied to the cabin.

In addition to the aim of ensuring passenger comfort, humidity levels must be controlled to prevent damage to electrical and electronic equipment due to excessive condensation. Humidity control also reduces the need for windscreen and window de-misting and anti-misting systems.

The fine mist of water droplets in the cold cabin inlet supply must be coalesced into large droplets that can then be trapped and drained away. Two types of water separator are in common use with air cycle refrigeration systems: a centrifugal device and a mechanical device. In the centrifugal devices a turbine is commonly used to swirl the moist air. The relatively heavy water droplets are forced to the sides of a tube, where the water and a small amount of air is trapped and drained away, thus reducing the water content of the air downstream of a water separator.

The mechanical water separator, which consists of a coalescer, a relief valve and a water collector, achieves the same result by forcing the moist air to flow through the coalescer where large droplets are formed and blown onto collector plates. The water runs down the plates and is then drained away. The relief valve opens to allow the air to bypass the water separator if ice forms.

Simple water collection devices can be used in vapour cycle refrigeration systems to reduce humidity levels since the air is cooled to its dew point as it flows through the evaporator. Water droplets collect on the heat exchanger surfaces and can be simply trapped and drained away.

Chemicals can also be used to reduce moisture content. In civil aircraft the air gaps between two plates of the passenger windows are commonly vented via an absorbent material such as silica gel to prevent condensation of moisture on the window. Moisture is condensed from the air as it flows through the gel, and the latent heat given up by the condensing moisture increases the air temperature.

Molecular sieves can also be used to remove moisture from air. These are absorbent materials which are used to sieve out the large water molecules from the air in the same way as the molecular sieve oxygen concentrators described later in this chapter remove the large gas molecules and impurities from air to leave almost pure oxygen.

7.7 The Inefficiency of Present Systems

In cooling down engine bleed air to temperatures low enough to provide adequate cooling capacity for aircrew, passengers and equipment, a great deal of heat and therefore potentially useful energy is rejected to atmosphere. Typically, the ratio of engine power used to heat load cooled in order to provide sufficient cooling for the total aircraft heat load is 10:1.

In addition, further engine power is required to overcome the drag caused by the ram air heat exchangers. This problem becomes worse, particularly on military aircraft which suffer a continually increasing avionics heat load; while the design requirements are to improve engine performance and reduce aircraft weight. The more avionics, the heavier the aircraft, not only due to the avionics equipment weight itself, but also due to the weight of the environmental control system equipment and the air distribution pipework. Furthermore, additional engine bleed air is required as the avionics heat load increases, but bleeding more air off the engines is detrimental to engine performance. More efficient cooling by closed loop systems would undoubtedly increase equipment reliability.

The increasing avionics heat load on military aircraft may lead to further developments of closed loop environmental control systems in the future, since there is potential to vastly reduce the amount of engine bleed required, and thus overcome the problems of detrimental effects of open loop systems on engine performance.

7.8 Air Distribution Systems

7.8.1 Avionics Cooling

In civil aircraft the total avionics heat load is low when compared with the many applications which have been, and continually are being, found in military aircraft. In civil aircraft it is often sufficient to draw cabin ambient air over the avionics equipment racks using fans. This will have the effect of increasing the overall cabin temperature but, since the total avionics heat load is not massive, the environmental control system has sufficient capacity to maintain cabin temperatures at acceptable levels.

However, on a military aircraft with a high avionics heat load, only a few items of the avionics equipment are located in the cabin. The majority are located in either conditioned or nonconditioned equipment bays, an installation decision which is made by taking into consideration such criteria as the effect of temperature on equipment reliability or damage, and the amount of engine bleed available for air conditioning.

Since the equipment can operate in ambient temperatures higher than humans can tolerate, the air used to condition it tends to be cabin exhaust air. There is usually very little space in equipment bays as they are tightly packed with equipment. There is little space left for the installation of cooling air ducts. Therefore, the equipment racking and air distribution system must be carefully designed to ensure an even temperature distribution.

7.8.2 Unconditioned Bays

Unconditioned bays may reach temperatures up to recovery temperature. However, air in these bays is not totally stagnant. The aircraft is usually designed to have a continuous venting flow through each equipment bay, only the pressure cabin is sealed. This ensures that there is no build up of differential pressure between bays, particularly during rapid climb and descent. The venting flow tends to be the conditioned bay outlet flow.

7.8.3 Conditioned Bays

Equipment can be cooled by a variety of methods, including the following;

- cooling by convection air blown over the outside walls of the equipment boxes (external air wash)
- air blown through the boxes and over the printed circuit boards (direct forced air)
- air blown through a cold wall heat exchanger inside the box (indirect forced air)
- fans installed in the box to draw a supply of cooling air from the box surroundings

The first method of cooling is adequate for equipment with low heat loads. As the heat load increases it tends to become very inefficient, requiring a lot more cooling air than the other three methods to achieve the same degree of cooling. It is very difficult to design an avionics equipment box with a high heat load to enable the efficient dissipation of heat by convection via the box walls. Local 'hot spots' inside the box will lead to component unreliability.

The other three methods of cooling are very much more efficient, but the boxes must have a good thermal design to ensure precious conditioning air is not wasted.

The second method is often the most efficient way of cooling. The box is thermally designed so that the component heat load is conducted to a cold wall heat sink. The cold wall acts as a small heat exchanger.

The final method is only acceptable in an equipment bay layout where there is no chance of re-ingestion of hot exhaust air from another box. This is not practical in a closely packed equipment bay.

Particular attention must be given to the cooling requirements of equipment whose correct operation is critical to the safety of the aircraft – such computers must be continuously fully conditioned, since failure of all computers would render the aircraft uncontrollable. Otherwise computers performing flight-critical functions must be designed to operate uncooled for the duration of a flight albeit in a limited bay environmental temperature. What suffers is long-term reliability of the computer.

Figure 7.19 shows a typical method of air distribution. The distribution ducting provides a supply of air into a plenum chamber which is built into a shelf on which the equipment is installed. The air is supplied directly into the equipment via orifices in the shelf and the equipment box. It is prevented from leaking away by a soft seal between the shelf and the box. The air exhausts from the box through louvres in the wall.

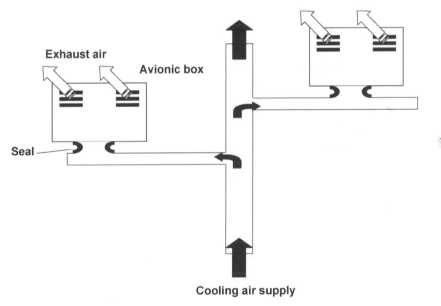

Figure 7.19 Typical air distribution scheme

7.8.4 Conditioned Bay Equipment Racking

In a closely packed equipment bay cooling the first three methods are often used side by side on shelves specially designed to accommodate the various

cool air interface. The standard for the equipment enclosures will specify the appropriate wall to wall gap to ensure correct cooling air flow between units. Figure 7.20a and b illustrate some methods of cooling equipment.

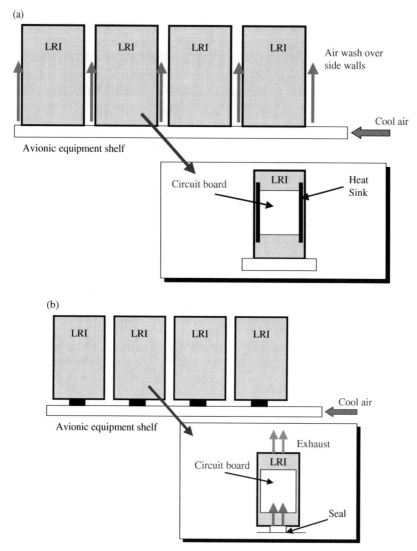

Figure 7.20a and b   Different methods of cooling avionics equipment.

7.8.5 Ground Cooling

For aircraft with separate equipment bays fans are provided which are often located in the undercarriage bays. These are used to provide ambient cooling air flow for the avionics bays when the aircraft is on the ground, and there is only enough bleed air flow from the engines in this case to provide cabin conditioning. The fans can also be used to cool the equipment if the environmental control system fails.

7.8.6 Cabin Distribution Systems

Cabin distribution systems on both civil and military aircraft are designed to provide as comfortable an environment as possible. The aircrew and passenger's body temperature should be kept to acceptable levels without hot spots, cold spots or draughts. Civil aircraft are designed to maintain good comfort levels throughout the cabin since passengers are free to move about. On some aircraft each passenger has personal control of flow and direction of local air from an air vent above the head (often known as a 'punkah louvre'), although on modern large aircraft total air conditioning is provided. The personal air vent is no longer provided, partly because of the better performance of air conditioning systems, and also because the increased height of passenger cabins means that passengers are no longer able to reach the vent while seated. There are usually additional vents which blow air into the region of the passengers' feet so that there is no temperature gradient between the head and feet. Figure 7.21 shows an example of a Boeing B777 air conditioning pack and an illustration of the way in which air enters at the roof and is extracted at floor level in a typical cabin. Air flows predominantly down from the roof vents across the front of each passenger, and is extracted at floor level. A proportion of the exhaust air (up to 50%) is recirculated by being first

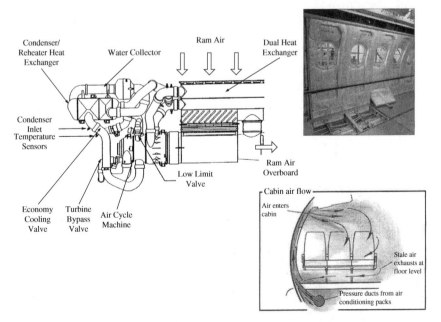

Figure 7.21   Example air conditioning and distribution system

'cleaned' in high efficiency filters to remove bacterial and viral particles, and then mixed with clean incoming air.

This form of ventilation can be more difficult to achieve for the pilot on a fighter aircraft, where his head receives the full effect of solar radiation through the transparency. The air velocities must be high and the air temperatures near freezing for the pilot to feel any effect through his clothing (including an immersion suit). The distribution system must also be designed so that the cold air jet picks up as little heat as possible from its surrounding environment before it reaches the subject to be cooled.

7.9 Cabin Noise

Aircraft are designed aerodynamically or structurally to keep externally generated noise levels to a minimum. At crew stations the noise levels should be such that the aircrew are able to communicate satisfactorily over a radio or intercom, or to operate direct voice input avionics systems. As in any other work environment noise levels must be kept to satisfactory levels to avoid damage to hearing. The noise levels in the passenger cabin of a civil aircraft are kept to a minimum to ensure passenger comfort since fare paying passengers are free to take their custom elsewhere.

Noise in the military aircraft cockpit can be distracting for the pilot and adds to other sources of noise to present a health and safety problem. Legislation is gradually lowering the threshold for noise for people at work, and it must be noted that aircrew and cabin crew are 'in the office' when they are working on an aircraft. If cabin noise approaches or exceeds the permitted thresholds or accumulated noise dose, then crew may be prohibited from flying for periods of time.

7.10 Cabin Pressurisation

Cabin pressurisation is achieved by a cabin pressure control valve which is installed in the cabin wall to control cabin pressure to the required value depending on the aircraft altitude by regulating the flow of air from the cabin.

For aircraft where oxygen is not used routinely, and where the crew and passengers are free to move around as in a long range passenger airliner, the cabin will be pressurised so that a cabin altitude of about 8000 ft is never exceeded. This leads to a high differential pressure between the cabin and the external environment. Typically for an airliner cruising at 35 000 ft with a cabin altitude of 8000 ft there will be a differential pressure of about 50 kpa (0.5 atmosphere) across the cabin wall. The crew is able to select a desired cabin altitude from the cockpit and cabin pressurisation will begin when the aircraft reaches this altitude. This will be maintained until the maximum design cabin differential pressure is reached. This is also true for large military aircraft such as surveillance platforms or air-to-air refuelling tankers.

For aircraft with the crew in fixed positions, using oxygen routinely as in a military aircraft, the pressurisation system is usually designed so that the cabin altitude does not exceed about 20 000 ft. Figure 7.22a shows a typical fighter aircraft automatic pressurisation schedule with tolerances plotting Cabin Altitude (y-axis) versus Aircraft Altitude (x-axis). The cabin pressure control

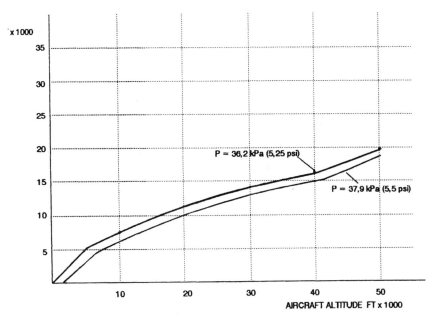

Figure 7.22a Fighter aircraft pressurisation schedule

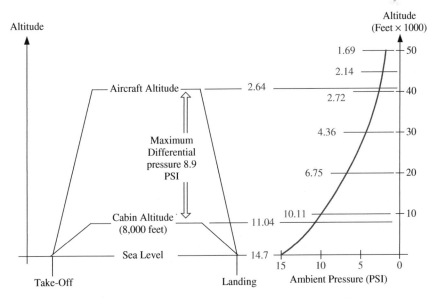

Figure 7.22b Typical commercial aircraft pressurisation schedule

valve is designed to automatically maintain the cabin altitude to this schedule depending on aircraft altitude without any intervention from the pilot.

The differential pressure is maintained high enough so that if the cabin pressurisation fails when the aircraft is at a high altitude there is sufficient time for the pilot to descend. For example, at 50 000 ft; the pressure will not leak away causing the cabin altitude to exceed a safe value before the pilot has had enough time to descend to a safe altitude.

Therefore, the cabin must be designed as a pressure vessel with minimum leakage. In the event of loss of pressurisation the cabin pressure control valve will close and the only leakage will be through the structure. Non-return valves are installed in the air distribution pipes where they pass through the cabin wall, so that when the air supply fails the air already in the cabin cannot leak back out through the pipes. A safety valve is installed in the cabin wall to relieve internal pressure if it increases above a certain value in the event of failure of the pressure control valve. The principles of operation are illustrated in Figure 7.23. An example of the Boeing B777 pressurisation system is shown in Figure 7.24, and typical cabin discharge valves shown in Figure 7.25.

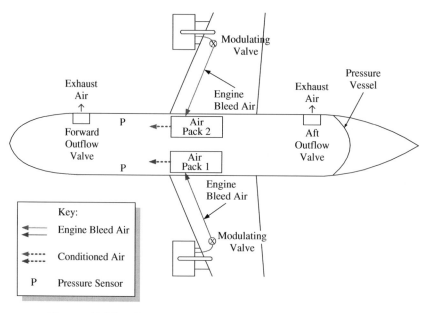

Figure 7.23 Cabin pressurisation control principles

Following the loss of the cabin pressurisation system and descent to a safe altitude, the pilot can select the opening of a valve to enable ram air to be forced into the distribution system by a scoop which faces into the external airflow. This system of purging with ram air can also be selected should the cabin be contaminated by fumes or smoke coming from the main environmental control system air supply.

It should be noted that military aircraft can suffer from rapid or explosive decompression if the canopy or the aircraft structure is penetrated by ordnance

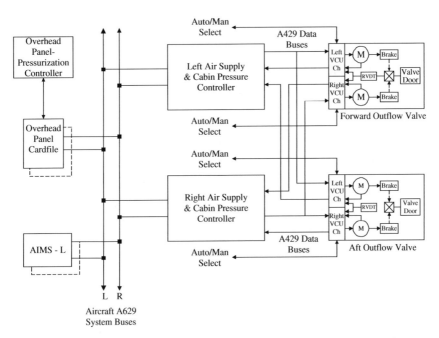

Figure 7.24   Boeing B777 pressurisation control example

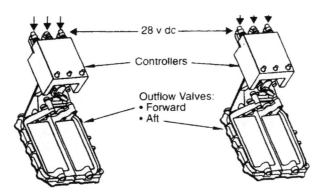

Figure 7.25   Boeing B777 discharge valves

fragments. The pilot will be protected by his clothing, helmet, visor and the use of pressure breathing. However, this circumstance must be taken into account to protect some avionic equipment from the rapid change of pressure, see Chapter 13.

7.11 Hypoxia

Oxygen is essential for the maintenance of life. If the oxygen supply to the brain is cut off, unconsciousness soon follows, and brain death is likely to occur within 4 to 5 minutes. Breathing air at reduced atmospheric pressure results in a reduction in alveolar oxygen pressure which in turn results in an oxygen supply deficiency to the body and brain tissues.

This condition is termed hypoxia.

The effects of hypoxia can be demonstrated by imagining a slow ascent by balloon. Up to 10 000 ft there will be no significant effects of altitude on the body. At 15 000 ft it will be markedly more difficult to perform physical tasks, and the ability to perform skilled tasks will be severely reduced, although this fact will probably go unnoticed. At 20 000 ft the performance of physical tasks will be grossly impaired, thinking will be slow, and calculating ability will be unreliable. However, the occupants of the balloon will be unaware of their deficiencies, and may become light-headed and over-confident. Any physical exertion may cause unconsciousness. Even a highly qualified and experienced pilot will be ain a totally unfit state to fly an aircraft. Above 20 000 ft loss of consciousness sets in [2].

During a rapid climb to altitude in a fighter aircraft, without any protection against hypoxia, rapid and sudden loss of consciousness will result without any of the symptoms of hypoxia appearing. The dangerous effects of breathing air at reduced atmospheric pressures can be alleviated by pressurising the cabin. Typically, on a civil aircraft with a maximum operating altitude of 25 000 ft, the cabin will be pressurised to maintain a cabin altitude of 8000 ft.

An alternative method of preventing hypoxia is to increase the concentration of oxygen in the cabin atmosphere. Civil aircraft only supply oxygen in cases of rapid cabin depressurisation or contamination of the cabin air by smoke or harmful gases. Emergency procedures require quick action from the pilot and crew, or an automatic system, in the event of a rapid depressurisation since hypoxia is much more severe when it is initiated by a sudden exposure to high altitude compared to a more gradual degradation of performance with gradually increasing altitude.

For both civil and military applications, an oxygen regulator is used to control the flow of breathing gas in response to the breathing action of the person requiring the supply of gas. The proportions of air to oxygen mixture supplied can be varied depending on the altitude. A mask is connected by hoses and connectors to the regulator output.

The source of breathing gas will be either from precharged or liquid oxygen bottles, or from a Molecular Sieve Oxygen Concentrator (MSOC) which produces breathable gas from engine bleed air.

7.12 Molecular Sieve Oxygen Concentrators

Until recently the only practical means of supplying oxygen during flight has been from a cylinder or a liquid oxygen bottle. This has several disadvantages, particularly for military aircraft. It limits sortie duration (fuel may not be the limiting factor if in-flight refuelling is used), the equipment is heavy and the bottles need replenishing frequently.

Molecular Sieve Oxygen Concentrators (MSOC) are currently being developed for military applications. The MSOCs use air taken from the environmental control systems as their source of gas. Most of the gases in air have larger molecules than oxygen. These molecules are sieved out of the air mixture until mostly oxygen remains. This means that a continuous supply of oxygen can be made available without needing to replenish the traditional oxygen storage system after each flight. The residual inert gases can be used for fuel tank pressurisation and inerting. A system designed specifically for the production of inert gases is known as On-Board Inert Gas Generating System (OBIGGS).

However, MSOCs have a major disadvantage. If the environmental air supply from the engines stops then so does the supply of oxygen. Therefore, small backup oxygen systems are required for emergency situations to enable the pilot to descend to altitudes where oxygen levels are high enough for breathing. Developments of MSOCs are watched with interest, and further systems may be efficient enough to provide oxygen enriched air for civil aircraft cabins.

In military aircraft which are typically designed to fly to altitudes in excess of 50 000 ft, both cabin pressurisation and oxygen systems are employed to help alleviate the effects of hypoxia. In cases where aircrew are exposed to altitudes greater than 40 000 ft, either due to cabin depressurisation or following escape from their aircraft, then additional protection is required. In the event of cabin depressurisation the pilot would normally initiate an emergency descent to a 'safe' altitude. However, short-term protection against the effects of high altitude is still required.

At altitudes up to 33 000 ft, the alveolar oxygen pressure can be increased up to its value at ground level by increasing the concentration of oxygen in the breathing gas. However, even when 100 per cent oxygen is breathed, the alveolar oxygen pressure begins to fall at altitudes above 33 000 ft. It is possible to overcome this problem by increasing the pressure in the lungs above the surrounding environmental pressure. This is called positive pressure breathing. At altitudes above 40 000 ft the rise in pressure in the lungs relative to the pressure external to the body seriously affects blood circulation round the body and makes breathing more difficult. Partial pressure suits are designed to apply pressure to parts of the body to counter the problems of pressure breathing for short durations above 40 000 ft.

A partial pressure suit typically includes a pressure helmet and a bladder garment which covers the entire trunk and the upper part of the thighs. The pressure garments are inflated when required by air taken from the environmental control system and are used in conjunction with an inflatable bladder in anti-g trousers which are used primarily to increase the tolerance of the aircrew to the effects of g.

Full pressure suits can be used to apply an increase in pressure over the entire surface of the body. This increases duration at altitude. For durations exceeding 10 minutes, however, other problems such as decompression sickness and the effects of exposure to the extremely low temperatures at altitude must be overcome. Limited altitude and g protection is afforded by the provision of anti-g trousers and a slightly increased breathing pressure. In this case, under high g, the pilot would still need to perform so called 'g straining manoeuvres' which are tiring.

A typical OBOG system is shown in Figure 7.26 in this case for a two-seat aircraft, although the architecture is the same for a single-seat aircraft, but with only one regulator. The following description is from a Honeywell Aerospace Yeovil paper (Yeoell & Kneebone, 2003) [3].

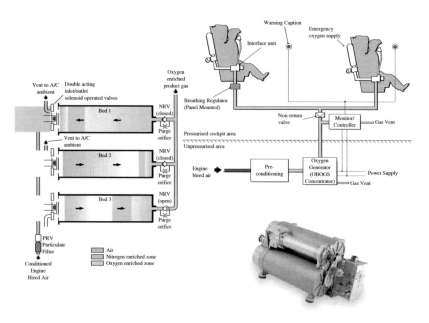

Figure 7.26 On-board oxygen generation system (Courtesy of Honeywell Aerospace Yeovil)

Engine bleed air enters the pre-conditioning system element where the temperature is reduced, ideally to less than 70 °C and water is removed as far as possible. In addition it is normal at this stage to use a combined particulate and coalescing filter to remove potential contaminants including free-water that may still be contained in the inlet air.

The OBOGS contains a pressure reducing valve to reduce the inlet air pressure of the air supply to that required by the OBOG generator, typically 35 psig.

The next system element is the oxygen generator, or more correctly, the OBOGS concentrator that uses multiple zeolite beds to produce the oxygen-rich product gas.

The switching between the zeolite beds is achieved using solenoid actuated pneumatic diaphragm valves controlled by the system monitor/controller.

These valves are 'wear-free' and allow the concentrator to be a 'fit and forget' system that requires no scheduled maintenance and exhibits high reliability.

The system monitor/controller is a solid state electronic device that monitors the PPO_2 level of the OBOGS concentrator product gas, and adjusts the cycling of the beds to produce the desired level of oxygen concentration for cockpit altitudes below 15 000 ft. This process is known as concentration control and means that no air-mix, or dilution, of the product gas is required at the regulator, hence preventing the ingress of any smoke or fumes from the cabin into the pilots breathing gas supply.

The breathing gas then passes to the pilots breathing regulator, in this case a panel mounted unit is shown; however, ejection seat and pilot mounted devices can also be used.

7.13 g Tolerance

For aircraft which are likely to perform frequent high g manoeuvres such as Typhoon, a 'relaxed g protection' system is beneficial. This consists of increased coverage g trousers and pressure breathing with g and altitude which requires a breathing gas regulator and mask capable of providing increased pressure gas, and a pressurised upper body garment to provide external counter pressure (a chest counter pressure garment). This enables the pilot to perform repeated high g manoeuvres without the need for g straining. It also provides altitude protection in the case of a cabin decompression in a manner similar to a full pressure suit.

Engineers strive constantly to improve the agility and combat performance of military aircraft. Indeed technology is such that it is now man who is the limiting factor and not the machine. Accelerations occur whenever there is a change in velocity or a change in direction of a body at uniform velocity. For a centripetal acceleration, towards the centre of rotation, a resultant centrifugal force will act to make the body feel heavier than normal, as illustrated in Figure 7.23.

Forces due to acceleration are measured in g. 1 g is the acceleration due to gravity, i.e. 9.81 m/s. A typical pilot is capable of performing aircraft manoeuvres up to 3 or 4 g, i.e. until he feels about three or four times his normal body weight. At g levels above this the heart becomes unable to maintain an adequate supply of oxygenated blood to the brain, which will result in blackout. This is a very dangerous condition, particularly in low-flying aircraft.

If the acceleration onset is gradual then the blood supply to the eyes is the first to reduce sufficiently to provide the symptoms of tunnelling of vision, before blackout and loss of consciousness occurs.

Anti-g trousers are used partially to alleviate the effects of excessive g on the body. The trousers consist of inflatable air bladders retained beneath a nonstretch belt and leggings. The trousers are inflated using air from the environmental control system. Inflation and deflation of the trousers is typically controlled by an inertial valve. The valve consists of a weight acting on a spring. At the onset of g, as the pilot is pushed down in his seat, the weight compresses the spring which acts to open the valve, thus allowing a supply of air to inflate the bladders in the trousers. The inflation action acts to restrict the flow of blood away from the brain. Using anti-g trousers a typical pilot can perform manoeuvres up to about 8 g. Positive pressure breathing also increases short term resistance to g.

Another method of increasing g tolerance is to recline the pilot's seat. This increases the ability of the heart to provide an adequate supply of blood to the brain under high g conditions. However, in practice the seat can only be slightly reclined because of cockpit design problems, pilot visibility and the need to provide a safe ejection pathway to ensure injury free emergency exit from the cockpit.

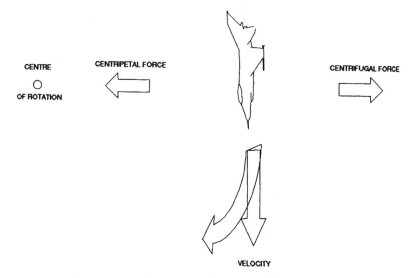

Figure 7.27   g forces in a combat aircraft

7.14 Rain Dispersal

A pilot must have clear vision through the windscreen under all weather conditions, particularly on approach to landing. The use of windscreen wipers can be effective up to high subsonic speeds, particularly on large screens. As on a car, wipers are used in conjunction with washing fluid to clean the screen of insect debris, dust, dirt and salt spray etc. However, wipers are not suitable for use with plastic windscreens since they tend to scratch the surface. They also have the disadvantage of increasing drag.

Hot air jets for rain dispersal can be used up to much higher speeds than wipers and are suitable for use on glass and plastic. The air is discharged at high velocity over the outside surface of the screen from a row of nozzles at the base.

The air discharged from the nozzles is supplied from the environmental control system at temperatures of at least 100 °C. Such high temperatures are required to evaporate the water. However, the nozzles must be designed so that the windscreen surface temperature is not increased to such an extent that damage occurs. This is particularly a problem with stretched acrylic windscreens which begin to shrink back to the cast acrylic state at temperatures above 120 °C.

Distortion of the surface of the acrylic at the locations where the air jet impinges on the screen has been known to occur.

The system can also be used for anti-icing and de-misting.

7.15 Anti-Misting and De-Misting

Misting will occur when the surface temperature of the transparency falls below the dew point of the surrounding air. Misting typically occurs when an aircraft which has been cruising at an altitude where air is cold and relatively dry. When the aircraft descends into a warmer and more humid atmosphere, misting will occur on the surfaces which have not had enough time to warm up to a temperature above the dew point of the air.

An anti-misting system can be provided to keep the surface temperature of the transparency above the dew point and thus prevent misting. A system of nozzles blowing air at about 100 °C over the canopy from its base can be used, or alternatively an electrically heated gold or metal oxide film can be deposited on the transparency surface or placed between laminations.

A transparency de-mist system can be provided to clear the transparency of mist should misting occur suddenly, or if the anti-mist system fails. This is particularly important on landing for aircraft where the pilot is tightly strapped into his seat and cannot clear the screen with his hand. The de-mist system consists of nozzles blowing environmental control system air at high flow rates across the transparency.

7.16 Aircraft Icing

Aircraft icing is an extremely important factor in the world of aviation and despite the best efforts of the industry there are still continuing accidents resulting from inadequate protection against the impact of icing on aircraft controllability.

Icing is caused either by the freezing on to aircraft surfaces of some from of precipitation, this usually occurs on the ground; or by supercooled liquid water droplets found in clouds or rain solidifying on impact with aircraft structure, which is at a sufficiently low temperature, during flight. This 'accretion' occurs on areas of the airframe where the airflow is near to stagnation, i.e. close to a rest, such as wing or tailplane leading edges, engine intakes or helicopter rotor blade leading edges. The factors determining rate of accretion are complex but include the temperature of the surface, its radius and sweep, the size and

temperature of the water droplets, the aircraft altitude and the intensity of
the icing conditions. The latter has various means of definition but the most
commonly used, particularly in civil aviation, such as those detailed in Table 7.1
below, are based on rate of accumulation or accretion.

Table 7.1 Definition of aircraft icing

Intensity	Ice accumulation
Trace	Ice becomes perceptible. Rate of accumulation slightly greater then rate of sublimation. It is not hazardous even though de-icing/anti-icing equipment is not utilised unless encountered for more than one hour.
Light	The rate of accumulation might create a problem if flight in this environment exceeds one hour. Occasional use of de-icing/anti-icing equipment removes /prevents accumulation.
Moderate	The rate of accumulation is such that even short encounters become potentially hazardous and use of de-icing/anti-icing equipment or diversion is necessary.
Severe	The rate of accumulation is such that de-icing/anti-icing equipment fails to reduce or control the hazard. Immediate diversion is necessary.

Other definitions such as FAR 25 are based on the cloud liquid water content
and droplet size and are used in the design and testing process.

For the ground icing hazard de-icing fluids are universally used to ensure the
aircraft surface is free of ice at take-off. The method of countering the airborne
icing hazard varies depending on the nature of the aircraft and its operational
requirements. These can be divided broadly into two categories, those aircraft
requiring ice protection and those not. The former category contains aircraft
which are clearly prohibited from flying in icing conditions and fast jet military
aircraft which have the operational flexibility to avoid such conditions, either
by achieving an 'ice free speed', i.e. accelerating to a speed where surface
temperatures are too high for icing to occur, or by flying around them. In
this case a limited duration clearance (typically five minutes) may be given
for takeoff and landing through icing conditions by analysis of the accretion
characteristics of the airframe utilising specialised icing prediction computer
programs, and testing in natural icing conditions or behind icing tankers. In
these cases the aircraft are often fitted with ice detectors to inform pilots of the
presence of icing conditions and the need to take action.

Where some form of ice protection is required there are various methods
utilised. A distinction is made between an anti-ice system where ice accretion is
prevented, and a de-ice system where a limited amount of accretion is allowed
before some action is taken to shed it. These systems are used in conjunction
with an ice detector.

Anti-ice systems can utilise:

- Hot bleed air where either continuously or when icing conditions are present, hot air is projected on the inside of a surface subject to ice accretion such as a wing leading edge or engine bullet
- Electrical heating where elements are embedded in the structure susceptible to icing to achieve a continuous surface temperature above freezing level
- Liquid ice protection where a freezing point depressant liquid is deposited on a surface or extruded through a porous surface to prevent freezing

De-ice systems can utilise:

- Pneumatic boots where a reinforced synthetic rubber layer is overlaid on the susceptible surface and periodically inflated in conditions of ice accretion thereby breaking and shedding the ice
- Electrical heating which can be used in a de-icing mode by switching on and off periodically during exposure to icing conditions
- Electro-expulsive system which utilises opposing magnetic fields or eddy currents induced by conductors embedded in a flexible surface to create relative movement and hence the breakage and shedding of the accreted ice
- Electro magnetic impulse de-icing which utilises coils inside the leading edge inducing eddy currents in metal skin with the result that the surface is deformed, breaking the ice

All of the above have their relative advantages and disadvantages the balance of which is dependent on the operational requirements and characteristics of the target aircraft. Extensive use is made of sophisticated computer icing prediction models to determine areas requiring protection and the requirements for the protection, followed by testing of sample areas in wind tunnels and icing tunnels, and flight testing in icing conditions. Flight testing with simulated icing shapes attached to vulnerable areas of the airframe is also performed to determine handling characteristics with ice accreted.

References

[1] Society of Aerospace Engineering (SAE) (1969) *Aerospace Applied Thermodynamics Manual*. Developed by SAE Committee, AC-9, Aircraft Environmental Systems.
[2] Ernsting, J. (ed.) (1988) *Aviation Medicine*, 2nd edition, Butterworths, London.
[3] Yeoell, Lawrence and Kneebone, Robert (2003) On-Board Oxygen Generation Systems (OBOGS). For In-service Military Aircraft – The Benefits and Challenges of Retro-Fitting.

8

Emergency Systems

8.1 Introduction

Despite the best efforts of designers, constructors and operators, there will always be the risk of failure or accident that impairs the continuing safe operation of the aircraft. Under such circumstances there is the possibility of damage to it, and the risk of injury and death to the occupants or to members of the public on the ground. Although it can never be possible to cover all eventualities and account for them in design, it is possible to predict certain failures or accidents. If the statistical probability of their occurrence is sufficiently high, and the consequences of such occurrences sufficiently severe, then the aircraft design will incorporate emergency systems to improve the survivability of the aircraft and its crew.

Because emergency systems may be the final means of survival for the aircraft, crew and passengers, then the integrity of these systems must be high. Hence there is a need to separate them from the aircraft primary systems so that failures are not propagated from the primary systems into the emergency systems. Emphasis is placed on separate sources of power, alternative methods of operation and clear emergency warning indications. This will ensure that the systems can be operated during or after an emergency, and, if necessary, by untrained operators such as passengers or rescuers at a crash site.

Often the systems are designed to operate once and only when there is an emergency. Because of this it is not possible to test the systems at the beginning of each flight – the systems are essentially 'dormant'. Reliance is placed on sound design to ensure that the system will work when it is needed, and on periodic or sample testing. Examples of such systems are parachutes, passenger escape slides and ejection seats.

Aircraft Systems: Mechanical, electrical, and avionics subsystems integration. 3rd Edition Ian Moir and Allan Seabridge
© 2008 John Wiley & Sons, Ltd

8.2 Warning Systems

Since many systems in a modern aircraft perform their functions automatically and in many instances take full control of the aircraft's flight and propulsion systems, it is essential that any detected malfunctions are instantly signalled to the crew.

In previous generations of aircraft, warnings were presented to the crew as individual warning lights, each with an engraved legend on the lamp lens or on the instrument panel. Such warnings were rarely placed together but tended to be sited on the cockpit panels near to the controls or indicators of the system to which they related, or even wherever there happened to be sufficient space. This is illustrated in the Spitfire aircraft cockpit shown in Figure 8.1.

Haphazard though this may seem, a traditional hierarchy of warnings and a philosophy of colour usage emerged. Red was used for failures requiring instant corrective action, amber was used for cautions with less need for an immediate response; blue, green or clear were used as advisory or status indications. This was developed further by grouping together warnings into a single area of the cockpit or flight deck in the form of a central warning panel or master caution panel, an example of which is shown in Figure 8.2.

The attention of the crew to the generation of a warning can be achieved by incorporating a flashing lamp or attention-getter in the direct vision of the pilot, and by using audible tones in the cockpit or on the crew headsets. Bells, buzzers, electronic warbles and tones are in use on many aircraft today. A hierarchy of tones is required to ensure unambiguous attention getting in circumstances where a number of warnings arise together.

A typical sequence of events for an immediate attention warning is as follows:

- A system warning is detected by a sensor or control unit
- A signal is sent to the central warning panel
- The attention-getters flash, an audible tone sounds in the pilot's headset, and a caption on the panel is illuminated
- The pilot presses the attention-getter to stop it flashing and to silence the tone
- The pilot reads the caption and takes the necessary corrective action

Any further warnings will start the sequence again. This can be a nuisance if an intermittent fault keeps repeating itself, in which case the attention getters will repeat. Care must be taken in the design of systems to prevent intermittent faults, or to filter out repeats. To ensure that the pilot takes the correct action, a set of flight reference cards is carried. The cards enable the pilot to locate the caption rapidly and to read from the cards a series of corrective actions. Aircraft being built today tend to use Multi-Function Displays (MFDs) units

Figure 8.1 Spitfire cockpit (Courtesy of Gordon G. Bartley)

for the presentation of aircraft data, and areas on the screen can be reserved for the display of warning messages (Figure 8.3). The use of voice is available as an alternative to audio tones, it allows multiple word messages to be generated in response to different failures. An incidental benefit of this method is that such messages will automatically be recorded on the cockpit voice recorder for analysis in the event of a crash.

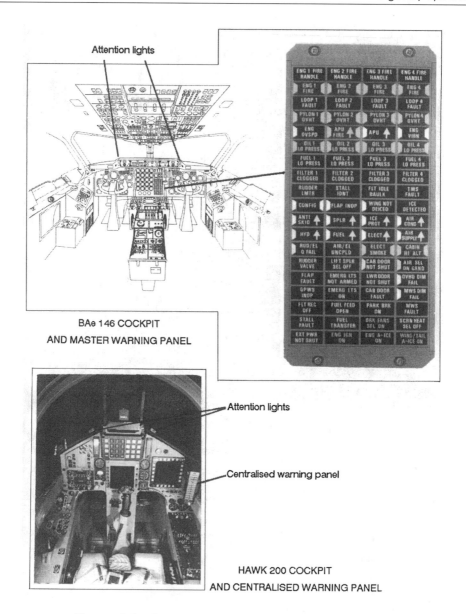

Figure 8.2 Examples of central warning panels

Multi-word visual and aural messages can be sufficiently explicit about the failure condition and do not leave the crew with the difficult task of trying to decipher a single word lamp legend together with systems indications in a stressful situation. In fact modern display systems can tell the crew what the system failure is and what actions they should take to recover to a safe condition. This electronic flight reference can be used to replace the flight reference cards. Further information on warning systems can be found in Institution of Mechanical Engineers (1991) Seminar S969 on the Philosophy of Warning Systems [1].

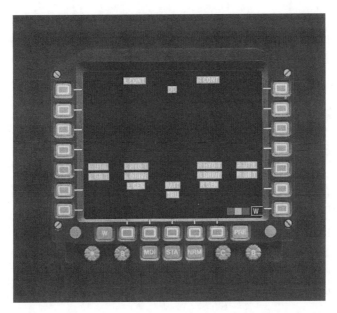

Figure 8.3   M FD warning and display system (Courtesy of Smiths Group – now GE Aviation)

8.3 Fire Detection and Suppression

The occurrence of a fire in an aircraft is an extremely serious event, since the structure is unlikely to remain sound in the continued presence of flame or hot gases. The most likely place for a fire to start is the engine compartment. Fires may occur as a result of mechanical damage leading to the engine breaking up or overheating, from pipe or casing ruptures leading to the escape of hot gases which may impinge on the structure, or from escaping fuel coming into contact with hot surfaces.

All the necessary ingredients for starting and maintaining a fire are readily available – plenty of fuel, plenty of air and hot surfaces. Needless to say, everything that can be done to prevent the escape of fluids and to reduce the risk of fire is done. Nevertheless it is prudent to install some means of detecting one and a means of extinguishing it.

Detection systems are usually installed in bays where the main and auxiliary power-plants are located (Figure 8.4 and 8.5). The intention is to monitor the temperature of the bays and to warn the crew when a predetermined temperature has been exceeded. The system consists of a temperature measuring mechanism, either discrete or continuous, a control unit and a connection to the aircraft warning displays, as shown in Figure 8.5. The temperature detection mechanism is usually installed in different zones of the engine bay so that fires can be localised to individual areas of the power-plant.

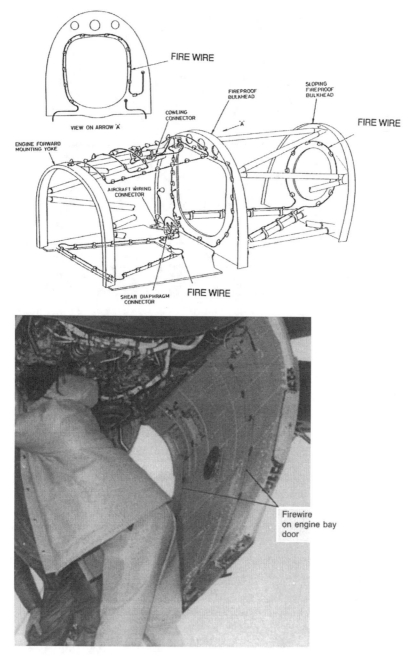

Figure 8.4 Fire detection system in an engine bay

Discrete temperature sensors usually take the form of bi-metallic strips constructed so that a contact is made up to a certain temperature, when the strips part. A number of sensors are placed at strategic locations in the engine compartment, and wired to cause the contacts to open, then the control unit

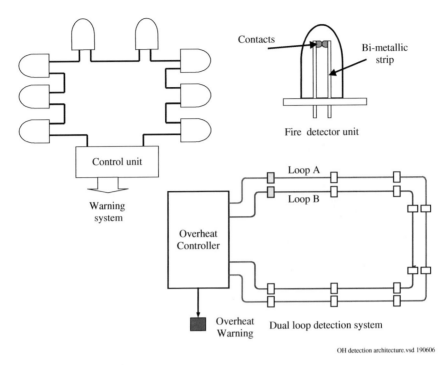

Contacts

Bi-metallic strip

Fire detector unit

Control unit

Warning system

Overheat Controller

Loop A

Loop B

Overheat Warning   Dual loop detection system

OH detection architecture.vsd 190606

Figure 8.5 Diagrams of discrete and continuous systems

detects the change in resistance of the series wiring and causes a warning to illuminate in the cockpit.

An alternative method is a continuous loop of tubular steel coaxial sensor which can be routed around the engine bay. This sensor changes its physical and electrical characteristics when subjected to heat. This change of characteristic is sensed by a control unit which causes a warning to light (Figure 8.6).

The Graviner FIREWIRE sensing element is a slim stainless-steel tube with a centrally located coaxial wire surrounded by a temperature-sensitive, semi-conductive material. This material has a negative temperature coefficient of resistance. The resistance measured between the centre wire and the outer sheath decreases with temperature, and is accompanied by a corresponding increase in capacitance. The resistance and capacitance of a loop is monitored continuously by a control unit. The control unit will provide a warning signal when the resistance reaches a predetermined value, as long as the capacitance is sufficiently high. Monitoring both parameters in this way reduces the potential for false recognition of fires resulting from damage or moisture contamination of the element.

The Kidde CFD system uses a ceramic-like thermistor surrounding two electrical conductors. The thermistor material has a high resistance at normal ambient temperature which reduces rapidly as the sensor is heated. A control unit senses the resistance and signals a warning when the value drops below a preset condition.

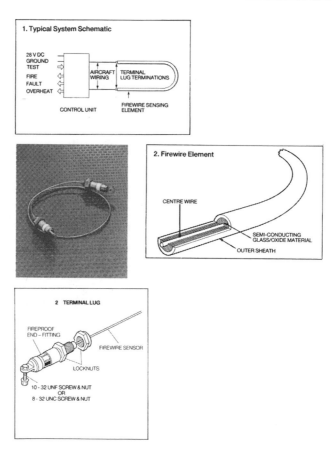

Figure 8.6 Construction of continuous firewire system (Courtesy of Kidde-Graviner)

Both discrete and continuous systems work as detectors of overheating or fire, but both are susceptible to damage by the very condition they are monitoring. The fire or jet of hot gas the leads to the temperature rise can easily burn through the wiring or the sensor. The system must be designed so that if this does occur, then the warning is not extinguished. Equally the system must be designed so that no warnings are given when there is no fire. Both these conditions are dangerous. The first because the crew may think that a fire has been extinguished, the second because a system which continually gives spurious warnings may be disregarded when a real fire occurs.

Once a fire warning is observed a formal drill is initiated by the crew to extinguish it (the fire). This will include shutting down the engine and isolating the fuel system at the engine fire wall by closing a cock in the fuel system, and then discharging extinguisher fluid into the bay.

This is done by pressing a switch in the cockpit (often a switch built into the fire warning lamp) which fires a cartridge built into a bottle containing a fluid such as BromoChloro-diFluoro-methane (BCF). This causes a spray of fluid to be directed into the engine bay. Usually the bottles are single shot. If, after discharging the bottle, the fire warning remains, the crew must decide if the warning is genuine. In a commercial aircraft this can be done by looking out of a window to see if flames can be seen in the engine nacelle, in a military aircraft by asking another aircraft to observe from behind. If a fire is confirmed then the aircraft must be landed as soon as possible or abandoned.

It should be noted that chlorofluorocarbons (CFC) endanger the ozone layer and are the subject of much debate calling for a limitation in their use. This has resulted in the development of new fluids for fire extinguishing, although some legacy aircraft may still contain CFC based fluids.

8.4 Emergency Power Sources

Modern commercial aircraft rely on multiple redundancy to achieve continued safe operation in the presence of single or even multiple failures in critical systems such as electrical or hydraulic power generation, engine or flight control. This redundancy may achieve levels as high as quadruple independent systems.

Military aircraft can rarely go to such levels and it is necessary to provide some form of emergency power source in some types. Notably these are aircraft with full authority electrical engine and flight control systems in which total power loss would result in loss of the aircraft. Very often this applies only to prototypes and test aircraft which are flown up to and beyond normal flight envelope restrictions.

An aircraft exploring high incidence boundaries is likely to depart into a stall or a spin, which may lead to such a disturbance of the engine intake air flow as to cause all engines to flame out. This will result in a total loss of engine generated power, such as electrical and hydraulic power, both of which may be required to attain the correct flight attitude and forward speed necessary to restart the engines.

Emergency power can be provided by a number of means including an Emergency Power Unit (EPU), an electro-hydraulic pump, or a Ram Air Turbine (RAT).

An emergency power unit consists of a turbine which is caused to rotate by the release of energy from a mono-fuel such as hydrazine. The hydrazine is stored in a sealed tank and isolated from the turbine by a shut-off cock. The shut-off cock is opened in emergency conditions, either manually by a pilot operated switch or automatically by a sensor which detects that the aircraft is in flight and that all engines are below a predetermined speed of rotation. The rotating turbine drives an aircraft gearbox which enables at least one hydraulic pump and one generator to be energised. A hydrazine EPU was used in Concorde prototypes and some Tornado prototypes.

An electro-hydraulic pump can be used to provide hydraulic power for aircraft in which the flight control system can be used without the need for electrical control. A manual or automatic operation can be used to initiate a one-shot or thermal battery to drive a hydraulic pump. This will provide power for a limited duration, sufficient to recover the aircraft and start the engines. Such a unit was used on the Jaguar prototype for spinning trials.

The Interdictor/Strike (IDS) Tornado emergency power system (EPS) provides hydraulic power following a double engine flame-out, a double generator failure or a double transformer rectifier unit failure.In this system a single shot battery is activated by an explosive device. This activation can be automatic or initiated by the pilot. As well as an hydraulic pump, the system also drives a fuel pump which can be supplied with power for up to 13 minutes as long as hydraulic demands are minimised. The cockpit controls are shown in Figure 8.7.

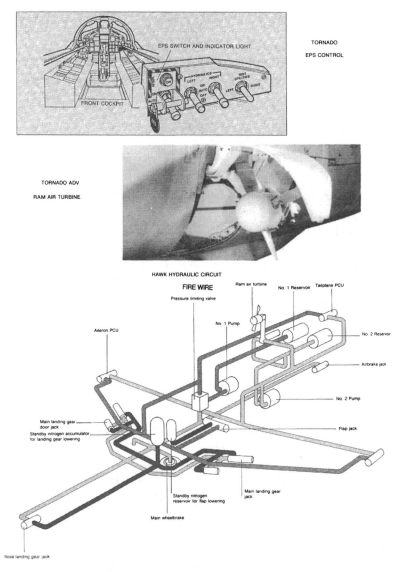

Figure 8.7 Examples of emergency power sources (Courtesy of BAE Systems)

A ram air turbine does not require a source of power other than that provided by forward movement of the aircraft. It is limited in the amount of power that can be provided. The multi-bladed unit drops from a stowed position in the aircraft and provides electrical power. The Air Defence Variant (ADV) Tornado is fitted with a RAT which is deployed automatically when both engine speeds fall below a prescribed level. The RAT maintains sufficient pressure in the No. 1 hydraulic system to provide adequate taileron control during engine re-light. The Tornado RAT is shown in Figure 8.7.

The Hawk aircraft also uses a RAT which extends into the airstream from the top fuselage following an engine failure, thereby providing power to the flying controls down to landing speed. The position of the RAT in the Hawk hydraulic circuit is shown in Figure 8.7.

8.5 Explosion Suppression

The volume above the surface of fuel in the tanks is referred to as the 'ullage'. This volume is a mixture of fuel vapour and air and must be considered as an explosion hazard. It is probably the only enclosed space on an aircraft that qualifies as Zone 1 of ATEX. The obvious way to prevent ignition of this explosive mixture is to design the fuel system to prevent that ever happening. However, fuel tank explosions have occurred in commercial aircraft. The causes of most have been explained as lightning strike, external fire, fire after refuelling and fires in heated fuel tanks. Military aircraft are at risk for similar reasons, with the added hazards of air-to-air refuelling and munitions damage in battle.

Electrical components that are installed in fuel tanks which are a potential source of ignition include:

- Fuel gauge probes
- Density measuring probes
- Level sensors
- Transfer pumps
- Boost pumps

When designing the fuel system it is important to consultant the most recent advice from FAR/JAR. In this volume, the FAA has proposed an Advisory Circular AC 25.981-1B which sets limits on energy (in microJoules) and current (in mA) that can be dissipated in components in fuel of the ullage [2]. The FAA has also defined a level of oxygen concentration of 12 % for commercial aircraft and 9% for military aircraft at sea level which should not be exceeded. These issues are discussed more fully in Chapter 3 – Fuel System.

Methods of reducing the risk of explosions include filling the ullage space with reticulated foam or with nitrogen gas (or air with a high concentration of nitrogen). Nitrogen gas is provided either from an external source of gaseous nitrogen, or may be generated on the aircraft using a molecular sieve which depletes air of oxygen. This is known as an On Board Inert Gas Generation System (OBIGGS).

8.6 Emergency Oxygen

Commercial aircraft operating above 10 000 feet pressurise the fuselage to an altitude condition that is comfortable for the crew and passengers. If there is any failure of the cabin pressurisation system then oxygen must be provided for the occupants. The aircrew are provided with face masks which they can fit rapidly to obtain oxygen from a pressurised bottle. Face masks for passengers are normally stowed in the racks above the seats, and fall automatically on depressurisation. Before each flight the cabin crew will brief passengers and demonstrate the use of the masks. The aircraft descends to an altitude where oxygen concentration in the air is sufficiently high to allow normal air breathing.

Most combat aircraft crews breathe oxygen throughout the flight using a face mask supplied with oxygen from a liquid oxygen (LOX) container which can be charged before each flight. One or two wire-wound cylinders are provided in the aircraft. The gas flows through a pressure regulating valve, and a regulator enables the pilot to select an oxygen–air mixture or pure oxygen. Two 1400 litre bottles provide sufficient oxygen for up to five hours with air-oxygen (Airmix) or up to three hours on 100 per cent oxygen for a sustained cruise at 35 000 feet.

A contents gauge and a doll's eye flow indicator are provided, as well as a failure warning light to enable the pilot to monitor the system.

If the normal oxygen supply should fail then the crew can change over to the oxygen bottle carried on their ejection seats. Although this will provide oxygen for a limited duration only, it will be sufficient to return to base. A cylinder of about 70 litres capacity is connected so that gas flow is routed through a seat mounted demand regulator. Selection of emergency oxygen automatically ensures a supply of 100 per cent oxygen irrespective of any previous crew selections. The bottle also provides an automatic supply of oxygen to the pilot upon ejection (Figure 8.8).

8.7 Passenger Evacuation

Commercial aircraft and military transports must provide a means of allowing all passengers to evacuate the aircraft in a certain time. Emergency exit doors are provided at strategic locations in the aircraft, and the doors are fitted with escape chutes so that passengers can slide to the ground. The chutes are designed to operate automatically or manually, and to inflate rapidly on command (Figure 8.9). Doors are designed to open outwards and are of sufficient width to allow passengers to exit rapidly. All doors and exits are identified with illuminated signs.

The Airbus A380 provides routes of escape to occupants from all doorways in the event of an emergency. The deployment of the emergency slides is powered by the aircraft's internal battery power. Typically the A380 will have around 555 seats in three classes although many airliners will have fewer than

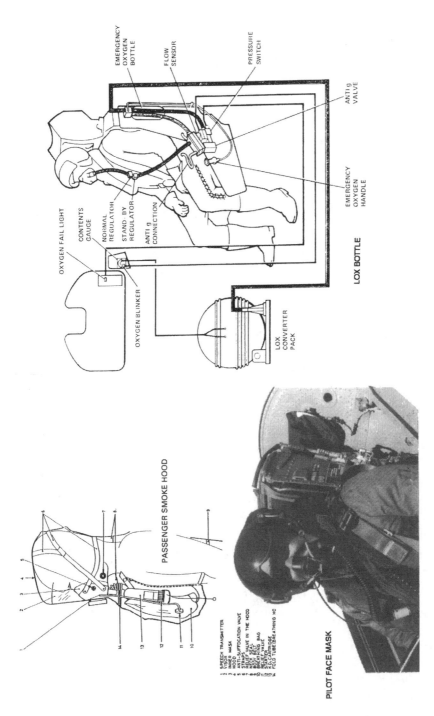

EMERGENCY
OXYGEN
BOTTLE

FLOW
SENSOR

PRESSURE
SWITCH

ANTI g
VALVE

EMERGENCY
OXYGEN
HANDLE

OXYGEN FAIL LIGHT

CONTENTS
GAUGE

NORMAL
REGULATOR

STAND - BY
REGULATOR

ANTI g
CONNECTION

OXYGEN BLINKER

LOX BOTTLE

LOX
CONVERTER
PACK

PASSENGER SMOKE HOOD

1 SPEECH TRANSMITTER
2 VISOR
3 INNER MASK
4 HOOD
5 STRAPS
6 ANTI-SUFFOCATION VALVE
7 RELIEF VALVE IN THE HOOD
8 NECK SEAL
9 RELIEF VALVE
10 BREATHING BAG
11 STARTER
12 KO2 CARTRIDGE
13 FOLD TUBE/BREATHING HO

PILOT FACE MASK

Figure 8.8 Example of a face mask, passenger hood and LOX bottle

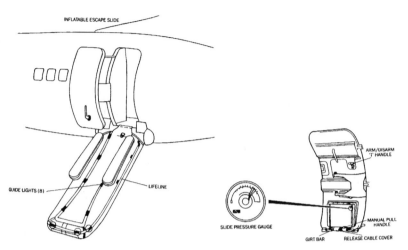

Figure 8.9 The BAE SYSTEMS ATP escape chute system (Courtesy of BAE Systems)

500. Sixteen emergency slides can be deployed at the same time from both upper and lower decks using only the aircraft's battery power.

Life vests are carried beneath the passenger seats, and the aircraft is equipped with life rafts and with locator beacons.

8.8 Crew Escape

The crew of a commercial aircraft can escape through the passenger emergency exits or by using an escape rope to slide down from the flight deck through the opening side windows.

Military crews in combat aircraft are provided with ejection seats which allow them to abandon their aircraft at all flight conditions ranging from high-speed, high-altitude to zero-speed and zero-height (zero–zero). The seat is provided with a full harness, restraints to avoid limb flailing and pull the legs and arms into the seat to avoid injury, a parachute, dinghy, an oxygen supply and locator beacon. The seat is mounted in the aircraft on a slide rail which permits the seat to travel in a controlled manner upwards and out of the cockpit. The design of the seat and the rail allows the seat and occupant to exit the aircraft with sufficient clearance between the cockpit panels and the pilot to avoid injury. The seat is operated by pulling a handle which initiates a rocket motor to propel the seat up the slide. The ejection system may be synchronised to allow the canopy to be explosively ejected or shattered before the seat reaches it, or the seat top will be designed to shatter the canopy. The canopy can be shattered by a pattern of miniature detonating cord embedded in the acrylic canopy (Figure 8.10).

The cord is a continuous loop of small diameter lead tubing filled with explosive material. The loop is bonded to the canopy transparency in a pattern that causes the canopy to fragment before the pilot leaves the aircraft. The

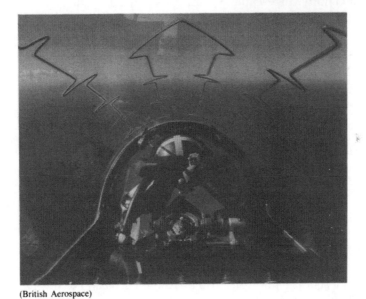

(British Aerospace)

Figure 8.10 Crew escape (Courtesy of BAE Systems/Martin Baker Engineering)

fragmentation system can be fired from outside the aircraft to allow rescuers to free the crew of a downed aircraft.

Some canopy materials such as polycarbonate will not shatter and explosive charges are placed in a pattern on the material in order to cut the canopy and use the air stream to lift the pieces of material clear of the aircraft.

The seat leaves the aircraft in a controlled manner to reduce the effect of acceleration on the crew member and a parachute is deployed to decelerate the seat and to stabilise its position. After an interval the seat detaches from the man and a personal parachute opens.

The Martin Baker Mk 10 ejection seat fitted to the Tornado has a zero–zero capability, enabling safe ejection from zero altitude and zero forward speed. This means that the crew can safely eject from an aircraft on the ground while it is stationary or taxiing. A fast, efficient ejection is absolutely essential for an aircraft designed to operate at low level and high speed. Operations in such conditions leave little time for crew escape in the event of a catastrophe. The crew can elect to eject by pulling the seat ejection handle. The escape sequence is then fully automatic, and takes about 2.5 seconds for the parachute to be fully deployed. A baro-static mechanism ensures that the seat detaches from the pilot automatically below 10 000 ft.

The Tornado is typical of present day, in-service systems. It is a two-crew aircraft with a fully automatic escape system, which needs a single input from either crewmember in order to initiate it. Each cockpit is provided with a Martin Baker Mark 10A ejection seat and both cockpits are covered by a single transparent canopy. Its escape system provides all of the following functions automatically:

- Primary canopy removal by jettison
- Secondary, backup canopy removal
- Jettison of night vision devices
- Ejection of the rear seat
- Ejection of the front seat
- Seat/occupant separation
- Personal locator beacon switch on
- Parachute deployment
- Lowering of survival aids container
- Inflation of lifejacket and liferaft on water entry

8.9 Computer-Controlled Seats

To operate the seat the pilot pulls on a handle between the thighs. This causes the canopy to be ejected or shattered and the pilot's legs and arms are restrained into the seat to stop the limbs flailing. A ballistic gas generator then ejects the seat up a pair of guide rails and out of the aircraft. At separation a rocket motor fires to continue the trajectory over the fin.

After ejection a stabilising drogue parachute is activated, the seat is separated from the pilot and a main parachute deploys. This process is controlled by an onboard multi-mode electronic sequencer and backed up by a barostatic pressure sensor.

The Mk 16 seat used on the Eurofighter Typhoon utilises a second-generation digital seat sequencer which continuously senses external environmental parameters. Under certain speed and altitude conditions the recovery timings at which the parachute is deployed are varied in order to optimise the terrain clearance.

The seat used in the Joint Strike Fighter F-35 is the US-16E which is common to all F-35 variants. The seat is modular and contains the following major assemblies:

- A seat bucket within which is located the survival aids container, a backrest and under-seat rocket motor
- A twin tube catapult with integral canopy penetrators; on the catapult is located an energy absorbing head pad, a drogue parachute, and inertial retraction device and a third-generation COTS electronic sequencer
- Side-mounted guide rails
- Fully integrated Life Support & Helmet Mounted Display equipment

The seat incorporates an auto eject function for the F-35 STOVL aircraft to be used in the event of lift fan failure. The auto ejection system utilises a signal from the FCS to initiate ejection.

8.10 Ejection System Timing

The required time delay between canopy jettison and seats is not significantly speed dependent. Improvements to low-level escape capability can be achieved by sensing the speed (and altitude) during the ejection to vary the timing of the ejection seat sequence of events. This is done on electronically controlled seats such as those for Typhoon and F-35 where the point at which the seat parachute is deployed is varied dependent on speed and altitude. Current escape system sequences have fixed time delays built into them, to ensure safe separation between the individual elements that are launched from the aircraft. For example, the Tornado has a fully automatic sequence to manage:

- Jettison of the canopy
- Ejection of the rear seat
- Ejection of the front seat

Two fixed timers are used to sequence these three elements such that there is a nominal delay of 0.30 seconds between the canopy and the rear seat and a nominal delay of 0.34 seconds between the front and rear seats. These delays are set to give safe separation across the whole escape envelope. They are also subject to production tolerances. The total delay deliberately introduced into the sequence is 0.79 seconds or approximately 80 % of the overall time taken for the canopy and both seats to separate from the aircraft.

Future improvements in low-level escape capability will come from the introduction of variable time delays, based on actual conditions rather than a single worst design case. An aircraft travelling at 450 knots in a 60-degree dive will descend through 520 ft during the 0.79 seconds delay of the Tornado

system. In a 30-degree dive, it will descend through 300 ft. As fast jet aircraft
routinely operate at or below such altitudes, any reduction in sequence delays
will reduce the height required at system initiation for a safe escape and
increase the probability of a survivable ejection. At 450 knots, the total time
delay could be reduced considerably, by 50 % or more, reducing the safe
ejection height for the Tornado front seat by some 200 ft or so. For higher sink
rates at ejection, the gain would be even greater.

In order to achieve variable sequence timings, technologies that allow position sensing and algorithms that can establish the appropriate timing for the
prevailing conditions will have to be developed. The introduction of computer
controlled sequencers onto the ejection seats will facilitate the development
and integration of these more intelligent overall system sequences.

8.11 High Speed Escape

Over the years, escape speeds have been slowly increasing. Table 8.1 shows
the percentage of ejections occurring at or above given speeds. The Hunter was
fitted with a Martin Baker MK4 ejection seat and was typical of the aircraft
in service from the mid 1950s to the mid 1970s. The Tornado is fitted with a
Martin Baker Mk10 ejection seat and is typical of aircraft in service from the
mid 1970s to today. These are predominantly peacetime ejections, in wartime
ejection speeds tend to increase overall.

Table 8.1 Comparison of ejection speeds

Aircraft	350 Kts	400 Kts	450 Kts
Hunter	19%	8%	4%
Tornado	36%	16%	8%

As ejection speeds increase, the potential for injury from air blast increases.
The face and limbs are particularly vulnerable to airblast damage.

In some multi-crew combat aircraft such as the General Dynamics F-111 the
crew escape in a module, the entire cockpit being designed to be jettisoned
and parachuted to the ground.

8.12 Crash Recorder

It is a mandatory requirement to carry a recorder in commercial aircraft and in
military aircraft operating in civilian airspace so that certain critical parameters
are continuously recorded for analysis after an accident. The recorder, variously
known as crash recorder, accident data recorder, flight recorder or, in the press,
black box recorder, is a crash survivable machine which may be ejected from
the aircraft after a crash and contains a radio and sonar locator to guide rescue
crews to its location.

The recorder is connected to the aircraft systems so that flight critical parameters are continuously recorded together with information about the aircraft's flight conditions. For example, control column and throttle positions, flight control surface positions, engine speed, pressure and temperature will be recorded together with altitude, airspeed, attitude, position and time. Analysis of this data after an accident will be used to determine the cause of the incident. Recording all crew conversations and communications with the outside world is also carried out, either on the same recorder or on a separate cockpit voice recorder (Figure 8.11).

Figure 8.11   Examples of crash recorders (Courtesy of BAE Systems)

8.13 Crash Switch

On many military aircraft it is accepted that an aircraft may have to be landed in a dangerous condition either wheels up or wheels down. The crew will have to exit the aircraft quickly and safely in these circumstances and the risk of fire must be reduced as far as possible. A crash switch is designed to do this by providing a single means of shutting down engines, closing fuel cocks, disconnecting the aircraft battery from the busbars and discharging the fire extinguishers into the engine bays.

These precautions can be provided manually or automatically. The manual method provides a number of switches in the cockpit which are linked by a bar so that a single action will operate all the switches. The pilot will do this immediately before or as soon as the aircraft hits the ground. The automatic method is provided by inertia switches that operate under crash conditions.

8.14 Emergency Landing

In the event of an emergency landing or an aborted take-off it is necessary to provide an alternative to onboard systems to stop a military aircraft. There are two methods in common use – an arrestor hook engaging on a wire across the runway, and a barrier net across the runway.

Arrestor gear is found at nearly all military aerodromes. The gear usually consists of a cable laid across the runway about 1500/2000 ft in from each end. When fully rigged, the cable is held off the runway a few inches by rubber 'doughnuts', to allow the aircraft hook to pick it up. The cable is connected to rotary hydraulic equipment that provides the retarding force when an aircraft engages the system. This airfield installation is known as Rotary Hydraulic Arresting Gear (RHAG).

Nearly all fast-jet (fighter type) aircraft have an arrestor hook for engaging these systems, which is normally retracted at the tail of the aircraft. In an emergency the aircraft would lower the hook and engage the cable. The hook is normally locked up and is released by a cockpit switch that operates a solenoid to release the up-lock mechanism.

Typical emergencies that require this type of action include aborted take-off, or landings with a known failure of brakes/hydraulics/flaps/slats/wing-sweep/engine or anything that increases the landing speed above normal or reduces the stopping power.

If an airfield does not have a RHAG, if the cable engagement fails or a fast-jet without a hook declares an emergency, then a barrier engagement is the only alternative. A barrier is a heavy duty net which can be raised hydraulically to stop the aircraft before it goes off the end of the runway. An additional deceleration device is a bed of energy absorbing material beyond the arresting devices. Airfields will have on or more of these mechanisms in series, and some will have all three as illustrated in Figure 8.12.

In the USA the arresting system has the designation MAK 2 or BAK 13. There is a BAK 14 system which can be recessed into the runway surface and is raised on demand. FAA Advisory Circular AC 25.981-1B (2001) describes these arresting systems [2].

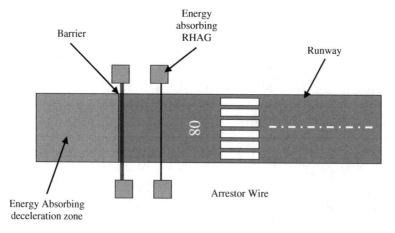

Figure 8.12 A runway equipped with Arrestor wire, barrier and deceleration zone

The arrestor wire is wound around large diameter drum equipped with paddles which is immersed in water. As the arrestor wire is engaged and extended by the aircraft it rotates the drum which provides retardation and absorbs the large amount of energy in the reservoir of water. Figure 8.13 shows an example of an arrestor wire and an aircraft engaging the wire.

Figure 8.13 F-15 Eagle engaging the arresting system (DoD photo by TSGT Edward Boyce)

8.15 Emergency System Testing

The emergency systems described in this chapter are crucial to the safety of the aircraft, crew and passengers. For this reason they must work when required to do so. Wherever possible a means of testing the systems prior to flight is made available so that the crew can have confidence in the ability of the system to provide its correct function. Proof of correct operation during design cannot for practical reasons embrace the flight testing of an ejection seat in a aircraft it is designed for. Ejection seats are tested on high speed ground tracks with dummies and flight tested on specially adapted test aircraft.

Some systems, however, cannot be tested on a pre-flight routine basis – it would obviously be impractical to test an ejection seat. There are other examples where the crew must depend on periodic testing or have confidence in the correct assembly of the system. This is a dilemma for designers and users – to establish a balance between confidence in design, and proof of design, and practical pre-flight testing.

References

[1] Institution of Mechanical Engineers (1991) Seminar S969 on the Philosophy of Warning Systems, March.
[2] FAA Advisory Circular AC 25.981–1B, 2001.
[3] Dept of the Air Force, Air Traffic Control Training Services, Equipment – Arresting Systems AT-G-02 (March 2004).

9

Rotary Wing Systems

9.1 Introduction

The helicopter was a late arrival on the aviation scene compared to more
conventional fixed wing aircraft. A number of designers experimented with
autogyros in the late 1920s and 1930s but it was not until the mid to late 1930s
that serious helicopter designs emerged. The Royal Air Force used an autogyro
which was a Cierva design, licence-built by Avro, and some Sikorsky Hoverfly
I and II examples were used for limited squadron service and evaluation
purposes. In general, the helicopter was regarded at the time as something
of an anachronism and it was not until the early 1940s that the Sikorsky R5
emerged.

In the UK, Bristol produced the Sycamore Type 171, which entered service
with the Royal Air Force in 1953. Bristol also produced the twin-rotor Type 173
which was developed for the military as the Type 192 and subsequently named
Belvedere, entering service in the 1960s. The development of helicopters in the
UK was in the main based upon UK derivatives of US designs of which the
Dragonfly, Whirlwind, Wessex and Sea King have been notable examples.

In the late 1960s and early 1970s Westland Helicopters became involved with
the joint design of a family of helicopters together with Aerospatiale of France.
This led to the development of the Gazelle, Lynx and Puma helicopters, all of
which have served with various branches of the UK Armed Forces.

The helicopter came of age as a fighting vehicle in the late 1960s and the US
involvement in the Korean War was probably the first large-scale conflict in
which it played a major part in a variety of roles. This pattern has been followed
by the British involvement in the Falklands Campaign where the shortage of
helicopters imposed severe operational limitations upon the ground troops.
More recently the role of the helicopter in the Gulf War has emphasised its
place in the order of battle – in particular the heavy battlefield attack machine
(Apache) and the missile-equipped helicopter (Lynx).

Aircraft Systems: Mechanical, electrical, and avionics subsystems integration. 3rd Edition   Ian Moir and Allan Seabridge
© 2008 John Wiley & Sons, Ltd

As their roles became more demanding so the helicopters became more sophisticated and complex. As the number of systems fitted increased to satisfy greater and more difficult tasks, the amount of propulsive power required and both the power and the number of engines fitted have increased to accommodate these needs. The Dragonfly of the 1950s required a single 550 hp engine to power the 5500 lb fully loaded helicopter. The EH101 of the 1990s (see Figure 9.1) has three T700 engines, each rated at 1437 shp to lift the helicopter with an all-up-weight of around 30 000 lb. As the size of the helicopter and engines has increased so has the complexity of the various systems. The amount of electrical power required by a large helicopter of this type equates to that needed for most jet fighters a few years ago. The EH101 also requires a complex autopilot and flight control system to provide the necessary handling characteristics so that the crew can devote their attention to the demands of the mission. Electrical and hydraulic systems also require higher levels of redundancy to support the mission requirements. Finally, the avionic equipment required to undertake a range of missions also places additional demands upon the baseline helicopter systems.

Figure 9.1 European Helicopter Industries EH 101 Merlin (Courtesy of AGUSTA WESTLAND)

9.2 Special Requirements of Helicopters

The unique nature of the helicopter compared to conventional fixed-wing aircraft deserves special consideration in relation to aircraft systems. Despite the fact that many of the same principles apply, the vertical take-off and landing features of the helicopter place a different emphasis upon their embodiment. Vertical take-off imposes a requirement for a high power to weight ratio. It is generally reckoned that for an aircraft to take-off vertically with an adequate control margin, a thrust to weight ratio of 1.25:1 is required. This ratio applies after various transmission losses have been taken into account. Here the excess thrust is that which allows the vehicle to accelerate vertically (in normal level flight – thrust = weight) and provide the ability to manoeuvre.

The means of controlling a helicopter is by its very nature totally different to the methods used by fixed wing aircraft. Also, due to unique properties such as hovering flight, and the ability to land vertically in confined areas, some system requirements are unusual. These led to the adoption of autopilot control modes such as auto-hover, which are not possible on fixed-wing aircraft. The ability to hover also dictates the need for winch systems and has led to the development of specialised autopilot modes. The need to land and remain tethered on ship decks in high seas has resulted in the introduction and use of deck locking systems.

9.3 Principles of Helicopter Flight

Whereas the lift force for a fixed wing system is produced by the passage of air over the wing aerofoil, the helicopter rotor rotating blades are aerofoils which generate the lift force to counteract the vehicle weight (see Figure 9.2). While it is more usual to have one rotor, there are a number of twin-rotor helicopters where the twin rotors may be located fore-and-aft in tandem, while others may have the rotors located side-by-side on either side of the fuselage. The rotors may comprise a number of blades which may vary between – usually between two and six though some designs may use 7 blades (CH-53E) and even 8 blades (Mi-26).

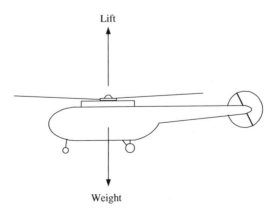

Figure 9.2   Helicopter lift forces

The fact that the helicopter lift force is generated by rotation of the rotor causes additional complication for the helicopter. As the helicopter propulsion system drives the rotor head in one direction, a Newtonian equal and opposite reaction tends to rotate the fuselage in the other direction and clearly this would be unacceptable for normal controlled flight (see Figure 9.3). This problem is overcome by using a tail rotor which applies a counter-acting force (effectively a horizontal 'lift' force) which prevents the helicopter fuselage from rotating. The tail rotor is driven by an extension of the gearbox, transmission system which couples the rotor head to the prime movers – the engines. An alternative method, called NOTAR, or NO Tail Rotor, has been developed recently and this is described later in this chapter. Twin rotors use counter-rotating rotors to help balance net torque, but it cannot be balanced exactly. Therefore, helicopters like the CH-47 etc. use differential cyclic to actually balance the torque.

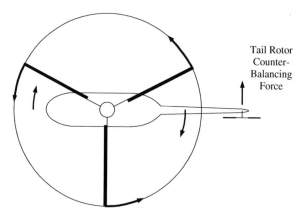

Tail Rotor
Counter-
Balancing
Force

Figure 9.3   Rotor torque effects and the need for a tail rotor

Tilting the rotor head provides the longitudinal (fore-and-aft) and lateral (side-to-side) forces necessary to give the helicopter horizontal movement. This is achieved by varying the cyclic pitch of the rotor head. Moving the pilot's stick forward alters the cyclic pitch such that the rotor tilts forward, thereby adding a forward component to the lift force and enabling the helicopter to move forwards. Moving the pilot's stick back causes the rotor to tilt backwards and the resulting aft component of the lift force makes the helicopter fly backwards. Figure 9.4. shows the effect of the pilot's controls on the rotor head and the subsequent helicopter motion.

Movement of the pilot's control column from side to side tilts the rotor accordingly and causes the helicopter to move laterally from left to right. Yaw control is by means of rudder pedals as for a fixed-wing aircraft. In the case of the helicopter, movement of the rudder pedals modifies the pitch of the tail rotor blades and therefore the thrust force generated by the tail rotor. Moving the rudder pedals to the left to initiate a yaw movement to the left increases the thrust of the tail rotor and causes the helicopter to rotate (yaw) to the left.

**Control
Movement**   **Helicopter
Motion**

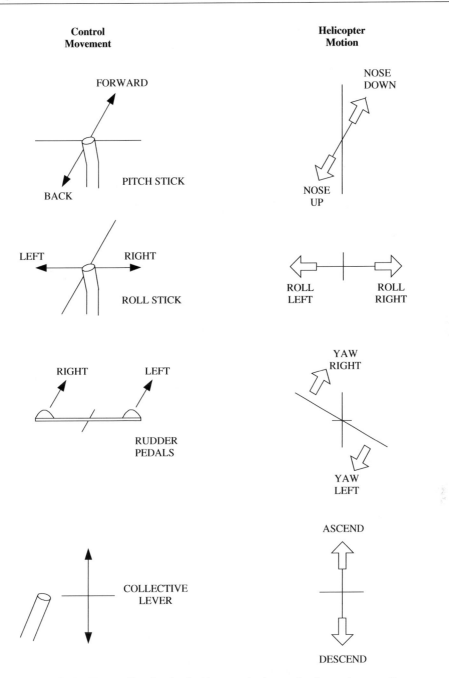

Figure 9.4 The effect of pilot's controls on helicopter motion

Moving the rudder pedals to the right causes a corresponding reduction in the tail rotor thrust and the helicopter yaws right. This is the situation for a nominal case with a counter-clockwise rotation when viewed in plane view; some helicopters use clockwise rotation in which case the reverse holds true.

Vertical movement of the helicopter is initiated by varying the pitch of all of the rotor blades and thereby increasing or decreasing total rotor lift by using the excess power available. Reducing rotor lift results in a resultant downward force which causes the helicopter to descend. Increasing rotor lift generates a resulting upward force which causes it to ascend. The pitch of the main rotor blades is varied by means of a collective pitch lever. The engine power, or torque, is controlled by a throttle twist-grip located at the end of the collective lever and is usually operated in conjunction with the collective pitch lever to cause the helicopter to climb smoothly or descend as required. Flying the helicopter is therefore achieved by a smooth coordination of pitch and lateral cyclic control, together with rudder pedals, power and collective pitch controls.

In general the helicopter is more unstable than its conventional fixed-wing counterpart. Furthermore, the secondary effects of some of the helicopter controls are more pronounced, thus requiring greater compensatory control corrections by the pilot. It follows that flying a helicopter is generally much more difficult than flying a fixed-wing aircraft, particularly when an attempt is made to execute precise tracking or positional tasks in gusty or turbulent conditions. For this reason, some sophisticated helicopters possess auto-stabilisation and multi-mode autopilot systems to minimise the effects of interreactions, thereby reducing pilot workload and thus enabling him to concentrate on crucial aspects of the flight or mission. For an easily digestible, but nonetheless comprehensive description of the helicopter and how it flies, see Fay (1987) [1].

9.4 Helicopter Flight Control

As has already been mentioned, the helicopter flight control system is more complex than that for fixed wing aircraft. Figure 9.5 depicts a typical flight control system including control runs which are of the push-rod and bell crank variety – similar to a fighter aircraft.

- The pitch and roll commands are input by means of a cyclic stick, much the same as the stick on a conventional aircraft
- Yaw commands are input by means of rudder pedals – again similar to a fixed wing aircraft and with the same sense of operation
- Thrust and lift demands are input using the collective stick, located to the left of the pilot. Raising and lowering the collective lever increases or decreases the collective pitch, thereby increasing or decreasing main rotor lift. The throttle is located at the end of the collective lever to increase or decrease the shaft horsepower or torque being extracted from the engine(s). The rotor thrust is often fashioned as a twist grip – as may be found on a motorcycle – and the combination or operating the collective lever and the throttle twist grip allows thrust and power to be coordinated using a single control

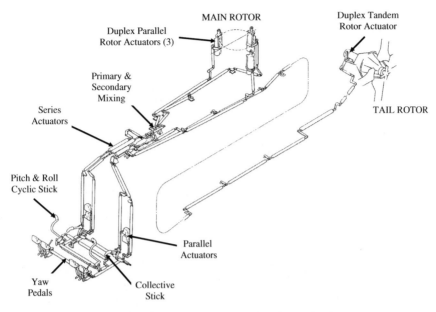

Figure 9.5 Helicopter flight controls

Other elements located within the control runs include:

- Series and parallel actuators to alter the control runs according to auto-stabilisation or autopilot commands
- Electrical trim actuators in pitch and roll activated by a stick top control
- Pitch and roll spring feel units
- Primary and secondary mixing units that mechanically sum the various channel demands to provide a harmonised response as the helicopter manoeuvres
- Rotor head actuators
- Tail rotor actuators

9.5 Primary Flight Control Actuation

On a helicopter pitch and roll manoeuvres are accomplished by tilting the rotor head using a combination of three equidistantly placed actuators 120° apart around the rotor head; one positioned to the rear on the centre-line and the others positioned forward left and right (see Figure 9.6). Extending the forward actuators while retracting the rear causes the rotor head to tilt aft with the result the helicopter will pitch up. Conversely, extending the rear and retracting the front actuators causes the rotor and therefore the helicopter to pitch forward. Differential extension or retraction across the centre-line causes the rotor and helicopter to roll right or left appropriately. The rotor disc is tilted by inducing flapping. In general, it is the case that the tilt of the swashplate plane parallels that of the rotor disk plane. Tilting the swashplate induces cyclic pitch, which

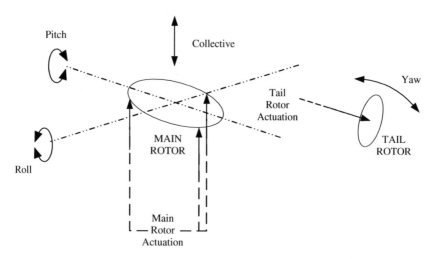

Figure 9.6 Main rotor and tail rotor actuation

leads to blade flapping, and blade flapping lags the cyclic pitch inputs by
something less than 90°. The exact relationship between the swashplate plane
and the rotor tip path plane (its tilt) depends on the mechanical design of the
rotor head, including the location of the flap hinge relative to the rotational
axis of the rotor.

The helicopter is yawed to the left or right by varying the pitch of the tail
rotor.

Figure 9.7 illustrates where the major flight control assemblies are located
on the Westland Agusta EH101 helicopter. The main rotor and tail rotor
hydraulic power actuation (as opposed to AFCS series and parallel electrical actuation that will be described later) is provided by three duplex
parallel and one duplex tandem actuator respectively. The EH101 is a highly
capable and sophisticated helicopter in service with a number of air forces
and navies. It has also been selected to provide the US Presidential helicopter as the US101. The aircraft is fitted with a Flight Control System (FCS)
that allows many features of such a system to be described in a rotary wing
context.

To demonstrate and outline the interaction of the various modes of flight
control and the associated actuators the EH101 system is described progressively as operating in three distinct modes:

- Manual control with no auto-stabilisation operative
- Manual control with auto-stabilisation engaged
- Full autopilot mode

9.5.1 Manual Control

The configuration of the primary flight control, hydraulically powered actuators is shown in Figure 9.8. Both main and tail rotor actuators receive duplex

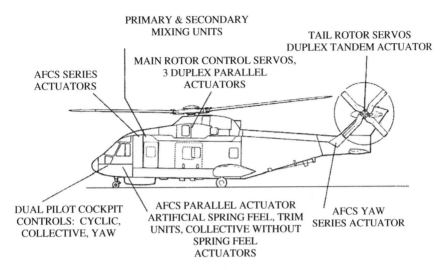

Figure 9.7 EH 101 flight controls – location of major assemblies

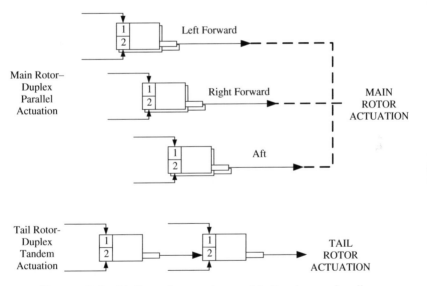

Figure 9.8 Helicopter main and tail rotor actuation

command inputs but the actuators are configured in parallel and tandem configurations as described below:

- *Main rotor actuation.* As has already been described actuation of the helicopter main rotor is achieved by three equidistant actuators. Each of the three main actuators receives duplex servo-valve commands and uses parallel actuation for each location, thereby providing redundancy in terms of signalling and actuation. As there are three actuators located around the rotor head additional redundancy is available in the event that a single actuator fails though the range of movement and rate of response may be limited

- *Tail rotor actuation.* The tail rotor actuators also receive duplex servo-valve commands but the actuators are located in a tandem (or series) configuration. The complete loss of one actuator may be tolerated assuming the actuator linear range of authority is sufficiently high

This arrangement provides redundancy for both main and tail rotor actuation. On an aircraft such as the EH101 three independent hydraulic systems supply power to this actuation system.

In the absence of auto-stabilisation and autopilot systems and with the pilot flying the helicopter manually, the functional layout of the EH101 FCS may be seen by reference to Figure 9.9. Pilot inputs in pitch and roll (cyclic stick), lift commands (collective stick) and yaw commands via the rudder pedals cause appropriate movement in the control runs. After primary and secondary mixing, outputs are fed to the main rotor and tail rotor actuators. With the exception of the pilot introducing electrically actuated pitch or roll trim commands by means of the four-way trim switch on top of the cyclic stick, all actuation is manually induced and hydraulically powered. However helicopters are inherently unstable machines, especially in gusty conditions and pilot workload can be unacceptably high when performing operational tasks. Accordingly, the FCS is augmented by additional systems to ease pilot workload and permit better execution of the operational task. The additional elements are:

- An auto-stabiliser to stabilise the helicopter and dampen out unwanted perturbations
- An autopilot to provide additional closed loop control to facilitate the mission

In a practical system these functions may be highly interrelated and inter-woven.

9.5.2 Auto-Stabilisation

In order to reduce pilot workload and improve handing qualities helicopters, in common with many conventional aircraft, employ an auto-stabilisation system. Auto-stabilisation systems use aircraft attitude, body rates and accelerations and air data inputs to perform additional calculations to modify the aircraft response: necessary control corrections are fed into the actuation system. The auto stabilisation system effectively closes a loop around the aircraft natural response to modify the handling characteristics. In order to make the necessary corrections this loop needs to operate with high bandwidth but relatively low control authority.

In the helicopter FCS this is achieved by means of an electrically driven series actuator. As there are four control channels: pitch, roll, yaw and collective then a series actuator is placed in each control run (see Figure 9.10). In the case of

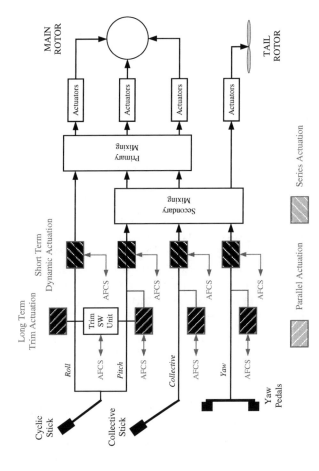

Figure 9.9 Typical helicopter – manual control

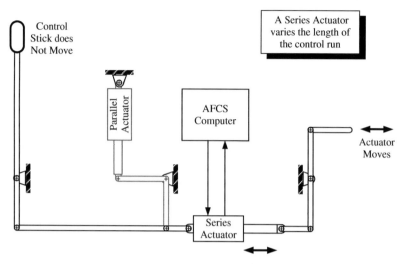

Figure 9.10 Series actuation – auto-stabilisation

series actuation, control inputs are not visible to the pilot as the relatively high frequency corrections are fed in downstream of the feel units.

The Automatic Flight Control System (AFCS) computer calculates the inputs for each control channel and feeds these to series actuator(s) receiving feedback information from the actuator. The series actuator introduces high frequency low amplitude inputs into the control run as the auto-stabiliser algorithms make the continuous corrections to maintain stability. When the auto-stabiliser function is engaged the actuator receives continual inputs of low magnitude but the control run upstream of the actuator not move and the pilot is unaware of what is happening; apart from the evident improvement in handling characteristics. Figure 9.11 illustrates the helicopter FCS with auto-stabiliser engaged.

9.5.3 Autopilot Modes

Whereas the auto-stabilisation function inputs small high frequency control inputs into the FCS the autopilot provides longer term, more slowly varying inputs to change a control variable gradually over time to satisfy the trajectory requirements of the helicopter. The principle of operation of a parallel actuator is shown in Figure 9.12. The parallel actuator receives demands from the AFCS and provides feedback upon the actuator position. The actuator is connected to the appropriate control run via a suitable linkage and as the parallel actuator extends or retracts according to the AFCS demand, so the position of the entire control run varies. A parallel input will cause the actuator servo-valve to move; it will also move the pilot's control stick.

Parallel actuators in the pitch, roll, yaw and collective channels therefore introduce a slow but long-term change in the control run and therefore in the appropriate helicopter attitude. A speed hold mode such as HEIGHT

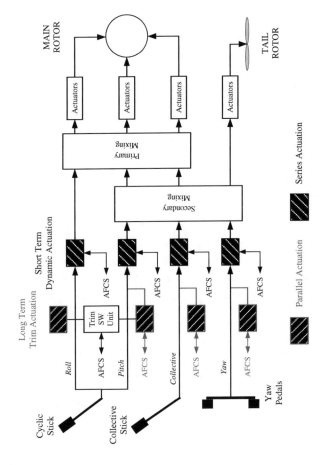

Figure 9.11 Typical helicopter control – auto-stabilisation mode

HOLD will use the collective parallel actuator (controlling thrust) and engine torque (controlling power) to maintain a given barometric or radio altimeter datum.

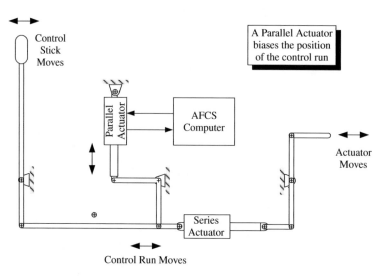

Figure 9.12   Parallel actuation – autopilot functions

In a practical AFCS both the auto-stabiliser and autopilot functions will be operating in conjunction to stabilise the helicopter and to provide the autopilot trajectory commands required by the pilot. Therefore the situation shown in Figure 9.13 will prevail with both series and parallel actuators receiving

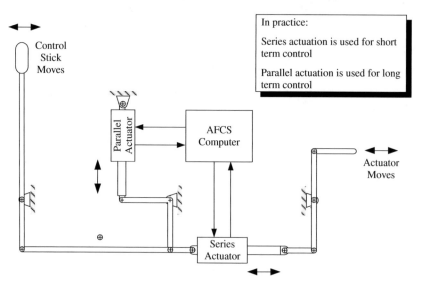

Figure 9.13   Series and parallel actuation – AFCS

AFCS commands: short-term demands for the series (auto-stabiliser) actuator and long-term demands for the parallel (autopilot) actuator. The control run bias inputs that the parallel actuators impart are analogous to trimming the aircraft; such that if the autopilot control authority is lost (e.g. autopilot drop-out) the vehicle is in the correct trim state for the prevailing flight conditions. Figure 9.14 illustrates the complete EH 101 FCS control arrangement when operating in full auto-stabilisation and autopilot modes. The control channels operate as follows:

- Roll (R) demands may be influenced by roll trim, roll auto-stabiliser or autopilot roll demands. Roll demands feed directly into the primary mixing unit and then to the main rotor actuators
- Pitch (P) demands may be influenced by pitch trim, pitch auto-stabiliser or autopilot pitch demands. Pitch demands feed directly into the secondary mixing unit and then primary mixing unit to the main rotor actuators
- Collective (C) demands may be influenced by collective auto-stabiliser or autopilot collective demands. Collective demands feed directly into the secondary mixing unit and then to the primary mixing unit to the main rotor actuators
- Yaw (Y) demands may be influenced by yaw auto-stabiliser or autopilot yaw demands. Yaw demands feed directly the secondary mixing unit and then downstream to the tail rotor actuator

Figure 9.14 combines the information from Figure 9.7 and 9.13 to show how the main pitch, roll, collective and yaw control channel demands are translated to the main and tail rotors to fly the aircraft:

- Roll, pitch and collective demands from the primary mixing units are fed proportionately into the left forward, right forward and aft main rotor duplex parallel actuators resulting in the main rotor performing roll, pitch and collective motion
- Yaw demands from the secondary mixing unit to the duplex tandem tail rotor actuator resulting in a yaw motion

This section should be read in conjunction with the description of the Smiths/GE Aviation AFDS later in this chapter.

9.6 Key Helicopter Systems

The basic principles of many helicopter systems are identical to similar systems in fixed-wing aircraft. However, the unique nature of the helicopter places a different emphasis upon how these systems are implemented and also introduces a requirement for some totally new systems. A range of these systems

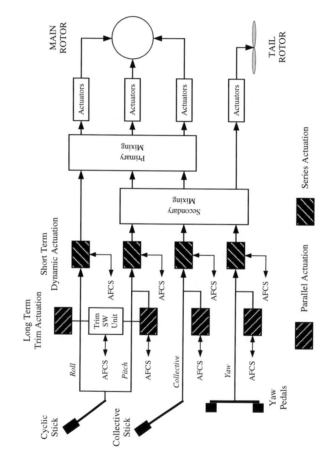

Figure 9.14 Typical helicopter – full AFCS control

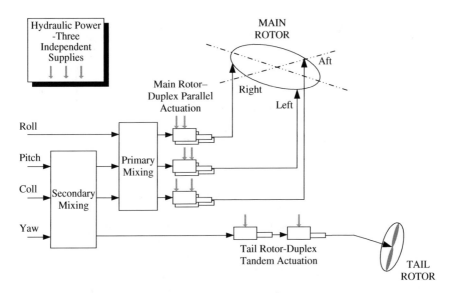

Figure 9.15 Helicopter hydraulic power actuation

is described so that a comparison might be made with the fixed-wing aircraft equivalent. They are:

- Engine and transmission system
- Hydraulic system
- Electrical system
- Health monitoring system
- Specialised helicopter systems

9.6.1 Engine and Transmission System

Many helicopters today have a number of engines to supply motive power to the rotor and transmission system. In fact, all but the smallest helicopters usually have two engines, and some larger ones have three. The need for multiple engines is obvious; helicopter lift is wholly dependent upon rotor speed, which in turn depends upon the power provided by the engines. In the event of engine failure it is still necessary to have power available to drive the rotor, therefore multiple engines are needed so that the remaining engine(s) can satisfy this requirement. Although it is possible to land a single-engined helicopter following engine failure, using a technique called auto-rotation, this mode of unpowered flight takes time to establish. If the helicopter is flying at around 500 ft or less then it is unlikely that safe auto-rotation recovery can be carried out. Engines are usually sized so that the aircraft can fly for a period of time with one engine failed, except in the most extreme flight conditions:when the helicopter is flying heavily loaded or 'hot and high' [2]. The EH 101 Merlin is fitted with a variant of the General Electric T700-GE-401 turbo-shaft engines in the naval variant while civil and military versions are powered by the General Electric CT7-6, a variant of the T700 developed specifically for the EH 101. Martin (1984) gives more detail regarding the development of the T700 family of engines [3].

A more recently developed engine available for this class of helicopter is the Rolls/Turbomeca RTM 322 which is designed to operate at 2100 shp (shaft horse power) and weighs around 530 lb. This engine is of a suitable size to power up-rated versions of the EH 101. It is being produced with a 50/50 work share by Rolls-Royce and Turbomeca and an indication of the engine configuration and work share is given in Figure 9.5. Bryanton (1985) and Buller and Lewis (1985) describe the development programme of the RTM 322 [4, 5].

The majority of new helicopters use gas turbines rather than internal combustion engines, for a variety of reasons. Most engines are electronically controlled using computers and over recent years control has become digital in nature, using Full Authority Digital Engine Control (FADEC). These units are usually configured with two lanes or channels of control, though, for a single-engined helicopter, a dual channel and a hydro-mechanical standby channel may be provided. Typical control laws which would be embodied are:

- *Acceleration control*. Acceleration control (of the gas generator) is for surge prevention and is done using either fuel flow scheduling using a control law such as:

$$\frac{W_F}{P_C} = f(N_H)$$

Where W_F = Fuel flow and P_C = Compressor discharge pressure
The ratio: $\frac{W_F}{P_C}$ is a powerful parameter since it approximates fuel/air ratio as long as the high pressure turbine is choked. Deceleration limiting to prevent engine flameout may be similarly implemented

- *NH control*. More recently control of 'N_H Dot' as a function of engine inlet conditions is used together with surge recovery algorithms. (Note: N_H Dot acceleration control is used by the venerable Adour engine using a dashpot that varies with altitude. This being implemented hydro-mechanically is not very sophisticated while modern N_H Dot system implemented in software can be much more complex without penalty.) N_H control is perhaps the most challenging issue in helicopter control since this is the control of power delivered to the rotor system. As the pilot makes sudden changes in collective, the gas turbine must respond immediately to deliver more gas horsepower in order to maintain an essentially constant rotor speed
- *Maximum/minimum fuel flow limiting*
- *Torque limiter*. Torque sharing between engines is also a common control requirement when isochronous rotor speed governing is employed

These control laws are complex and detailed description is outside the scope of this book. Saunders (1983) gives a fuller description of these control laws and their implementation [6]. A system comprising more than one engine/FADEC may also incorporate features whereby one will be accelerated to maximum power if one of the other engines fails or the thrust drops below a predetermined level. Such a system is likely to apply power more quickly than the pilot when operating in a critical flight mode such as the hover.

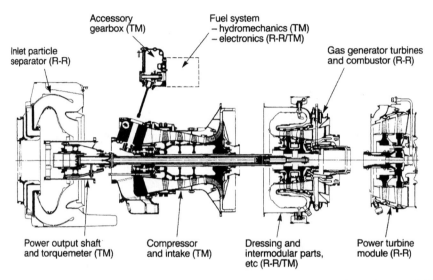

Figure 9.16 RTM 322 configuration and work share (Courtesy of Rolls Royce/Turbomeca)

An idea of the complexity of the transmission system needed for a three-engined helicopter may be gained from Figure 9.17. This depicts how each of the three engines drive though a series of reduction gears to the third stage collector gear. The collector gear drives the rotor at 210 rpm through a sun and planet gear. The tail rotor shaft is driven off the collector gear at 3312 rpm. The accessory gearbox is also driven off the collector gear, however when the rotor is stationary it is possible to drive the accessory gearbox by the APU or from No. 1 engine by pilot selection. The accessory gearbox drives two of the three hydraulic pumps and the two AC generators. The third hydraulic pump is driven directly off the main gearbox. The main gearbox lubrication system comprises two independent lubrication circuits, each with its own oil pump filter and cooler.

The EH 101 main gearbox and engine installation are shown in Figure 9.18. The nose of the helicopter is to the left of this diagram. The three engines can be clearly seen as can the APU which is to the rear of the main gearbox and just above the tail rotordrive shaft. The accessory gearbox is located

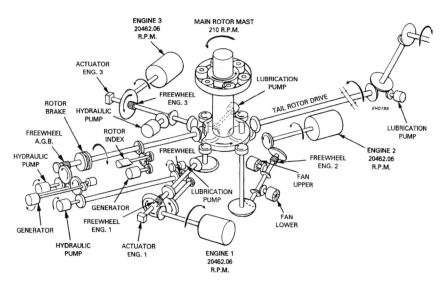

Figure 9.17   EH 101 Merlin main transmission system (Courtesy of AGUSTA WESTLAND)

on the front of the main gearbox and the main rotor drive rises vertically from the main gearbox. Due to the obvious importance of the transmission system a considerable degree of monitoring is in-built to detect failures at an incipient stage. Typical parameters which are monitored are oil pressures and temperatures, bearing temperatures, wear, and in some cases accelerations. The role of the health and usage monitoring system on board helicopters is assuming paramount importance and will be discussed later in this chapter.

9.6.2 Hydraulic Systems

For helicopters, the hydraulic systems are a major source of power for the flying controls as for various other ancillary services. A typical large helicopter, such as the EH 101, has three hydraulic systems, though smaller vehicles may not be so well-endowed. The number of hydraulic systems will depend upon integrity requirements and helicopter handling following loss of hydraulic power.

The main hydraulic loads supplied are:

- *Powered flying controls:*
  - 3 dual main rotor jacks
  - 1 dual tail rotor jack

- *Ancillary services:*
  - landing gear
  - steering
  - wheel brakes
  - rotor brake
  - winch (if needed)

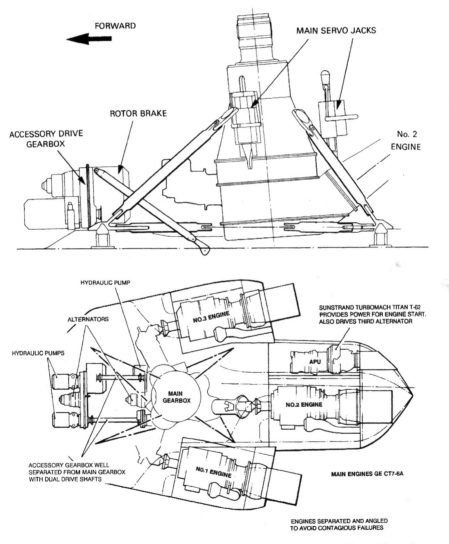

Figure 9.18 EH 101 main gearbox and engine installation (Courtesy of AGUSTA WESTLAND)

9.6.3 Electrical System

The EH 101 electrical system is shown in the simplified block schematic in Figure 9.19 and is typical of the electrical system of this size of helicopter. AC generation is supplied by two main generators each of 45 kVA capacity driven by the accessory gearbox. An emergency AC generator is driven directly off the main gearbox. The arrangement of the main generator tie contactors and the bus tie contactors, controlled by the two generator control units, is typical of a system of this configuration. In the event of an under-voltage condition

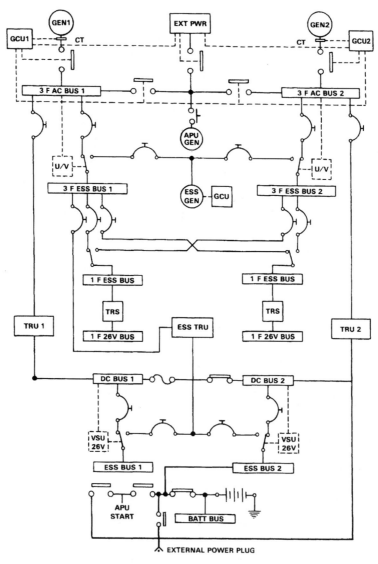

Figure 9.19 EH 101 simplified electrical system (Courtesy of AGUSTA WESTLAND)

being sensed, bus transfer relays switch the output of the essential ac generator onto No. 1 and No. 2 essential buses as appropriate. These feed, in turn, single phase essential buses and the essential TRU. In normal conditions No. 1 and No. 2 TRUs feed dc buses No. 1 and No. 2 respectively. A battery is provided, mainly to start the APU, however this can provide short duration emergency power in the event of a triple electrical systems failure.

9.6.4 Health Monitoring System

The importance of the health monitoring system has already been briefly mentioned in the section on the engine and transmission system. The importance of Health and Usage Monitoring Systems, or HUMS as they are also known (in abbreviated parlance) are now considered to be so important that the UK Civil Aviation Authority (CAA) now specifies the equipment as mandatory for all helicopters certified in the UK.

There are two notable aspects to the use of HUMS. The first relates to criticality and flight safety, the second to cost savings. If the correct critical parameters in an engine and transmission system are monitored then it is possible to identify deterioration of components before a critical failure occurs. This is done by establishing a time-history of the parameter during normal operation of the aircraft, and carrying out trend analysis using computers and data reduction techniques. The tendency for a parameter to exceed set thresholds on either an occasional or regular basis can be readily identified as may a steadily rising trend in a component vibration measurement. Such trends may be identified as heralding a gearbox failure – possibly an impending gear tooth failure – or increasing torque levels in a transmission shaft which might indicate that the component is being over-stressed and may fail in a catastrophic fashion. Many such failures in a helicopter gearbox and transmission system could cause the loss of the helicopter and occupants.

With regard to cost savings HUMS helps to avoid the expense of a major failure and the significant engine damage and expense which this entails. As has been shown, a multi-engined helicopter is well capable of flying and landing with two or even one remaining engine, so the flight hazard is of a lower order. However, the expense of overhauling an engine after a major failure is considerable. It therefore makes sound economic sense to monitor key engine parameters and forestall the problem by removing the engine for overhaul when certain critical exceedances have been attained. The ability to monitor the consumption of component life may be used to modify the way in which the helicopter is operated or maintained. If it is apparent that operating the aircraft in a certain way consumes component life in an excessive manner then the pilots may be instructed to modify the flight envelope to avoid the flight condition responsible. From a maintenance standpoint it may be possible to extend the life of certain components if more information is available regarding true degradation or wear. In some cases it may be possible to dispense with a rigid component lifing policy and replace units in a more intelligent way based upon component condition.

The parameters which may be monitored are extensive and may depend to some degree upon the precise engine/gearbox/rotor configuration. Listed below is a range of typical parameters together with the reason for their use:

- *Speed probes and tachometer generators*: the measurement of speed is of importance to ensure that a rotating component does not exceed limits with the risk of being over-stressed
- *Temperature measurement*: the exceedance of temperature limits or a tendency to run hot is often a prelude to a major component or system failure
- *Pressure measurement*: a tendency to over-pressure or low pressure may be an indication of impending failure or a loss of vital system fluids
- *Acceleration*: higher acceleration readings than normal may indicate that a component has been over-stressed or that abnormal wear is occurring. The use of low-cycle fatigue algorithms may indicate blade fatigue which could result in blade failure
- *Particle detection*: metal particle detection may indicate higher than normal metal contamination in an engine or gearbox oil system resulting from abnormal or excessive wear of a bearing which could fail if left unchecked

Most HUMS systems continuously monitor and log the above mentioned parameters and would only indicate to the pilot when an exceedance had occurred. The data accumulated is regularly downloaded from the aircraft using a data transfer unit. The data is then transferred to a ground-based computer and replay facility which performs the necessary data reduction and performance/trend algorithms, as well as providing a means of displaying the data. In this way it is possible to maintain a record of every helicopter in the fleet and to take the necessary actions when any exceedances or unhealthy trends have been identified. Astridge and Roe (1984) describes the health and usage monitoring system of the Westland 30 helicopter [7].

9.6.5 Specialised Helicopter Systems

A number of systems to be found on a helicopter are specific to the nature of its mode of operation and would find no equivalent application on a fixed-wing aircraft. Two such systems are the winch and the deck-locking system.

Winch System

The helicopter's ability to hover, when coupled with the provision of a winch system, clearly enhances its flexibility in a range of roles such as the lifting and handling of loads or the recovery of personnel from the ground in an emergency situation. The winch may be either electrically or hydraulically operated and some aircraft may offer both. The winch operates by using either source of power to drive a reversible motor which pays out or retrieves the winch cable. The winch control system has the ability to lock the cable at any position while under load. Winch power may be controlled by the pilots using a control unit in the cockpit. However, it is usual to provide a control station adjacent to the cargo door where the winch may be controlled by a dedicated cable operator. The system may include a guillotine arrangement whereby it may be severed should be winch operation endanger the helicopter. This could occur if the winch hook became entangled with an object on the ground or if the helicopter suffered an engine failure or power loss while lifting a heavy load.

Deck-Locking System

The deck-locking system enables a helicopter to land on, and remain secured to, the desk of a heaving ship in gale force winds up to 50 knots. The principle has been in use since the early 1960s when a rudimentary system was tested by the Royal Aircraft Establishment (now the Defence Research Agency), using a Dragonfly helicopter. The system allows the pilot to 'capture' the deck, either for a final recovery landing or to re-arm or refuel prior to an additional sortie.

The deck lock system is developed and produced by Claverham – now Hamilton Sundstrand – and is in use for the recovery of helicopters up to 20 000 lb. Later systems under development for use with the EH 101 will enable operation with helicopters up to 30 000 lb. The ship deck has a grid into which a helicopter mounted harpoon arrangement may engage. The helicopter hovers above the deck as the pilot 'arms' the system. This causes the deck-lock to be lowered from the stowage bay into an extended position. By judging the movement of the ship, the pilot elects to touchdown and activates the system by pressing a switch located on his collective lever. This enables the engagement beak and jaws to engage the deck grid and secure the helicopter to the deck. If for any reason the beak misses the grid, or encounters solid deck, the system automatically recycles and the pilot may re-attempt engagement. The engagement sequence is complete with 1.5 seconds. The deck lock system for the EH 101 is shown in Figure 9.20.

9.7 Helicopter Auto-Flight Control

9.7.1 EH 101 Flight Control System

The means by which the FCS provides auto-stabilisation and autopilot control has already been described earlier in the chapter. Handling a large helicopter such as the EH 101 requires a great deal of effort and concentration by pilots who have other considerable demands placed upon them, for instance by Air Traffic Control or mission requirements. The need for an advanced AFCS providing a wide range of autopilot modes is paramount

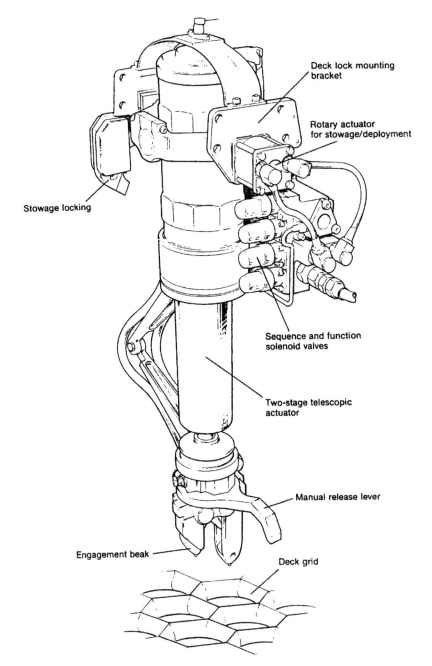

Deck lock mounting bracket

Rotary actuator for stowage/deployment

Stowage locking

Sequence and function solenoid valves

Two-stage telescopic actuator

Manual release lever

Engagement beak

Deck grid

Figure 9.20 EH 101 deck-lock system (Courtesy of Claverham/Hamilton Standard)

and the system developed by Smiths Group – now GE Aviation – and OMI Agusta provides the necessary automatic flight control. Within the EH101 AFCS, the auto-stabilisation and autopilot functions are partitioned into two main areas as follows:

- *Auto-stabilisation functions:*
  - Pitch, roll and yaw auto-stabilisation
  - Pitch and roll attitude hold
  - Heading hold
  - Turn coordination
  - Auto-trim

- *Autopilot functions:*
  - Barometric altitude hold
  - Radar altitude hold
  - Airspeed hold
  - Heading acquire
  - Vertical speed acquire
  - Navigation mode
  - Approach mode
  - Back course
  - Go-around
  - Hover hold
  - Hover trim
  - Cable hover
  - Transition up/down

The AFCS developed by Smiths Group – now GE Aviation/OMI Agusta is based upon a dual duplex architecture. Dissimilar microprocessors and software are utilised to meet the high integrity requirements. The simplified AFCS architecture is shown in Figure 9.21. At the heart of the system are the two Flight Control Computers (FCCs). Each FCC receives sensor information from the sensor unit as well as discrete and digital information from the aircraft sensors and systems. Both FCCs communicate with the other via digital and hardwired links. Both FCCs also communicate with the Pilots Control Unit shown in Figure 9.22. The control unit conveys to the pilot the status of the system and enables the pilot to monitor hover and radar altimeter altitude in feet and helicopter airspeed in knots.

Both FCCs output information to the aircraft management computer and to the Electronic Flight Instruments System (EFIS) displays. FCC 1 feeds lane 1 commands to the pitch, roll, yaw and collective series actuators. FCC 1 also supplies the parallel actuator pitch and roll commands. FCC 2 supplies lane 2 commands to the pitch, roll, yaw and collective series actuators. FCC 2 also supplies the yaw and collective parallel actuator commands.

To guard against failures and single point design errors several layers of protection are inherent in the AFDS implementation, these are:

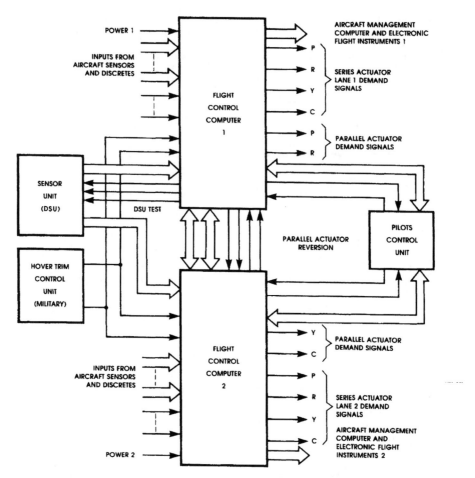

Figure 9.21 EH 101 Merlin AFCS architecture (Courtesy of AGUSTA WESTLAND)

- Dual redundancy of FCCs – FFC 1 and FCC 2
- Duplex implementation of control algorithms: pitch, roll, collective and yaw
- Monitored computation
- Dissimilar hardware – INTEL 80286 and Motorola 68000 – processors
- Dissimilar software defined and coded by two independent software teams
- Triplex PSUs in each FCC

9.7.2 NOTAR Method of Yaw Control

The helicopter systems described so far have been controlled in yaw by means of conventional use of the tail rotor. Boeing (formerly the McDonnell Douglas Helicopter Company) of Mesa, Arizona, introduced on some of their smaller helicopters an alternative method of yaw control called NOTAR (short for **NO TA**il **R**otor). This method replaces the variable pitch tail rotor and the rotating

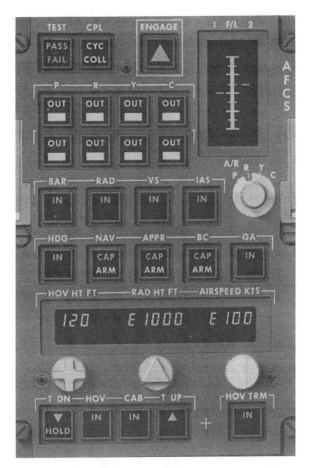

Figure 9.22 EH 101 AFCS pilot's control unit (Courtesy of Smiths Group – now GE Aviation)

drive shaft which has to pass the length of the tail boom to drive the tail rotor gearbox.

The NOTAR principle uses blown air to counteract the main rotor torque effect and it does this by employing two different means. Instead of a conventional tail boom structure the NOTAR tail comprises a hollow tube down which air is blown by a variable 13 blade 22 inch diameter fan. At the end of the boom, air is vented through direct jets which counteract the rotor torque. In addition downwash from the rotor passes externally over the boom causing a sideways anti-torque force very similar to the way in which an aircraft wing works. The airflow down the right hand side of the boom is encouraged to adhere to the boom by means of air bled out of thin longitudinal slots in the boom. The resulting forces induce a counter torque moment due to the Coanda effect. Measurements have indicated that approximately two-thirds of the counter torque force of the NOTAR concept is produced by the Coanda effect; the remaining third is generated by the low pressure air exhausting from the rear of the boom. See Figure 9.23.

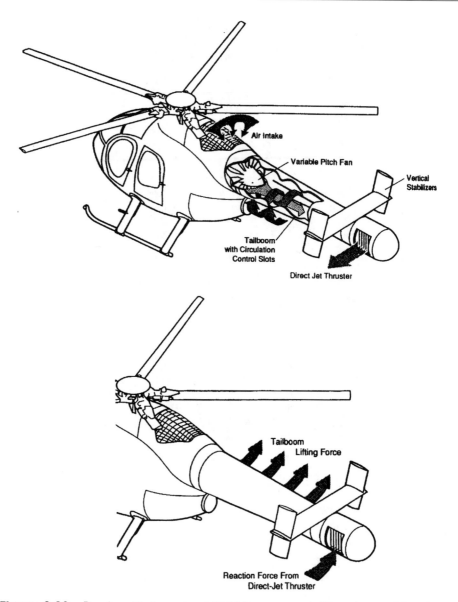

Figure 9.23   Boeing Helicopter NOTAR concept (Courtesy of Boeing)

The advantage of NOTAR is that it is relatively simple compared to the conventional tail rotor. The only moving parts are the fan and significant weight savings are achieved. The NOTAR concept has demonstrated up to 40 knots sideways motion using this principle and it is claimed that turns are much easier to coordinate, particularly in gusty conditions. Another advantage is that the concept is largely self-correcting with increases in power; as power is increased so does the rotor torque effect; however, so too does the rotor downwash and the Coanda effect and the counteracting force. A further benefit is the absence of rotating parts at the end of the tail boom which reduces the hazard to personnel on the ground and to the aircraft while manoeuvring close to trees in a combat situation.

McDonnell Douglas and their Superteam partner Bell included the NOTAR design in their submission for the US Army light helicopter (LH) proposal. This is the next generation of lightweight helicopters for the US Army. The contract award was, however, given to a Boeing/Sikorsky grouping. The US Army has reportedly modified 36 H-6-530 helicopters (US Army version of the MD530) to the NOTAR configuration. The modification is said to save 20 per cent of the airframe weight and it is expected that handling will be improved, noise reduced and power savings made.

9.8 Active Control Technology

Active Control Technology (ACT) is the term used in the UK to describe full authority manoeuvre demand flight control systems. Such a system would be fly-by-wire using electrical, or possibly fibre-optic signalling, instead of the conventional rod and lever flight control runs of the type already described for the EH 101 helicopter. The most obvious advantages of the ACT as applied to a helicopter are weight savings due to the removal of the mechanical control runs and pilot workload improvement due to enhanced handling characteristics. Future battlefield helicopters will need to be extremely agile compared to those of today and ACT is seen as vital in providing the necessary carefree manoeuvring capabilities. Richards [8] and Wyatt [9] address ACT and the helicopter.

The key issues relating to ACT and the helicopter are:

- The level of redundancy, i.e. triplex versus quadruplex lane architecture required to meet the integrity levels specified. This decision depends upon whether the helicopter requirements are military or civil and upon the effectiveness of BIT coverage and in-lane monitoring. Present thinking appears to favour a triplex implementation provided the monitoring and dissimilarity issues are addressed in a satisfactory manner
- The degree of dissimilarity between the processing 'strings' is important for high integrity. There is a general fear of the probability of a single catastrophic failure in the electronic computing elements or associated input/output which could cause a common mode failure of all lanes. This concern has become more prevalent due to the proliferation of commercial VLSI microelectronic chips, where it is almost impossible to conduct a Failure Modes and Effects Analysis (FMEA) with a high level of confidence. Equally there is concern regarding latent common mode software failures. The main way of solving these problems is by introducing hardware and software dissimilarity. It will be recalled that such a scheme is utilised in the EH 101 AFCS which is not fly-by-wire
- The signalling and transmission medium is a further consideration. The use of serial data buses offers great attractions: the main area of debate is whether

signalling should be by electrical or fibre-optic transmission. The appeal of the fibre-optic medium is an improved resistance to electro-magnetic interference (EMI)

9.9 Advanced Battlefield Helicopter

The most capable battlefield helicopter – arguably in the world – has recently had a chance to prove its combat effectiveness. The McDonnell Douglas AH-64A Apache was used to great effect during 'Desert Storm', in the Middle East, when its night capability and fearsome firepower was amply demonstrated. Due to the success of the helicopter during that conflict it is considered topical to outline some of the key characteristics in this chapter.

The Apache helicopter was originally designed by the Hughes Helicopter Company which was later acquired by the McDonnell Douglas Corporation, the company being renamed McDonnell Douglas Helicopter Company (MDHC). MDHC later became part of Boeing as has already been mentioned. The first Apache prototype flew in September 1975 and the first production aircraft in AH-64A/B was delivered in January 1984. By the end of 1987 some 300 aircraft had been delivered. The US Army subsequently significantly improved the capability of the aircraft by introducing the AH-64 C/D variants with increased electrical power, increased avionics bay capacity and the introduction of a mast mounted millimetric fire control radar (Longbow radar) on the –D variant. The basic AH-64A Apache configuration is shown in Figure 9.24. The aircraft for the –C/D variant were in the main remanufactured –A/B variants. The aircraft was also manufactured in the UK as the WAH-64D, all UK variants having the fire control radar.

The helicopter has a four blade articulated main rotor and a four blade tail rotor. The helicopter is powered by two 1696 shp General Electric T700/701 turbo-shaft engines which have an engine-out rating of 1723 shp. The hydraulic system comprises dual 3000 psi systems which power dual actuators for main and tail rotors. In the event that both systems fail a reversionary electrical link provides backup control. The AH-64A has two 35 kVA AC generators. A Garrett APU is provided for engine starting and to provide electrical power for maintenance. The helicopter is operated by two crew: a pilot sitting aft and a Co-Pilot Gunner (CPG) in the front cockpit.

Perhaps the two most striking features of the AH-64A are the night vision system and the extensive range of armaments which may be carried.

9.9.1 Target Acquisition and Designator System (TADS)/Pilots Night Vision System (PNVS)

The night vision system is called the Target Acquisition and Designator System/Pilots Night Vision System or TADS/PNVS for short. The systems, comprising two separate elements, are located in the bulbous protrusions on

Figure 9.24 McDonnell Douglas AH-64A Apache (Courtesy of Boeing)

the aircraft nose. Figure 9.25 shows the TADS/PNVS installation with some of the relevant fields of view.

The target acquisition and designator system (TADS) comprises the following facilities:

- Direct vision optics to enhance daylight long range target recognition; an optical relay tube transmits the direct view optics to the CPG
- Forward-looking infrared (FLIR)
- TV
- Laser designator/range finder
- Laser tracker

The Pilot Night Vision System (PNVS) gives the pilot a FLIR image over a 30° by 40° field of view which can be slaved to the direction the pilot is looking by means of an integrated helmet display.

The combination of TADS and PNVS capabilities gives the Apache a potent system which has demonstrated maturity and durability during the Gulf War. For more information relating to the Apache helicopter and its capabilities see Rorke [10] and Green [11].

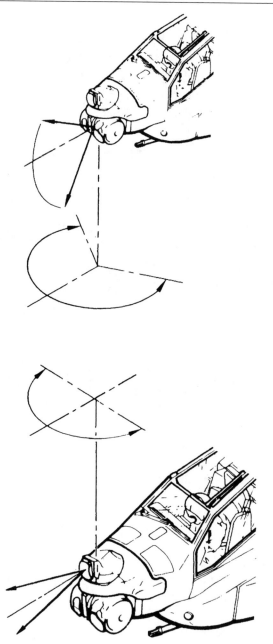

Figure 9.25 Apache TADS/PNVS installation (Courtesy of Boeing)

The aircraft can carry a variety of weapons and missiles. In addition a 30 mm chain cannon is fitted as standard. The weapons which can be carried include the following:

- 70 mm rockets
- Hellfire anti-tank missiles
- Stinger air-to-air missiles

The M230 chain gun is fitted under the forward fuselage as shown in Figure 9.26. The gun is a 30 mm cannon with a firing rate of 10 rounds/second. The gun may either fire a single shot or 10, 20 or 50 round bursts; it can be slaved to the pilot's helmet system and can therefore traverse over the field of view of the pilot to engage the target. A typical load of ammunition would be 1200 rounds.

9.9.2 AH-64 C/D Longbow Apache

In 1989 McDonnell Douglas embarked upon a development programme to upgrade the Apache helicopter to a configuration called Longbow Apache. The Longbow prefix related mainly to the addition of a millimetric, mast mounted, fire control radar previously called the Airborne Adverse Weather Weapon System (AAWWS). This advanced radar system allows the Apache to hover in a screened position behind a tree or ridge line with the radar illuminating and identifying targets out to several kilometres range. The helicopter system may then rise above the ridge line, launch missiles at several predesignated targets and then drop out of sight of the defending forces. This greatly improves the capability of the helicopter while reducing its vulnerability. The radar also operates well in conditions not suited to the use of the electro-optic TAD/PNVS. See Figure 9.27.

One of the improved system features of the Longbow Apache is a new electrical system called the Electrical Power Management System (EPMS). During the development of the Longbow configuration it was found necessary to increase the rating of the main AC generators from 35 kVA to 45 kVA to provide more power to feed the new systems being fitted to the aircraft. Improvements in system architecture were also made to eradicate certain single point failures in the system as well as to provide better reliability and maintainability. A circuit-breaker panel was removed from the left upper part of the pilot's cockpit thereby improving the field of view for air-to-air operations. Most circuit breakers were removed from the cockpit and a significant proportion of the aircraft electrical loads are now remotely switched from the cockpit using the touch-sensitive screens.

An early version of the EPMS developed and manufactured by Smiths Group – now GE Aviation – is shown in Figure 9.28 [12]. The system has progressively undergone a series of upgrades, improving packaging and increasing system functionality.

The prototype system comprised a total of nine Line Replaceable Units (LRUs). Six LRUs are high power switching modules which contain the primary power switching contactors and the aircraft primary 115 VAC, three-phase 400 Hz, 28 VDC and battery bus bars. These LRUs may be quickly replaced following failure and are partitioned and monitored using built-in test (BIT) such that electrical faults may be quickly traced to the correct module. Certain contactors have I^2t trip characteristics which enable fault conditions to be identified and removed within tighter tolerances than previously possible, thereby enabling reduction in the size of the busbars with consequent weight

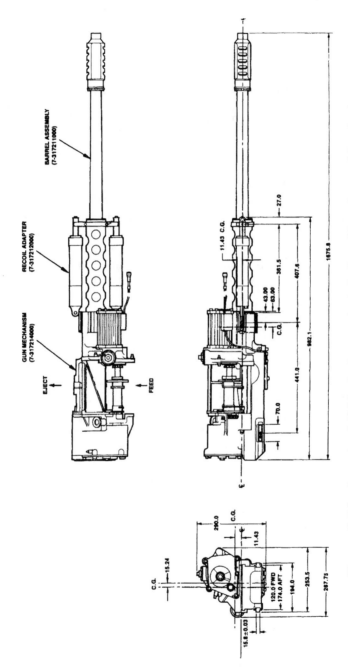

Figure 9.26 Apache M230 chain gun installation (Courtesy of Boeing)

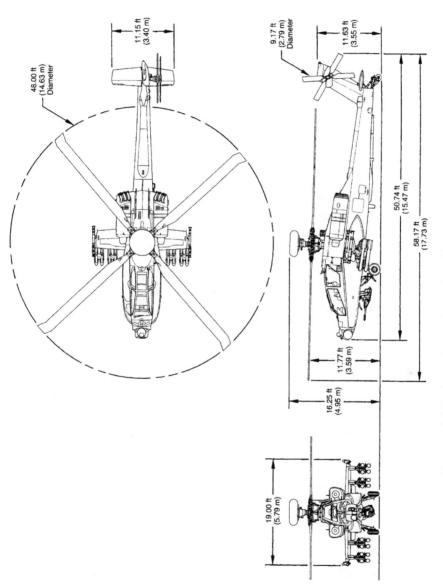

Figure 9.27 Longbow Apache (Courtesy of Boeing)

48.00 ft (14.63 m) Diameter

11.15 ft (3.40 m)

9.17 ft (2.79 m) Diameter

11.63 ft (3.55 m)

50.74 ft (15.47 m)

58.17 ft (17.73 m)

11.77 ft (3.59 m)

16.25 ft (4.95 m)

19.00 ft (5.79 m)

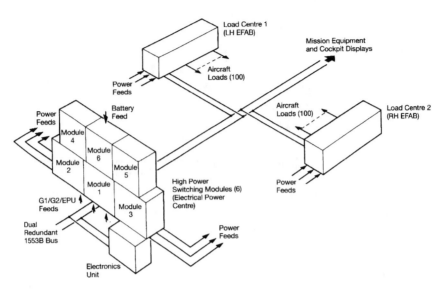

Figure 9.28 Longbow Apache Prototype EPMS (Courtesy of Smiths Group – now GE Aviation)

savings. The high power modules receive 115 VAC power from either main generator or the external power unit and DC power from the transformer rectifiers (TRs) or the battery. High power 115 VAC, 28 VDC, and battery feeds run forward to two load centres which control all aircraft secondary loads; those loads which are less than 20 amps.

The electronic unit has hardwired connections to the six high power modules enabling the primary contactors to be switched and status monitored as necessary. A large amount of system monitoring of current and voltage is possible. The electronic unit is connected to the aircraft 1553B dual redundant avionics bus. The six high power switching modules and the electronics unit are mounted on a bulkhead in the transmission bay, just behind the pilot's seat.

The two load centres are mounted in the forward avionics bays. The primary power fed from the high power switching modules is distributed to all the secondary loads and protected within the load centres. Each load centre protects and feeds around 100 secondary loads of which approximately 30–35 are remotely switched from the cockpit via the touch-sensitive displays and the 1553B avionics data bus. The load centres are supplied with conditioned air to remove excess heat.

The repackaging exercise for initial production aircraft rationalised the system to a total of four LRUs and introduced more processing capability. Subsequent modifications have included greater functionality to include the control of utilities systems.

9.10 Tilt Rotor Systems

9.10.1 Tilt Rotor Concept and Development

The tilt rotor concept as demonstrated by the Bell/Boeing V-22 is a concept that has been under development for almost half a century with a view to combining the most advantageous characteristics of both fixed and rotary wing aircraft. Fixed wing aircraft are efficient in the cruise configuration when the propulsion and aerodynamic configurations are operating somewhere near to optimum conditions. The disadvantage of fixed wing aircraft is that their flight characteristics are far from optimum during the low speed take-off and landing special provisions such as flaps, slats and other similar high lift devices are required to reduce aircraft speed to acceptable levels during these critical stages of flight.

Helicopters suffer from the reverse problem, being extremely effective for low or zero speed vertical landing and take-off but very limited for high speed cruise, due to rotor tip stall and other features of the rotary wing. Helicopters and other vertical take-off aircraft require a very high ratio of thrust to weight ratio necessitating relatively complex mechanical means. It therefore follows that a vehicle capable of exploiting both the characteristics of vertical take-off and landing and conventional flight should have a lot of advantages to offer even though it might not compete with the optimum machine designed for either regime. While aircraft such as the British Aerospace Harrier and the Soviet Yakolev YAK-36 (Forger) aircraft represent solutions to the problem when approached from the 'conventional' viewpoint, they both have limitations in that they are highly specialised military aircraft which can offer no possibility of adoption in the commercial aircraft arena.

A potential solution to this problem – starting from a helicopter baseline – was initially explored by the Bell Helicopter Company as long ago as 1944. It resulted in the development of the Bell XV-3 which first flew in 1953. The principle employed was that of the tilt rotor with twin engines located at the extremities of a conventional wing. During take-off the rotors were positioned with axes such that the aircraft operated as a conventional helicopter, albeit a twin rotor helicopter. As the aircraft transitioned into the cruise the rotor tilted forward until eventually both rotors acted as conventional propellers or airscrews pulling the aircraft forward in the normal way. For landing the situation was reversed with the rotors being tilted aft until the aircraft was flying in the helicopter mode once again. The XV-3 was powered by a single 450 hp piston engine which transmitted power to the rotors via a complex mechanical arrangement. In 1956 the aircraft suffered a serious crash which halted the development. It appears that the fundamental problem was a lack of structural rigidity due to rotor pylon coupling which led to a catastrophic failure while in the hover. Nevertheless the Bell XV-3 flew and demonstrated the concept of the tilt rotor with the transition to and from the contrasting conventional and helicopter modes of flight. Bell was followed by the Vought Corporation (later Ling Temco Vought or

LTV) who produced and flew several prototypes of the XC-142. This aircraft was a multiple-engined machine which was not fail-safe in the hover mode and was also very complex mechanically – this programme was eventually terminated.

However, interest persisted in the NASA organisation and eventually the XV-15 programme was initiated in 1971 with Bell selected in preference to Boeing Vertol; the twin-engined Bell XV-15 first flew in 1977 and executed a successful flight test programme including the demonstration of an engine-out capability. This aircraft benefited from turbine engines rated as 1550 shp with a higher thrust to weight ratio than had been possible on the XV-3.

Eventually, after many problems of funding, the concept was revived and the programme got fully under way in 1985 with US Navy and US Marine sponsorship as the V-22 Osprey. The US Army and US Air Force also showed an interest in limited quantities of the aircraft for specialised 'special forces' roles. At the time of the programme launch in 1985 the joint Bell/Boeing Vertol team saw prospects for the production of over 1000 aircraft for all four major US Services with the Marines being by far the largest customer. After a troubled and lengthy development process that is well documented elsewhere; the initial MV-22 US Marine Corps version has just entered service and at the time of writing is on the point of being deployed into the Afghanistan theatre.

9.10.2 V-22 OSPREY

The V-22 configuration is shown in Figure 9.29. The aircraft is powered by two 6000 shp Allison T-406 turbine engines, each of which is contained within the tilting nacelles. It is interesting to note that each engine/nacelle combination weighs about 5000 lb which is almost the same as the total weight of the original Bell XV-3. The total production aircraft weight is in the region of 32 000 lb empty. To minimise structural weight and maximise the payload the aircraft makes extensive use of composite materials.

The systems of particular interest on the V-22 are the following:

- Propulsion drive system
- Fuel system
- Flight control system

V-22 Propulsion Drive System

Figure 9.30 shows the mechanical drive system interconnection between the nacelles. Each engine drives a prop rotor gearbox located in the nacelles from which each rotor is driven in the opposing direction to the other thereby counter-balancing torque effects. Each prop rotor gearbox also drives through a tilt axis gearbox and mechanical linkage running through the wing to a midwing gearbox. This effectively interconnects the two systems

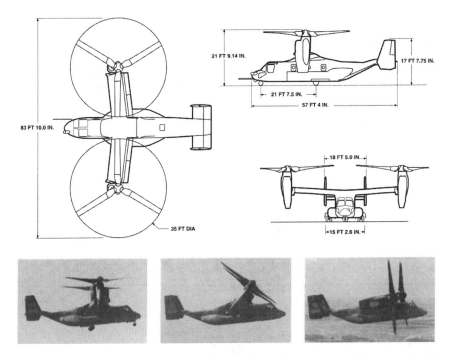

Figure 9.29 V-22 Osprey configuration (Courtesy of Boeing/Bell)

and also acts as the main aircraft accessory gearbox driving the constant frequency AC generators (two per aircraft, each rated at 40 kVA) and variable frequency AC generators (two per aircraft, each rated at 50/80 kVA) and 5000 psi hydraulic systems pumps. The APU also has the capability of driving the accessory gearbox which also drives the environmental system compressor.

V-22 Fuel System

The V-22 fuel system is much more complex than most helicopter systems. A number of possible configurations exist. Fuel is carried in tanks in the forward sections of the left and right sponsons (lower fuselage fairings) and in feed tanks just inboard of the nacelles.

The flexibility of the V-22 Osprey Tiltrotor is exemplified by the various fuel system options available with increasing capacity and therefore range capability as outlined in Table 9.1 and described below:

- The baseline configuration (MV-22) comprises wing feed tanks, forward sponson and rear sponson tanks
- The extended range configuration (CV-22) adds 8 wing tanks to the baseline configuration
- The ferry capability adds up to three cabin auxiliary tanks to whichever aircraft fuel configuration is embodied

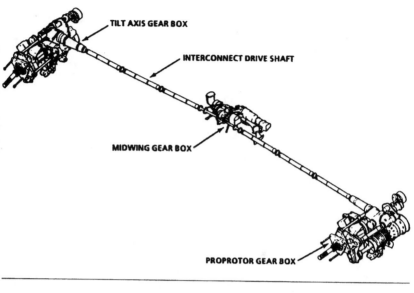

TILT AXIS GEAR BOX

INTERCONNECT DRIVE SHAFT

MIDWING GEAR BOX

PROPROTOR GEAR BOX

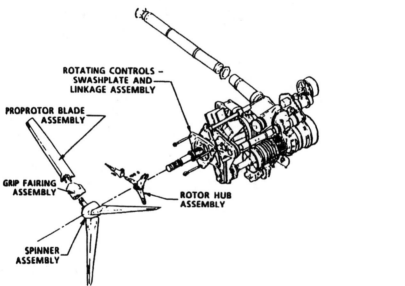

ROTATING CONTROLS –
SWASHPLATE AND
LINKAGE ASSEMBLY

PROPROTOR BLADE
ASSEMBLY

GRIP FAIRING
ASSEMBLY

ROTOR HUB
ASSEMBLY

SPINNER
ASSEMBLY

Figure 9.30 V-22 Osprey drive system (Courtesy of Boeing/Bell)

The total ferry load of ~ 30 000 lbs represents about 50% of the aircraft AUW.

Fuel transfer may be manually or automatically controlled. Fuel is transferred from the left and right wing tanks to the adjacent engine. Other tanks transfer fuel into these tanks and may also be cross-fed to the opposite fuel tank. Fuel transfer is carried out according to the following sequence: internal tanks; right aft sponson tank; wing tanks and then left and right forward sponson tanks. The feed tanks hold sufficient fuel for 30 minutes of flying time should there be a total failure of the fuel management and gauging system.

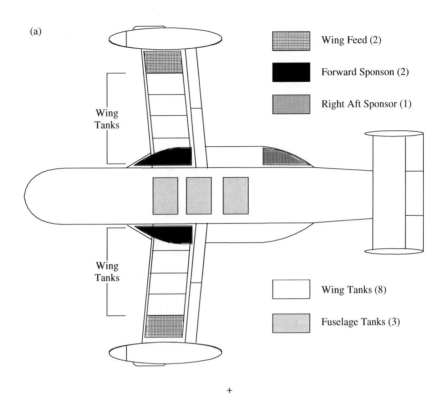

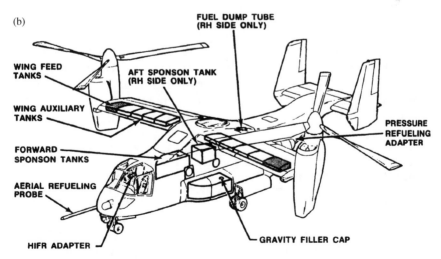

Figure 9.31 a and b   V-22 fuel system

Table 9.1 V-22 fuel configurations

	Number of tanks	Capacity (US gal)	Weight (lbs)
Typical Marine Configuration (MV-22):			
Wing Feed Tanks (L & R)	2	176	1200
Forward Sponson Tanks (L & R)	2	956	6500
Aft Right Sponson Tank	1	316	2150
Grand Total	**5**	**1448**	**9870**
Air Force Variant (CV-22):			
Wing Tanks (4 x L; 4 x R)	8	596	4000
Grand Total	**13**	**2044**	**13870**
Ferry Configuration:			
Cabin Auxiliary Tanks	3	2400	16350
Grand Total	**16**	**4444**	**30222**

Note: Nominal fuel characteristics assumed; fuel density alters with temperature

To control this system (excluding the ferry tanks) requires a total of 17 motorised valves and 9 fuel pumps and a miscellany of other valves. This is more in line with the complexity of a high performance aircraft than a normal fixed rotor helicopter.

To enhance flexibility of operation the aircraft has three possible methods of refuelling:

• Normal ground pressure refuelling
• Gravity feed through the forward left sponson tank
• Air-to-air refuelling using the refuelling probe

To enhance the aircraft survivability in a combat environment an On-Board Inert Gas Generation System (OBIGGS) is provided to supply nitrogen enriched air to the air space above the fuel surface. The inert gas displaces fuel vapour and reduces the possibility of fire and explosions after ballistic impact. Other fire suppression or fire reduction techniques include the use of rigid foam above and below the wing tanks and the sponson forward and aft bulkheads. The use of fire detector and chemical gas generating dispensing units also provide active explosion suppression.

V-22 Flight Control System

The tilt rotor concept offers huge operational advantages and flexibility; however the flight system capable of controlling an aircraft capable of conventional flight and helicopter-like Vertical Take-Off and Landing (VTOL) and mastering the transitions between these modes necessitates a complex flight control system.

The V-22 flight control system actuator configuration is summarised in Table 9.2. The various actuators are assigned the following functions:

Table 9.2 V-22 flight control actuators

Left Swash Plate (3)		Right Swash Plate (3)
Left Tilt Conversion (1)		Right Tilt Conversion (1)
Left Flaperon (4)	Right Flaperon (4)	
Left Rudder (1)	Right Rudder (1)	
	Elevators (3)	

- *Swash plate actuators* (6). Three swash plate actuators in each nacelle perform a function similar to the main rotor actuators on a conventional helicopter. They are located equally distant around the swash plate and allow each rotor to be tilted in the 'pitch' and 'roll' axes. Because there are two tilt rotors that may be tilted fore and aft either in synchronism or differentially the resulting vehicle motion is more complex. For example, demanding the rotors to tilt forward (left rotor) and aft (right rotor) causes the vehicle to yaw in a clockwise direction. These manoeuvring modes will be described later
- *Tilt conversion actuators* (2). Each tilt conversion actuator causes the appropriate nacelle to rotate between the VTOL and normal flight conditions
- *Flaperon actuators* (8). Four flaperon sections (combined flaps and ailerons), two on each wing, are powered by two flaperon actuators each
- *Rudder actuators* (2). The left and right rudder surfaces each have a single actuator
- *Elevator actuators* (3). There are three elevator sections, each powered by its own actuator

V-22 VTOL Modes of Flight

The VTOL modes of flight are illustrated in Figure 9.32 and are described below:

- *Pitch.* Forward movement of the cyclic stick causes both proprotor discs to tilt forward. The aircraft assumes a nose-down attitude and the airspeed increases in the same manner as a helicopter. Aft movement of the cyclic stick causes the proprotor discs to tilt aft, increasing nose-up attitude and decreasing speed
- *Roll.* Movement of the cyclic stick left and right demands differential application of collective pitch and lateral cyclic cause both proprotors to tilt in the appropriate direction causing the vehicle to roll. If rolling left the right proprotor increases collective pitch while the right proprotor decreases collective pitch. Both proprotor discs tilt to the left and the aircraft rolls left
- *Yaw.* Movement of the rudder bars causes differential fore and aft cyclic to be applied to the proprotors. If yawing left the left proprotor disc tilts aft and the right proprotor disc tilts forward causing a couple that yaws the aircraft left

- *Thrust.* Operation of the Thrust Control Lever (TCL) applies cyclic to both rotors causing the lift to increase and the aircraft to climb therefore increasing or decreasing altitude
- *Lateral translation.* Lateral operation of the inching control on top of the cyclic stick applies coordinated cyclic to both rotors causing the aircraft to translate left or right

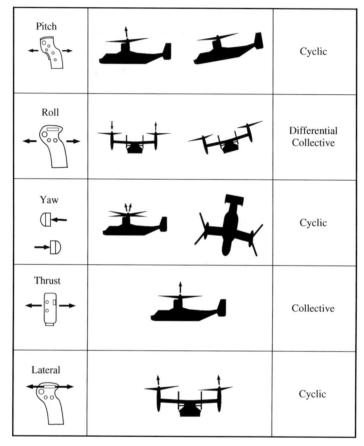

Figure 9.32 V-22 VTOL FCS modes

V-22 Conventional Modes of Flight

The conventional flight modes depicted in Figure 9.33. are controlled as follows:

- *Pitch.* Forward movement of the control stick causes the aircraft to pitch nose-down due to the operation of the elevators; altitude decreases and airspeed increases. Aft movement of the control stick causes the aircraft to pitch nose-up, altitude increases and airspeed decreases. The Angle of Attack (AoA) is monitored and limited

- *Roll.* Lateral movement of the control stick causes the aircraft to roll using the flaperons as ailerons. If rolling left, the left flaperons deflect up while right flaperons deflect down and the aircraft rolls left
- *Yaw.* Movement of the rudder pedals cause the aircraft to yaw using the twin rudders in a conventional aircraft sense
- *Thrust.* Operation of the thrust control lever alters proprotor blade pitch and control engine speed thrust and therefore forward speed

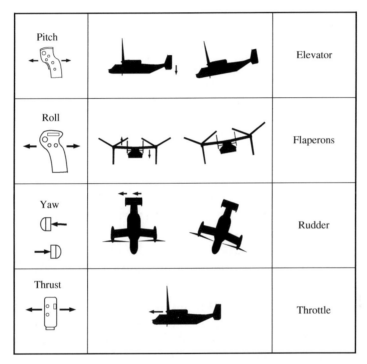

Figure 9.33   Aircraft FCS modes

The flight control system that accomplishes these tasks is shown in Figure 9.34. Flight control algorithms are performed by triple redundant Fight Control Computers (FCCs) that interface with MIL-STD-1553B data buses. These data buses provide the links with the avionics system, air data system, left and right Full Authority Digital Engine Control units (FADECs), and primary and secondary attitude data from the Inertial Navigation System (INS) and Secondary Attitude and Heading Reference System (SAHRS). Angle of Attack (AoA) data is also fed into the FCCs.

Hardwired interfaces from the left and right rotor transducer set provide rotor related data to the FCCs. Each FCC has hardwired interfaces to the actuator set to provide commands and receive feedback data. In common with many other flight control systems the actuators are provided with hydraulic power from three independent aircraft hydraulic system.

Further detail on the V-22 flight control system may be found in McManus (1987) [13].

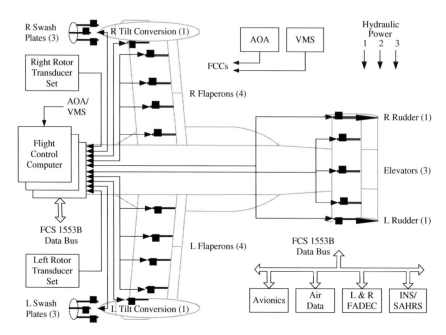

Figure 9.34 V-22 flight control system architecture

A further interesting and probably unique feature of the V-22 is the rotor and wing stowage facility. The need to stow the aircraft onboard aircraft carriers and amphibious assault ships dictates severe stowage constraints. The rotor/wing stowage occurs in the sequence shown in Figure 9.35. First the rotor blades are folded inboard to align with the wing. Then the nacelles are tilted forwards to place the rotor blades parallel with the wing leading edges. Finally, the whole wing is rotated 90° clockwise to be positioned along the top of the fuselage. For articles which further detail the V-22 Osprey see references [14, 15].

9.10.3 Civil Tilt Rotor

While the attractions of the tilt rotor vehicle are obvious in a military context, a development is under way by Bell, the original pioneer in this concept over 50 years ago. The Bell 609 is a civil tilt rotor aircraft that has flown and is entering full-scale development. The Bell 609 is a smaller aircraft designed to carry between 6 and 9 passengers. While the overall implementation is very similar to the V-22 the Bell 609 has no rudder, achieving yaw control in the aircraft

Figure 9.35 V-22 rotor and wing stowage sequence (Courtesy of Boeing/Bell)

mode by use of differential collective blade pitch on the rotors (while acting as propellers). In the VTOL mode the 609 uses differential cyclic to yaw the vehicle. Another key difference is that the 609 has no lateral cyclic actuators and therefore is unable to perform the lateral translation manoeuvres that are possible with the V-22. See Figure 9.36.

Figure 9.36 BA609 Civil Tiltrotor undergoing flight trials (Courtesy Bell Helicopter)

The paper 'Design and Development of the BA609 Civil Tiltrotor Hydraulic System Architecture', Bell Helicopter Textron Inc, Fort Worth, Texas, describes in detail the triple hydraulic system used on the vehicle and how the necessary high integrity is achieved.

References

[1] Fay, J. (1987) *The Helicopter, History, Piloting and How it Flies*. 4th edition, David & Charles.
[2] Hague, C.W. (1984) 'EH 101', *Aerospace*, July/August.
[3] Martin, E.E. (1984) 'T-700 – A Program Designed for Early Maturity and Growth Potential', *Tenth European Rotorcraft Forum*, The Hague, Netherlands.
[4] Bryanton, R. (1985) 'RTM 322 – Europe's newest helicopter engine', *Aerospace*, May.
[5] Buller, M.J., Lewis, D. (1985) 'The conception and development of a family of small engines for the 1990s', *Eleventh European Rotorcraft Forum*, London.
[6] Saunders, A.F. (1983) 'An Advanced Helicopter Engine Control System', *Aircraft Engineering*, March.
[7] Astridge, D.G., Roe, J.D. (1984) 'The Health and Usage Monitoring System of the Westland 30 Series 300 Helicopter', *Tenth European Rotorcraft Forum*, The Hague, Netherlands.
[8] Richards, W.R., 'ACT Applied to Helicopter Flight Control', *AGARD Conference Proceedings, 384*.

[9] Wyatt, G.C.F., 'The Evolution of Active Control Technology for the 1990s Helicopter', *AGARD Conference Proceedings*, 384.

[10] Rorke, J.B. (1987) 'Apache for the Battlefield of Today and the 21st Century', *AGARD Conference Proceedings*, (June) 423.

[11] Green, D.L. (1985) 'Flying in the Army's Latest Warrior – The Hughes AH-64A Apache', *Rotor & Wing International*, April.

[12] Moir, I., Filler, T. (1992) 'The Longbow Apache Electrical Power Management System', *Aerotech '92*, January.

[13] McManus, B.L. (1987) 'V-22 Tilt Rotor Fly-By-Wire Flight Control System', *NAECON*.

[14] Mark, Dr H. (1986) 'Aircraft without Airports. The Tilt-Rotor Concept and VTOL Aviation', *75th Wilbur and Orville Wright Lecture*, Royal Aeronautical Society, December.

[15] Moxon, J. (1988) 'V-22 Osprey Changing the Way Man Flies', *Flight International*, 14 May.

10

Advanced Systems

10.1 Introduction

This advanced systems chapter addresses some of those systems which broach new areas, having been either recently developed or under development. In many cases the concepts may have been under study for a number of years and recent developments in technology may have given the impetus and the means of implementation.

Some of these developments relate to the improved integration of aircraft systems to achieve hitherto unattainable benefits. Others embrace low-observability or 'stealth' technology. The following range of developments are addressed in this chapter:

10.1.1 STOL Manoeuvre Technology Demonstrator (SMTD)

The US Air Force SMTD F-15 upon which Integrated Flight and Propulsion Control (IFPC) allows closer integration of the aircraft flight control and engine control systems. Flight control systems are virtually all fly-by-wire in the modern fighter aircraft of today; the benefits being weight reduction and improved handling characteristics. New engines are likewise adopting Full Authority Digital Engine Control (FADEC) for the benefits offered by digital control. On aircraft such as the US Air Force SMTD F-15 these systems are being integrated to evaluate new control techniques applied to a modified F-15. This type of system could find application on the new generation of V/STOL aircraft to replace the Harrier in the early twenty-first century. In the event the Joint Strike Fighter (F-35) employed many of these technologies in its final design.

Aircraft Systems: Mechanical, electrical, and avionics subsystems integration. 3rd Edition   Ian Moir and Allan Seabridge
© 2008 John Wiley & Sons, Ltd

10.1.2 Vehicle Management Systems (VMS)

Vehicle management systems carry this integration still further, combining flight control and propulsion control with the control of utility and power management. This further improves the control of the aircraft systems and permits the integration of functions such as thermal management which will be vital to the performance of fighter aircraft cruising for extended periods at Mach 1.6 which is a requirement for the US Air Force F-22 Raptor. Thermal management is presently spread across several aircraft subsystems and these boundaries will need to be revised if the problem is to be properly tackled.

10.1.3 More-Electric Aircraft

For a number of years the concept of the 'All Electric Aircraft' has been espoused. The Bristol Brabazon utilised a great number of electrical systems and the Vickers Valiant V-Bomber was also highly electrical in nature. At the time – mid 1950s – the concept did not fully catch on, though over the years there has been a great deal of debate relating to the advantages of electrical versus other forms of secondary power, such as hydraulics or high pressure bleed air systems. The dialogue may be summarised by referring to the papers produced by Mike Cronin, formerly the Chief Electrical Engineer on the Brabazon Project and an employee of the Lockheed Aeronautical Systems Company for many years prior to his retirement in 1990. These merely represent a summary of Cronin's work on the subject and are referred to later in this chapter.

Over the past decade, examination of the benefits of the all-electric aircraft has been promoted by a number of aeronautical agencies in the US. In the early '80s NASA funded a number of studies addressing the Integrated Digital Electrical Airplane (IDEA). The IDEA concept studies embraced a range of technologies which could improve the efficiency of a 250–300 seater replacement for an aircraft such as the Lockheed L1011 (Tristar). The areas covered were:

- *Flight Control Technology* – relaxed stability augmentation leading to a reduction in trim drag with consequent down-sizing of the tailplane and fuel savings
- *Wing Technology* – use of efficient high aspect ratio wings using gust alleviation modes of the FCS to improve range and fuel consumption and reduce wing bending moments
- *Engine Power Extraction* – the reduction of engine power extraction losses by minimising the use of high pressure bleed air and hydraulic power and maximising the use of more efficient electrical power extraction techniques. See also more-electric engine
- *Flight Control Actuation* – the use of electro-mechanical actuation in lieu of hydro-mechanical actuation systems

- *Advanced Electrical Power Systems* – the development of new systems to generate and distribute electrical power as an adjunct to more efficient engine power extraction

Flight control system and flight control actuation developments are already underway or are embodied in major civil programmes as evidenced by systems on the Airbus A380 and Boeing 787 aircraft. The A380 and B787 also use novel more-electric features as will be described elsewhere in this book.

10.1.4 More-Electric Engine

The engine also benefits from the adoption of more-electric technology to address the following troublesome issues:

- *Reduction of bleed air offtakes* – As engine bypass ratios increase so does the burden on the central engine core, reducing engine efficiency and increasing fuel consumption. The reduction of engine HP air offtakes and the use of more-electric techniques has a considerable amelioration effect upon these adverse effects
- *Removal of the accessory gearbox* – Engine accessory gearboxes are becoming increasingly complex as the number of drives and power offtakes increase
- *Oil-less engine* – The engine oil system is complex on many engines, usually comprising a number of oil pumps, filter assemblies, coolers etc. The generation/conversion losses from the aircraft electrical generators reject heat into the engine. Great savings could be made if the oil system could be replaced with an alternative form of supporting the rotating engine assemblies. Electromagnetic bearing technology has been demonstrated on both sides of the Atlantic. However in order to be totally practicable, additional technologies have to be developed which permit the removal of the accessory gearbox and its associated power offtakes from the engine. Active Magnetic Bearings (AMB) technology is not yet able to support the big fans engine bearings during the blade-out condition which must be demonstrated as part of certification
- *IGV/VSV control* – Many engines use Variable Inlet-Guide Vanes (VIGVs), and Variable Stator Vanes (VSVs), to control the airflow into the engine central core. These may be variously powered by hydraulics, pneumatics (bleed air) or by fueldraulic means. Programmes are underway to examine the feasibility of using electrical actuation techniques to replace the fluidic power media
- *Distributed engine control* – Present primary engine control is by means of a Full Authority Digital Engine Control (FADEC) which is normally located on the engine fan casing. However there are many features of engine control which are distributed around the engine – such as reverse thrust, presently pneumatically actuated – which would need to be actuated by alternative means in a more-electric engine. This leads to the possibility of using distributed engine control

- *Electrically driven fuel pump* – Engine fuel is pressurised by means of a shaft driven High Pressure (HP) pump. The HP fuel pump is typically sized for engine starting and when driven directly by the engine (via a reduction gearbox) operation at high engine rotational speeds produces excess fuel flow that must be spilled back to the pump inlet. This is aggravated at cruise altitudes where the fuel required by the engine is about five times lower resulting in even more wasted pumping energy. In an all-electric engine the HP pump would be electrically driven at the optimum speed for the prevailing operating condition

The realisation of these technologies is reaching fruition in demonstration programmes as described later in the chapter.

10.2 Stealth

The development of 'low observable' aircraft has been given a high priority by the US Air Force in particular in the last two decades as a way of improving the combat effectiveness of the combat vehicle. The Lockheed F-117A 'stealth fighter', Northrop B-2 'stealth bomber' and the former Advanced Tactical Fighter (ATF) Dem/Val YF-22A and YF-23A projects were designed with this feature in mind. The selected F-22 Raptor underwent a protracted development and has entered service with the US Air Force in significant numbers over the past two years. Subsequent reports graphically indicated the benefits of this technology during the Gulf War; both the F-117A and the B-2 bomber were deployed during the 1999 Kosovo conflict. The F-117A has undergone a standard configuration fleet modification to standardise the low-observable coatings used across the fleet – previously a number of different techniques are utilised which evolved during the development and production phases. Recent reports of modifications to the B-2 bomber fleet have suggested the stealth technology, while operationally highly effective, does have a maintenance penalty.

10.2.1 Joint Strike Fighter (JSF)

The Joint Strike Fighter (JSF) was contested by competing teams from Boeing (incorporating the former McDonnell Douglas) with the X-32 and Lockheed Martin with the X-35. Both teams flew demonstration models with a down-selection to an overall winner in 2001. These designs also embody stealth technology. The aircraft are designed to meet the requirements of four Services: the US Air Force; US Navy; US Marines and British Royal Navy. Three main vehicle configurations are being developed:

- Conventional Take Off and Landing (CTOL) for the US Air Force
- Carrier Vehicle (CV) for the US Navy
- Short Take-Off Vertical Landing (STOVL) for the US Marines and Royal Navy

10.3 Integrated Flight and Propulsion Control (IFPC)

As avionics technologies have developed in the last decade, it has become commonplace for the control of major systems to be vested in electronic implementations; such systems may have previously been solely mechanically or electro-mechanically controlled. Moreover, the availability and maturity of the technologies required to satisfy avionics system integration have proved equally appealing in satisfying the requirements of more basic aircraft systems. The benefits of digital electronic control of mechanical systems are evident in greater precision and an ability to measure or predict performance degradation and incipient failure. Typical examples of this are digital implementations of flight control or fly-by-wire and digital engine control, or Full Authority Digital Engine Control (FADEC). As substantial benefits of improved performance and reliability are realised, e.g. weight reduction and other improvements in system integration and data flow, so the level of systems integration becomes correspondingly more ambitious.

It is therefore a logical progression that the demonstrated benefits of digital flight control and engine control systems has instigated development programmes which are examining the next level of integration – that of Integrated Flight and Propulsion Control (IFPC). IFPC is actively being developed in the US. The vehicle for this US Air Force funded programme is the F-15 STOL/Manoeuvre Technology Demonstrator (SMTD), a highly modified F-15B which has been flying for some time from Edwards Air Force Base. Other aims of the technology demonstrators were to show that a high performance fighter could land upon a roughly constructed (or repaired) concrete strip 1500 ft × 50 ft. This requires a sophisticated guidance system and an IFPC system to improve the aircraft response and therefore the precision with which the pilot can fly the aircraft during the approach. The configuration of the F-15 SMTD aircraft is shown in Figures 10.1a and 10.1b.

Of particular interest are the multiple effectors utilised on the SMTD aircraft which may be summarised as follows:

- Collective/differential canards
- Collective/differential flaps and ailerons
- Collective/differential stabilators
- Collective/differential rudders
- Variable geometry inlets
- Engine control
- Two-dimensional (2-D) vectoring/reversing nozzles

The collective/differential flight control surfaces allow a significant enhancement of the aircraft performance over and above that normally possible in an F-15 in the approach configuration. In addition normal control modes and the use of collective flight control surfaces should offer direct translational flight; that is, operation of those control surfaces should allow the aircraft to move, say vertically, without altering the pitch vector or attitude. The thrust vectoring control adds an additional facility and the aircraft has been flying with 2-D nozzles operational since May 1989. These may be operated in a thrust reverser mode. The F-15 SMTD has been under test since 1988 and has demonstrated operation of the thrust reversers in flight.

Figure 10.1a   F-15 SMTD (Courtesy of Boeing)

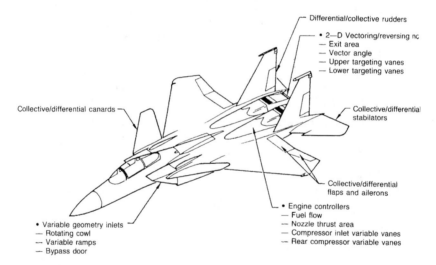

Figure 10.1b   F-15 SMTD flight control configuration

In order to gain some idea of the complexity of the IFPC, the following summarises the number of sensors and effectors associated with the system:

- *Flight control:*
    - 11 quadruplex sensors
    - 6 quadruplex actuators
    - 7 dual redundant actuators

- *Intake control (per engine):*
    - 3 sensors
    - 3 actuators

- *Engine control (per engine):*

  - 8 sensors (4 dual redundant)
  - 6 actuators

- *Nozzle control (per nozzle):*

  - 8 actuators

See Figure 10.2. For a detailed paper describing the IFPC fault tolerant design see Tuttle, Kisslinger and Ritzema (1990) [1].

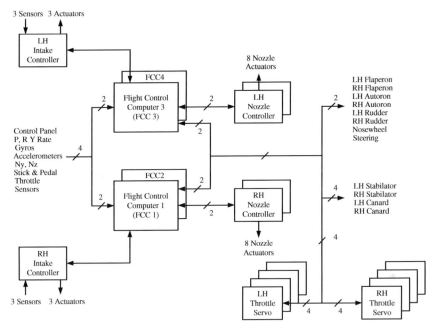

Figure 10.2 Simplified F-15 SMTD IFPC architecture

10.4 Vehicle Management System

The Integrated Flight and Propulsion Control (IFPC) described above is an integration of two main aircraft control systems into one. Vehicle Management Systems (VMS) relate to a higher level of system integration, that is the combination of flight control, propulsion control, and utilities/power management. One reason for combining these systems into a VMS is that the aircraft performance demands an improvement in the integration of these major systems. For example, thirty years ago no fighter aircraft would have been fitted with a full-authority fly-by-wire system. Stability augmentation systems were used as a matter of course but the flight control system was implemented using the push-pull rod systems of the type outlined in Chapter 1 – Flight

Control Systems. Nowadays, virtually all front line fighters routinely employ fly-by-wire systems: they offer artificial stability if the aircraft is unstable, or may merely improve aircraft handling. In either case, they improve handling and performance from the pilot's point of view. Fly-by-wire systems also save weight and can greatly ease or limit structural loading by curtailing demands where necessary. Of course this has all been made possible by advances in microelectronics and actuation techniques. The point is that these techniques have become the stock-in-trade of implementing flight control, as is shown by the extensive use of such systems in the new generation of stealth aircraft described later in this chapter.

In recent years engine control has moved toward Full-Authority Digital Engine Control solutions (FADEC) and the F-15 SMTD programme, already covered, shows how intake and nozzle control may need to be more closely integrated with digital engine control to satisfy some requirements. The air intake or inlet must be correctly matched to the engine or optimum performance will never be achieved, especially for supersonic aircraft. The F-15 SMTD two-dimensional nozzles require a total of six actuators to control the thrust vectoring in the vertical plane and reverse thrust modes for each engine. The nozzle control is normally an integral part of the FADEC which controls both the gas generator fuel flow and the afterburner (reheat) fuel flow.

Whereas pure raw performance may be the objective for some applications, others may seek to improve performance in more subtle ways. An F-117A stealth fighter seeks low observability as a primary mission goal, not the utmost in speed or excess thrust. The technical solutions adopted to achieve the primary goal of stealth mostly directly detract from performance; the means used to reduce the temperature and size of the exhaust plume reducing propulsive power. In this situation more elegant control methods may be required to ensure that these losses are not prohibitive.

Many aircraft systems, such as utilities management and electrical power management, require better control to meet more demanding problem statements. Systems such as fuel, hydraulics, secondary power, environmental control and electrical power systems are being improved by the use of digital control techniques. The UK Experimental Aircraft Programme (EAP) employed a Utilities Management System (UMS) which fully integrated many of these control functions into four dedicated control units as shown in Figure 10.3. This system first flew August 1986 and a similar system – Utility Control System (UCS) is fitted to Eurofighter Typhoon and a USM is fitted to the BAE Systems Nimrod MRA4. For more detail on the EAP system see Moir and Seabridge (1986) and Lowry, Moir and Seabridge (1987) [2, 3]. The Boeing AH-64C/D Longbow Apache employs an integrated electrical power management system (EPMS) to improve the control and distribution of the primary electrical system an advanced battlefield attack helicopter. See Chapter 9 – Helicopter Systems.

The VMS concept seeks to integrate all these major systems into one system responsible for controlling the air vehicle or aircraft. All of the systems utilise digital computer control and data buses which allow them to communicate

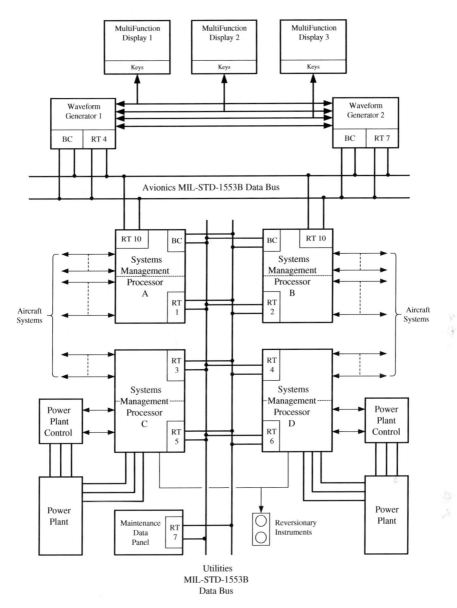

Figure 10.3 EAP utilities system management control units (Courtesy of Smiths Group – now GE Aviation)

with each other and with the remaining aircraft systems. This leads to the possibility of integrating the VMS using a series of data buses and one such architecture is shown in Figure 10.4. A major difference between the EAP and Eurofighter Utilities Management Systems and the VMS proposed for future aircraft is that high rate, closed loop servo systems have been included in the control concept.

This generic architecture shows a number of control units associated with flight control, engine control and utilities/power management. This allows the units to be closely tied to each other and to the sensors and actuators associated with the control task. In this scheme certain computers have responsibility for interfacing the VMS as a whole to the avionics system and to the pilot. This type of closely coupled control permits modes of operation that would be much more difficult to control if the systems were not integrated into a VMS. For example the fuel management system on a fighter can be used to control the aircraft CG. The position of the CG in relation to the centre of lift determines the aircraft stability and trim drag. For optimum cruise the CG could be positioned at or near the neutral point to minimise trim drag. For combat the CG could be moved aft to make the aircraft more manoeuvrable. Therefore in this example there is an inter-reaction between flight control and utility control which allows optimum modes to be selected for various phases of flight.

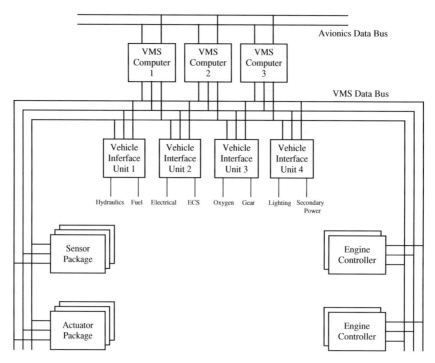

Figure 10.4   Generic VMS architecture

Thermal management is an area which is becoming more important in combat aircraft such as the F-22 Raptor which is designed for 'persistent supersonic cruise' operation. That is, the aircraft is designed to cruise for long periods at speeds of Mach 1.6 whereas previous fighters could only operate at such speeds during a short 'supersonic dash'. This leads to the problem of where to sink all the thermal energy generated during high speed cruise. The inter-reaction of the fuel system (fuel being used as a heat sink) and the environmental control system, is of great importance in solving the problem. More energy-efficient methods of extracting and utilising power from the engines can also help and is one of the reasons for studying the all-electric aircraft concept which is described in detail elsewhere in this chapter. Technology demonstration programs associated with the Joint Strike Fighter (JSF) made major advances in this area as will be described later in this chapter.

The US Air Force has embraced the VMS on recent programs in order that these improvements may be realised. Though the precise architectures may vary by programme depending upon the maturity of the various technologies, it is clear that many of the necessary technologies and building blocks are available and that such systems may be embodied without significant risk.

10.5 More-Electric Aircraft

10.5.1 Engine Power Offtakes

For the past few decades the way in which aircraft have extracted power from the engine has changed little though long standing studies exist which examine more – electric means – see references [4 to 13]. The three key methods or extracting energy from the engine have been:

- Electrical power by means of an accessory gearbox driven generator
- Hydraulic power by means of Engine Driven Pumps (EDPs) also run off the accessory gearbox but also by electrical and air driven means
- Pneumatic power achieved by bleeding air off the intermediate or HP compressor to provide energy for the environmental control system, cabin pressurisation and wing anti-icing system among others. High pressure air has also provided the means by which the engine is started with the air taken from a ground air start trolley, APU or another engine already running

While the engine is in effect a highly optimised gas generator, there are penalties in extracting bleed air which are disproportionate when compared to the power being extracted. This becomes more acute as the bypass ratio increases: original turbofans had relatively low bypass ratios of $\sim$1.4 (bypass) to 1 (engine core); more recent designs $\sim$4:1 and next generation turbofans such as the GE GEnex and Rolls-Royce Trent 1000 are close to 10:1. Modern engines have pressure ratios of the order of 30 to 35:1 and are more sensitive to the extraction of bleed air from an increasingly smaller and much more highly tuned engine central core.

The outcome is that to realise fully the benefits of emerging engine technology, a different and more efficient means of extracting power or energy for the aircraft systems becomes necessary. Efficient energy extraction for the aircraft without adversely affecting the performance of the engine core and the engine as a whole becomes an imperative reason for changing the architectures and technology utilised. Figure 10.5 illustrates the differences between conventional power extraction using bleed air on the left versus a more-electric version on the right. These architectures broadly represent the difference between the Boeing 767 (left) and its successor, the Boeing 787 (right). The main differences between the more-electric and conventional configurations are:

- *Reduced bleed air offtake*: the only bleed air offtake for the B787 is for engine cowl anti-icing – this can be fan air that may be used with much lower penalty than that extracted from the engine compressor
- *Increased electrical power generation*. The B787 system generates 500 kVA per channel instead of 120 kVA (B767-400). This increased electrical power is required in the main to provide energy to those systems no longer powered by bleed air
- *Electric engine start*: The B787 uses electric start since bleed air is no longer available for this purpose

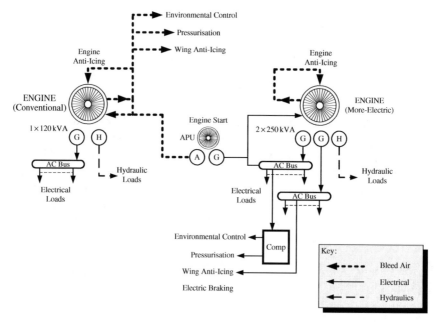

Figure 10.5 Comparison of conventional and more-electric power offtakes

10.5.2 Boeing 787 (More-Electric) Electrical System

The B787 electrical power system is portrayed at a top-level in Figure 10.6. A key feature is the adoption of three-phase 230 VAC electric power compared with the conventional three-phase 115 VAC solution universally adopted by the Industry to date. The increase in voltage by a factor of 2:1 decreases feeder losses in the electrical distribution system and allows significant wiring weight reduction. The use of higher 230 VAC phase voltage, or 400 VAC line-to-line, does require considerable care during design to avoid the possible effects of partial discharge, otherwise known as 'corona'.

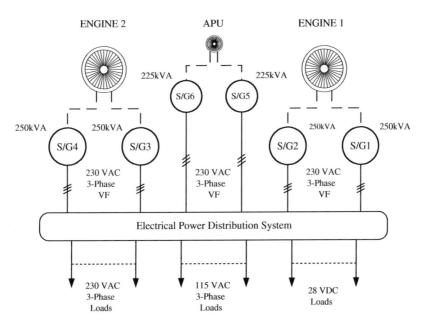

Figure 10.6 B787 (More-Electric) electrical system

A brief description of the B787 electrical system is given in Chapter 5; however, it is worth repeating the salient features here to help set the More-Electric developments in context. The salient features of the B787 electrical power system are:

- 2 × 250 kVA starter/generators per engine, resulting in 500 kVA of generated power per channel The generators are Variable Frequency (VF) reflecting recent industry trends in moving away from constant frequency (CF) 400 Hz power
- 2 × 225 kVA APU starter/generators, each starter/generator starting the APU and then acting as a generator during normal operation
- Each main generator feeds its own 230 VAC main bus before being fed into the power distribution system. As well as powering 230 VAC loads, electrical power is converted into 115 VAC and 28 VDC power to feed many of the legacy subsystems that require these more conventional supplies

The key features of the B787 electrical loads are given in Figure 10.7. As bleed air is no longer used within the airframe there are no air feeds to the environmental control system, cabin pressurisation system, wing anti-icing system as well as other air-powered subsystems. The only bleed air taken from the engine is low pressure fan air used to perform an anti-icing function for the engine cowl. Tapping bleed air off the engine compressor is extremely wasteful, especially as engine pressure ratios and bypass ratios increase on modern engines such as the General Electric GeNex and Rolls-Royce Trent 1000. An additional saving is removal of the overhead of providing large ducts throughout the airframe to transport the air; typically 8 inch diameter air ducts are required between engine and airframe and 7 inch ducts between

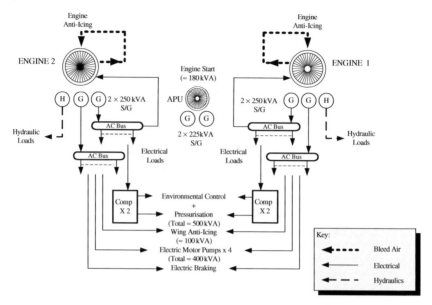

Figure 10.7   Boeing 787 electrical loads

APU and airframe and in the Air Driven Pump (ADP) feed. In some parts of the airframe the presence of these large bleed air ducts necessitate overheat detection systems to warn the flight crew of hot gas leaks.

The main more-electric loads in the B787 system are:

- *Environmental Control System (ECS) and pressurisation.* The removal of bleed air means that air for the ECS and pressurisation systems needs to be pressurised by electrical means; on the B787 four large electrically driven compressors are required drawing total electrical power in the region of 500 kVA
- *Wing anti-icing.* Non-availability of bleed air means that wing anti-icing has to be provided by electrical heating mats embedded in the wing leading edge. Wing anti-icing requires in the order of 100 kVA of electrical power
- *Electric motor pumps.* Some of the aircraft hydraulic Engine Driven Pumps (EDPs) are replaced by electrically driven pumps. The four new electrical motor pumps require ~100 kVA each giving a total load requirement of 400 Kva

A further outcome of the adoption of the 'bleedless engine' is that the aircraft engines cannot be started by the conventional means: high pressure air. The engines use the in-built starter/generators for this purpose and require ~180 kVA to start the engine.

10.5.3 More-Electric Hydraulic System

The effects on the hydraulic system of adopting more-electric concepts may be seen by comparing the hydraulic system configurations for the Boeing 767 (conventional Boeing wide body) versus the more-electric Boeing 787 as shown in Figure 10.8.

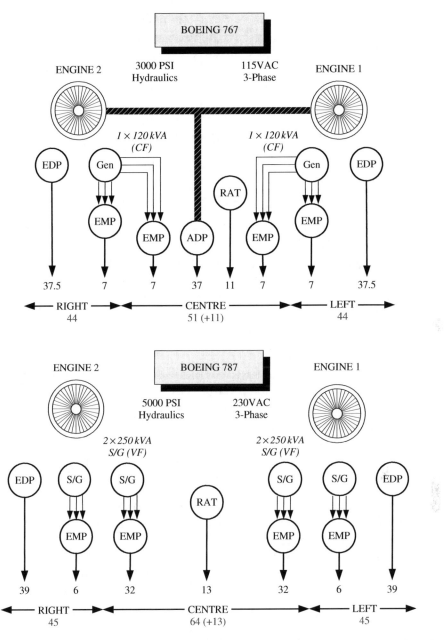

Figure 10.8 Comparison of conventional and more-electric hydraulic systems

Boeing have been more conservative regarding the use of centralised aircraft hydraulic systems on the Boeing 787 as opposed to the use of more de-centralised systems on the Airbus A380 and certainly Lockheed Martin F-35/JSF. Boeing also use conventional hydraulic actuation in general whereas the Airbus A380 makes considerable use of Electro-Hydrostatic

Actuators (EHA), and Electric Backup Hydrostatic Actuators (EBHAs) for primary flight control as described in Chapter 1. Nevertheless Figure 10.8 presents a valid comparison as it effectively contrasts conventional and more-electric hydraulic system architectures with one another. Furthermore, it is also a valid size comparison as the 787 family is the direct market successor to the 767.

Both 767 and 787 architectures use the Boeing Left (L), Centre (C), Right (R) hydraulic channel philosophy. The key differences are:

- Engine bleed air is removed with deletion of the Air Driven Pump (ADP)
- The use of 5000 psi rather than 3000 psi hydraulics system
- The adoption of 230 VAC, three-phase, VF primary power rather than 115 VAC, three-phase 400 Hz CF
- The use of starter/generators versus generators to facilitate electric engine start
- Use of larger Electric Motor Pumps (EMPs), around four times that of previous units

    Generally there are also increased levels of electrical power with the primary channels increasing from 120 kVA to 500 kVA. The levels of power for the Ram Air Turbine (RAT) and the Electric Motor Pumps (EMPs) have also increased dramatically

The three channel hydraulic system philosophy is more conservative than the '2H + 2E' philosophy adopted on the Airbus A380. On the A380 the blue hydraulic channel has effectively been replaced by a channel using distributed electrically powered actuation using EHAs and EBHAs. Both aircraft utilise 5000 psi hydraulic systems.

10.5.4 More-Electric Environmental Control System

The abolition of bleed air means that electrically driven compressors must be used to pressurise the cabin and provide a source of air for the environmental control system. See Figure 10.9.

In common with most aircraft of this size, the B787 is fitted with two air-conditioning packs, the difference being that they are electrically powered. Each pack has two electrically driven motor compressors each controlled by a motor controller located in the aft EE bays. Each permanent magnet motor requires ~125 kVA of electric power to drive it.

The outputs from these compressors enter a common manifold before being fed through primary and secondary heat exchangers, cooled by external ram air as would be the case in a conventionally driven Air Cycle Machine (ACM). The resulting cold air is mixed with recirculation air to maintain the desired cabin temperature.

Although the power required by the electric ECS is considerable, the key advantage is that air is not being extracted from the engine's central core. More importantly, the temperature and pressure of the delivered air is considerably lower, refer to Figure 10.10.

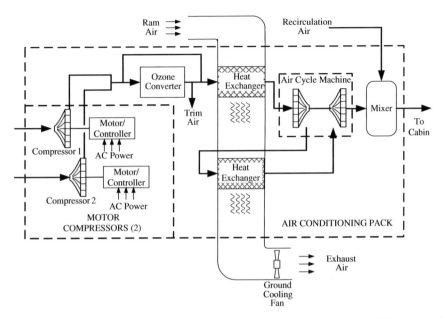

Figure 10.9   Electrically driven ECS system (Boeing 787)

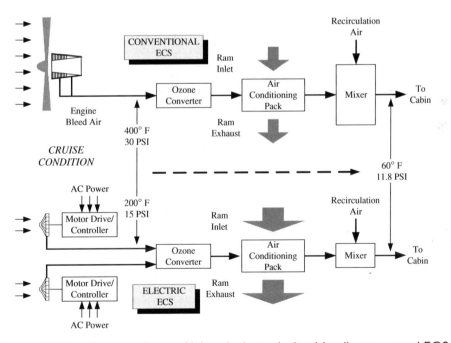

Figure 10.10   Comparison of bleed air and electrically powered ECS

Figure 10.10. portrays the differences between a conventional and an electrical ECS. Engine bleed air typically enters the conventional bleed air supplied ECS at ~400 °F and at about 30 psi. After being cooled by the air cycle machine the typical output will be ~60 °F and about 11.8 psi – this latter pressure equates to a cabin altitude of about 6000 ft.

The electrically driven motor compressors deliver air at ~200 °F and a lower pressure of 15 psi, as before this is reduced to 60 °F at 11.8 psi suitable for mixing with the warm recirculation air. It can be seen the energy expended on the cabin charge air differs greatly between the two methods; the difference represents energy loss and waste.

As has been described elsewhere, the high power distribution associated with these more-electric systems requires new technologies:

- The electrical power distribution cabinets that are located in the forward and aft EE bays weigh ~1000 lbs each
- Large loads require dedicated motor controllers
- In many cases ±270 VDC power is used within these cabinets
- The cabinets are liquid cooled due to the significant heat dissipation

10.6 More-Electric Actuation

10.6.1 Electro-Hydrostatic Actuators (EHA)

The principle of operation of the EHA is given in Chapter 1. The following is a selection of EHA applications on existing programmes or demonstrators:

- EHAs for A380 ailerons (4) and elevator (4)
- EHAs for F-35/JSF primary flight control surfaces (5)
- A380 rudder EBHAs (4) and spoiler EBHAs (4)

10.6.2 Electro-Mechanical Actuators (EMA)

The principle of operation of the EMA is given in Chapter 1. The following is a selection of EMA applications on existing programmes or demonstrators:

- EMAs for UK more-electric helicopter demonstator programme (HEAT)
- EMA for refuelling boom ruddervator
- EMA for UAV flight controls (Barracuda)

10.6.3 Electric Braking

Conventional braking systems are hydraulically powered using distributed aircraft hydraulic systems necessitating long pipe runs. Hydraulic leaks and fires can occur. Electrical braking is more efficient and is being introduced on the Boeing 787. Airbus are also about to demonstrate the concept on an Airbus A330.

10.7 More-Electric Engine

A major more-electric aircraft/more-electric engine technology demonstration programme called Power Optimised Aircraft (POA) has been underway for a number of years using European Union (EU) funding. The culmination of this programme in late 2006 is the ground running and demonstration of a more-electric engine in an engine cell in Spain. A top-level comparison of a conventional and a more-electric engine is shown in Figure 10.11.

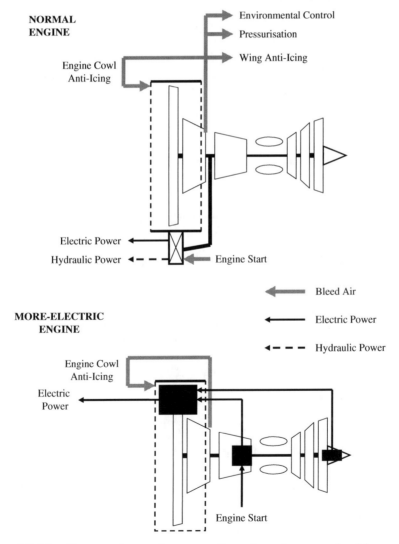

Figure 10.11 Comparison of conventional and more-electric engine

10.7.1 Conventional Engine Characteristics

In a normal engine bleed air is extracted from the engine and used for the following purposes:

- Engine anti-icing
- Wing anti-icing
- Environmental control and pressurisation

Engine power offtakes are by means a shaft powered accessory gearbox mounted on the lower part of the fan casing. Aircraft electrical generators and hydraulic pumps supply power to centralised systems as already described in the more-electric aircraft description.

10.7.2 More-Electric Engine Characteristics

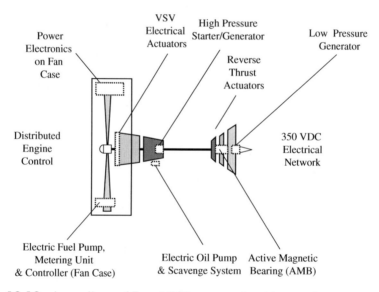

Figure 10.12   Location of Trent 500 more-electric engine components

In a more electric engine fan bleed air is used to anti-ice the engine; no bleed air is fed to the aircraft systems as described above. The more-electric engine Trent 500 shown in Figure 10.12 is being demonstrated by the European POA consortium. The engine has the following features:

- HP Starter Generator (HPSG) providing 150 kVA (Permanent Magnet)
- LP Fan Shaft Drive Generator (FSDG) providing 150 kVA (Switched Reluctance)
- Power Electronics Module (PEM) providing 350 VDC to engine and aircraft ME components; the PEM is located on engine fan case

- Electric Fuel Pump and Metering System (EFPMS) comprising an electric motor, pump and electronics drawing ~75 KW in total; this new concept permits extremely precise flow measurement compared to contemporary systems; perhaps more importantly, the pump provides only the required fuel flow hence no wasted pumping power with its attendant heat rejection problems; this is very significant at high altitude cruise conditions
- Electric oil pump
- Electric actuators for a range of applications:

   - Variable Stator Vanes (VSVs) actuators: using EMA actuation; replaces fueldraulic actuation; employing two physically identical actuators in a dual-redundant master/slave configuration
   - reverse thrust actuators: screw type linear actuators; flexible transmission shafts; drawing 35 kW; developed jointly by Hispano-Suiza (electrical and electronics units and wiring) and Honeywell mechanical components and integration); the A380 has pioneered electro-mechanical thrust reversing actuation

- Active Magnetic Bearing (AMB)
- 350 VDC electrical network

The main electrical power generation components are shown in Figure 10.13.

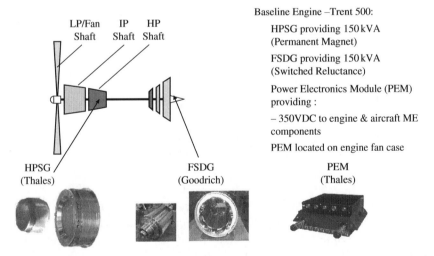

Baseline Engine – Trent 500:

HPSG providing 150 kVA (Permanent Magnet)

FSDG providing 150 kVA (Switched Reluctance)

Power Electronics Module (PEM) providing :

– 350VDC to engine & aircraft ME components

PEM located on engine fan case

Figure 10.13 More-Electric Trent 500 major electrical components

The HPSG located within the HP shaft is a permanent magnet machine producing 150 kVA of electrical power that is conditioned by a Power Electronics Module (PEM) located on the fan casing. This provides 350 VDC for use by the aircraft and engine more-electric systems.

The FSDG is a switched reluctance machine located on the fan shaft within the tailcone. This also produces 150 kVA but has significant advantages of supplying considerable quantities of power in emergency situations. In a wind-milling situation the engine fan shaft will continue to rotate at ~8% of full engine rpm, using the flexibility of the switched reluctance machine as described in Chapter 5 it is still possible to extract considerable amounts of electrical power from the FSDG. The FSDG therefore provides a viable alternative to a RAT for the provision of power in an emergency with the added advantage that it is integral with the engine. A further advantage is that FSDG is always in use whereas the RAT is a one-shot emergency system that conceivably may not work when called upon.

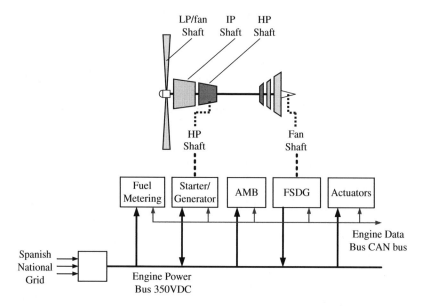

Figure 10.14   Overview of More-Electric Trent 500 electrical architecture

The key electrical attributes of the engine apart from the increased electrical actuation already described are portrayed in a simple form in Figure 10.14. These are:

- 350 VDC engine electrical bus. The HPSG receives 350 VDC from an external source in order to start the engine. In the demonstration scenario this supply comes from the Spanish national grid. Once the engine is running the HPSG provides 350 VDC to the engine bus via the PEM for use by the other subsystems: fuel metering, AMB and actuation. Once the engine is running the FSDG is also becomes a primary source of 350 VDC power
- Distributed engine control using deterministic CAN buses – high speed (up to 1 Mbits/sec and low speed (125 kbits/sec) used to integrate the control functions as shown in the figure. The high speed CANbus is used to integrate the primary engine control functions such as VSVs, fuel metering, electronic engine control and the oil system. Low speed CANbus is used to control ancillary control functions such as the LP generator, AMB, oil scavenge pump and oil breather

10.8 Impact of Stealth Design

Over the past fifteen years or so the term 'stealth' has become a common expression in relation to new combat aircraft programmes, particularly regarding recent developments in the US. The term 'stealth' relates to the ability of an aircraft to remain undetected and hence deny an adversary the opportunity to engage in combat. The main aircraft detection techniques involve the use of radar or infra-red thermal detection principles. It follows that stealth techniques aim to reduce radar and infra-red 'signature' emissions from the aircraft; this being what the use of stealth, or 'low observability' is all about. Though not totally new in principle, a range of new military aircraft developments by the US has, in recent years, given further impetus to the application of stealth techniques, to the point where military aircraft design, construction and manufacture, and operations are ruled by the stealth or low-observability requirements.

This principle is perhaps best illustrated by a simple example. The radar range equation governs the parameters which dictate the distance at which an aircraft will be detected. One of the key factors is the reflecting area of the target or aircraft. Typically for a fighter aircraft a radar reflecting area may be of the order of 10 square metres. For a stealth aircraft it may be assumed that this is reduced to 0.1 square metres – that is reduced by a factor of 100. The range at which an aircraft may be detected is proportional to the fourth root of the radar reflecting area. The fourth root of 100 is 3.16 and therefore the maximum detection range would have been reduced by this value. A radar previously able to detect a conventional target at 158 miles would now only be able to detect a stealthy target at 158/3.16 or 50 miles.

Detail of precisely how small the radar signature can be made is highly classified and it is likely to be much smaller than that given in the example. If the equivalent radar area were reduced by 10 000 rather than the factor of 100 used above, then the radar range would be reduced by a factor of 10 rather than 3.16 and the detection range would be reduced to 15.8 miles which would mean that the aircraft would be detected almost too late to engage successfully. The difficulty in rear aspect radar detection is almost certainly linked to the reduction in infrared or IR signature upon which many missile terminal guidance systems are based. The combination of significant reductions of both radar and IR signatures must make a stealthy aircraft very difficult to detect and engage by conventional means, herein lies the attraction.

The suppression of these two signatures has an impact upon aircraft design in the following areas:

- Most aircraft reflections are from the engine intake and exhausts and therefore considerable efforts may be expended to avoid these orifices acting as radar reflectors. The F-117 uses an inlet grid to minimise intake reflections; other aircraft use serpentine ducts to trap reflecting rf energy
- Intakes and jet pipes apart, angular corners or large plane reflecting surfaces should be avoided. Even straight edges such as wing leading or trailing edges may increase the reflecting area for some aircraft aspects

- Aircraft metal skins offer a good reflecting surface for radar emissions and the use of radar absorbent materials may also be considered. Straight edges on structure, doors and panels that are a right angles to the centre-line should be serrated to avoid coordinated refections back towards an irradiating radar
- To suppress the aircraft IR signature, efforts may be made to reduce the temperature of the jet plume issuing from the jet pipes by shielding the emissions or by diffusing cooler air into the jet exhaust to reduce the temperature

None of the techniques outlined above may be applied without accompanying penalties and it is interesting to contrast the differing stealth designs flying today as solutions to the problem, though the relative performance gains or losses must be purely a matter for speculation. The aircraft currently known today are:

- Lockheed/Martin F-117A stealth fighter
- Northrop B-2 stealth bomber
- Lockheed/Martin F-22 Raptor
- Lockheed Martin F-35 Lightning II

The Lockheed SR-71 Blackbird also made considerable use of stealth techniques.

10.8.1 Lockheed F-117A Nighthawk

The F-117A programme was commenced in 1978 and the aircraft first flew in 1981, though the US Air Force did not admit to its existence until November 1988 when the aircraft had already entered service. The general planform of the aircraft is depicted in Figure 10.15a. from which it can be seen that it has a highly angular almost prismatic construction comprising relatively few facets; the wings and fins are highly swept such that any incident radar energy which is reflected does not scatter in an organised fashion. The relatively simple polyhedron approach of the F-117A was presumably easier to model during early assessment of the low observability features of the design. It is also believed that the planar facets would have facilitated aircraft manufacture using radar absorbent material.

The aircraft is of subsonic performance, powered by the same General Electric 404 engines used on the McDonnell Douglas F/A-18 Hornet though no reheat is provided for the F-117A. The aircraft uses the same 40/45 kVA VSCF cycloconverter that was used on the F/A-18C/D. The engine air inlets are covered with grilles, supposedly using composite materials for a 0.75 in x 1.25 in mesh which prevents any reflections from the engine inlet turbine blades. The engine exhaust is diffused with cool air after exiting the engine and is spread by vanes to exhaust through wide shallow apertures across the entire inboard trailing edges of both wings – Figure 10.15b. The aircraft has a fly-by-wire control system though it is not known whether the aircraft is dynamically unstable. It is more likely that the fly-by-wire system is employed primarily to reduce weight and improve handling qualities.

Weapons are carried internally to preserve the low radar signature as is the case on all other stealth aircraft. Otherwise the aircraft systems are believed to be relatively conventional, some being purloined from other aircraft. The fuel system is certainly conventional if the in-flight refuelling photographs are anything to judge by – see Figure 10.15c. The aircraft was used operationally during the US intervention in Panama in 1990 and a number of aircraft were deployed to the Gulf in 1990 as part of the US response to that crisis. All the reports of the performance of the F-117A during Desert Storm suggest that the aircraft was extremely effective in terms of stealth and as a weapon delivery platform. A total of 59 aircraft were built under the US Force F-117A procurement contract. The aircraft has more recently performed creditably in the Bosnian conflict and the second Gulf War.

Figure 10.15a   F-117A Configuration (Courtesy of US Air Force)

Figure 10.15b   F-117A engine exhaust ducts (Courtesy of US Air Force)

Figure 10.15c   F-117A in-flight refuelling (Courtesy of US Air Force)

10.8.2 Northrop B-2 Spirit

The B-2 stealth bomber programme was publicly acknowledged before the US Air Force finally lifted the security veil in November 1988 at the aircraft roll-out. It is produced by Northrop with the Boeing Company as a major subcontractor. The flying wing design had been anticipated; however, what was unexpected was the angular wing platform with totally straight leading edge and the now customary zig-zag trailing edge. The aircraft also differed considerably from the previously unveiled F-117A in the degree of smooth fuselage/wing blended contours that are in stark contrast to the stealth fighter's polyhedral, planar faceted features. See Figure 10.16.

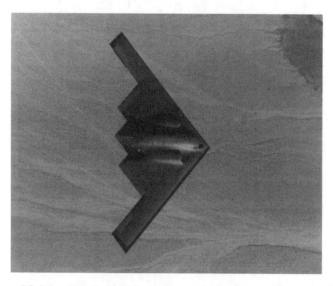

Figure 10.16   B-2 Spirit (Courtesy of Northrop Corporation)

The aircraft owes its pedigree to the Northrop flying wing designs of the immediate post-war era. One of them, the Northrop YB-49, was developed to the stage of having two flying prototypes. One crashed and the other was destroyed on take-off; the main difficulty being that of maintaining longitudinal stability. The B-2 uses a highly sophisticated quadruplex computer controlled fly-by-wire flight control system to provide stability. Unlike the F-117A the B-2 bomber is smoothly contoured with blended wing fuselage so that there are no abrupt changes of form. This probably offers a better or lower radar signature than the F-117A though it is probably correspondingly more difficult to manufacture. It has been reported in the aviation press that the prototypes have been manufactured to very precise production tooling standards and this may be a prerequisite to the smooth contouring of the aircraft.

Figure 10.17 B-2 refuelling in-flight (Courtesy of US Air Force)

The aircraft is controlled entirely by flying control surfaces along the wing trailing edge. Yaw is controlled by means of split ailerons on the outboard section of each wing. These have upper and lower surfaces that may be opened independently like airbrakes. Differential operation of the split ailerons allows differential drag to be applied to the aircraft allowing control in yaw.

See Figure 10.17. The centre rear portion of the fuselage, called the 'beaver's tail', is also believed to move vertically in a limited fashion and may permit trimming of the aircraft in pitch. The engine intakes and exhausts are situated on the upper surface of the wing where they are shielded from ground-based radars. Most of the fuel is believed to be carried in the outboard sections of the wing. The aircraft indicated a conventional in-flight refuelling capability at an early stage in the flight test programme as shown in Figure 10.17. The centre and inboard wing sections house the engines, intakes and exhausts and the internal weapon bay as the B-2 carries its weapons internally in common with the other stealth aircraft. During a much-publicised fault at an early stage in the flight test programme it was revealed that the aircraft was experiencing oil leaks from the AMAD – aircraft mounted accessory drives or gearboxes. This suggests that the aircraft is fairly conventional in terms of hydraulic and electrical systems.

The B-2 has been the subject of intense political debate due to the high programme costs and extremely high unit production costs of several hundred million dollars per aircraft. Congress eventually permitted production of 21 aircraft as opposed to the 132 that the US Air Force originally wished to purchase.

Despite the small numbers, the aircraft has made a major contribution to the air-launched component of the US Air Force during recent conflicts. In 1999 B-2 bombers were deployed directly from the US to bomb Yugoslavia using precision guided munitions during the Kosovo crisis. The aircraft was also deployed to great effect during the second Gulf War of 2003.

B-2 Flight Control System

The B-2 Spirit stealth bomber when unveiled in the late 1980s showed the Northrop flying wing heritage established by the XB-39 and YB-45 flown during the late 1940s. The flying wing poses interesting flight control problems, not the least being the provision of yaw control in an aircraft that has no vertical stabilisers or fins. The unique aircraft configuration together with the missions that the aircraft performs – typically 20 hour flights with air-to-air refuelling makes the B-2 FCS a topic worthy of further study.

The Northrop B-2 has the following flight control surfaces:

- Four elevons; two on each wing at the mid-section
- Two flaperons; one on each wing inboard of the elevons
- Beaver tail at the trailing edge of the centre section
- Split ailerons at the wing extremities

Refer to Figure 10.18 that shows the layout of the primary flight control surfaces on the B-2.

Pitch control is provided by the elevons together with the flaperons. The beaver tail provides trim and gust alleviation. The elevons provide roll control.

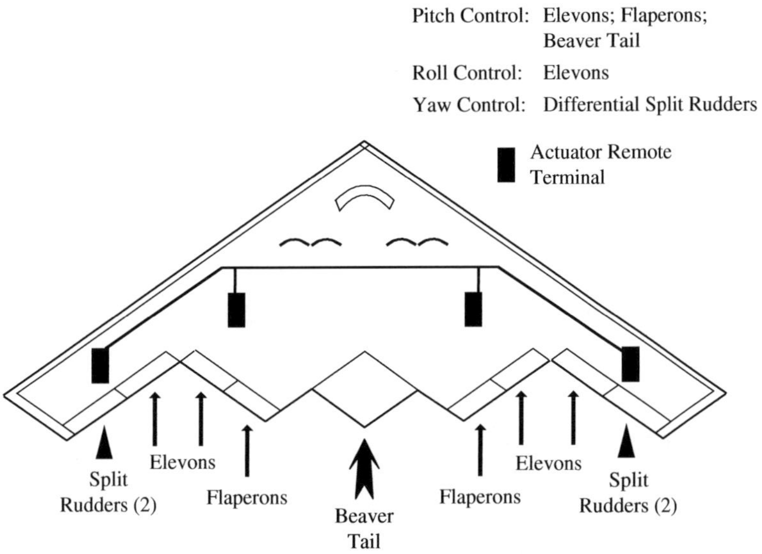

Pitch Control: Elevons; Flaperons;
                Beaver Tail

Roll Control: Elevons

Yaw Control: Differential Split Rudders

■ Actuator Remote
   Terminal

Figure 10.18   B-2 control surfaces

Differential operation of the split ailerons can also cause the aircraft to yaw; when deployed in synchronism they can act as an airbrake.

The flight control system is quadruplex redundant in several respects to assure full functionality and graceful degradation following multiple failures. The top-level architecture is depicted in Figure 10.19 with the following key attributes:

- Quadruplex-redundant flight control computers A, B, C, D receiving redundant air data, attitude and body rate data from the aircraft sensor sets
- Quadruplex MIL-STD-1553 data buses to interface the flight control computers to the four Actuator Remote Terminals (ARTs) in each wing. These actuator remote terminals each interface demands to the flight control surfaces via branches A, B, C, D respectively
- Hydraulic power is provided to the flight control architectures from four independent hydraulic systems

The quadruplex redundancy is continued down the control chain as shown in Figure 10.20.

Each ART branch feeds a control input to a pair of actuators for each flight control surface, therefore ART A feeds channel A of actuator 1 and actuator 2; ART B feeds the B channels and so on. Thus signalling to each of the flight control surface actuators is performed in quadruplex as well as there being two actuators per control surface. At the actuator level the hydraulic power supplies are redundant such that systems A or B may feed actuator 1 while systems C or D may feed actuator 2.

This system therefore incorporates quadruplex redundancy at multiple levels. The aircraft flight control system, as for its civil counterparts, has

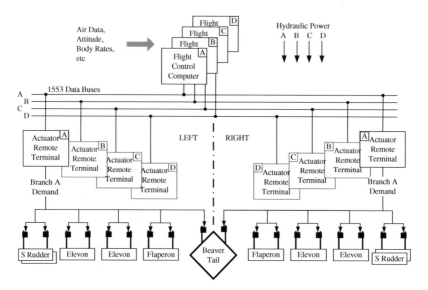

Figure 10.19   B-2 FCS architecture

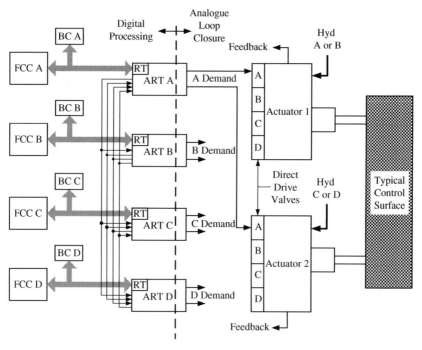

Figure 10.20   B-2 Actuator configuration

probably been designed to exceed catastrophic failures better than 1×10^{-9} or one per billion flight hours. The aircraft regularly flies 20 hour sorties on operational missions, i.e. an exposure period of 20 hours, which probably means that the design requirements were set at 1×10^{-11} per flight hour or better.

Schaefer, Inderhees and Moynes give an excellent overview of the B-2 flight control system [14].

10.8.3 Joint Strike Fighter – F-35 Lightning II

The latest fighter aircraft development programme is the Joint Strike Fighter (JSF) in which two competing teams developed flying demonstrators to prove the respective technologies and operating concepts. The winner of this competition was Lockheed Martin with the F-35 recently named the Lightning II, after the legendary Lockheed P-38 Lightning of WWII and the English Electric Lightning built in the UK. The F-35 aircraft addresses the requirements of the following customers:

- US Air Force: The US Air Force or Conventional Take-Off and Landing (CTOL) version has a fairly conventional set of requirements which include internal and external weapons carriage and a multi-role supersonic capability
- US Navy: The Carrier Vehicle (CV) variant has similar characteristics to the Air Force version but requires additional structural strength to accommodate the additional stresses associated with deck landings. Other key requirements are identical to those of the Air Force
- US Marines: The Marines version has similar requirements to the Air Force and Navy variants but mandates a Short Take-Off and Vertical Landing (STOVL) capability. This leads to the need for a direct lift propulsion system
- UK Royal Navy: The UK Navy requirement is directly equivalent to that for the US Marines

In addition, as the F-35 is intended as a possible replacement for F-16, AV-8B, Sea Harrier/Harrier and other present front-line fighter aircraft. Several nations have recently signed up as partners/participants in the programme: at the time of writing, these are:

- United States
- Unitex Kingdom
- Denmark
- Italy
- Netherlands
- Norway
- Australia
- Canada
- Turkey

The engine system selected for the F-35 is a derivative of the Pratt & Whitney F119 which is the engine well into development for the F-22 Raptor and has several thousands of ground test experience plus the flying experience gathered so far in the F-22 flight test programme. Recently an alternate engine development – the F136 produced by General Electric in conjunction with Rolls-Royce – had its funding reinstated.

Some of the requirements of the four sponsoring Services appear to be diametrically opposing in terms of achieving a final solution and present severe challenges. Nevertheless, if a high degree of commonality can be maintained between the competing variants and/or requirements then the US military authorities will have achieved a degree of standardisation which will doubtless yield significant benefits of scale: both to the operational Services and the taxpayer on both sides of the Atlantic

10.9 Technology Developments/Demonstrators

Supporting the F-35 flight demonstration programme is the JSF Integrated Subsystems Technology (J/IST) Demonstrator Program. Key among the aircraft systems related demonstrations are:

- Fault Tolerant 270VDC electrical power generation system
- Thermal and Energy Management Module (T/EMM)
- AFTI F-16 EHA Demonstration

10.9.1 Fault Tolerant 270VDC Electrical Power Generation System

The J/IST electrical power generation and distribution system as fitted to the NASA Dryden Advanced Fighter Technology Integration (AFTI) F-16 is based upon an 270 VDC 80 KW switched reluctance starter/generator incorporating a dual channel converter/controller supplied by Sundstrand. The aircraft also has a 270 VDC 15 KW emergency generator. This system provides flight critical power by means of two independent 270 VDC aircraft buses as shown in Figure 10.21. Each 270 VDC bus feeds one half of a Power Drive Electronics unit (PDE) of which there is one per primary flight control surface. The PDE in turn controls one half of the Parker Aerospace dual tandem 270 VDC EHA. All of the flight control actuator EHAs are supplied by a consortium of Parker Aerospace and Moog.

Five main flight control actuators so powered are:

- Left flaperon
- Right flaperon
- Left horizontal tail
- Right horizontal tail
- Rudder

10.9.2 Thermal and Energy Management Module

The Thermal and Energy Management Module (T/EMM) combines the function of a traditional APU, emergency power unit and environmental control

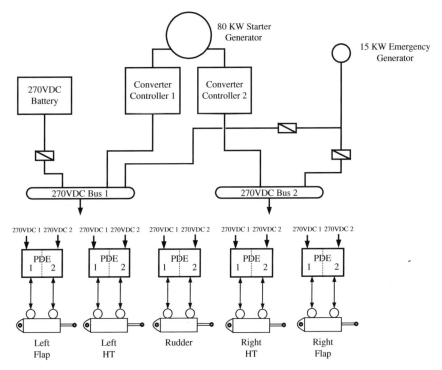

Figure 10.21 AFTI F-16 simplified 270 VDC system

system. This allows the conventional AMAD to be removed as is the aircraft central hydraulics system. The engine fan duct air is used as the heat sink thereby removing the usual heat exchangers and associated ducting. Extensive operation of the Honeywell (Allied Signal) supplied T/EMM has been accomplished during subsystem rig testing prior to engine and T/EMM integration in early 2000.

10.9.3 AFTI F-16 Flight Demonstration

The AFTI F-16 is the flight test bed for the flying elements of the J/IST demonstration programme. The aircraft has been modified to accommodate the 270 VDC architecture shown in Figure 10.22. The five PDEs each drive a dual-tandem actuator supplied by Parker Aerospace; one for each flight control surface as already mentioned. PDE channels 1 and 2 each drive a brushless DC motor which in turn powers half of the actuator package; a PDE channel performs loop closure around its respective components. Each half of the actuator comprises a motor, pump, local fluid reservoir and a valve assembly. As the name suggests, normally both channels operate in tandem. The valve assemblies ensure that each channel can drive the actuator ram if the other channel fails. In the event of both channels failing, aerodynamic pressure drives the control surface to central position where it becomes hydraulically locked. A simplified schematic of the dual-tandem actuator is at Figure 10.22.

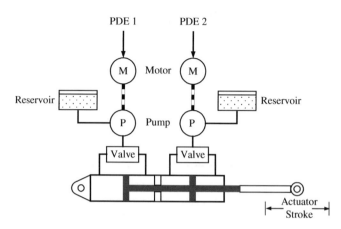

Figure 10.22 Simplified schematic of J/IST dual-tandem actuator

The control side of the implementation posed the problem of interfacing the existing quadruple-redundant Digital Flight Control Computer (DFCC), with the five new PDE actuator drive packages. This was achieved by the introduction of a new triple-redundant control electronics unit which interfaces the 'old' digital flight control system with the 'new' PDEs and actuators. For a comprehensive overview of the AFTI F-16 system see Schley and Kotalik [15]. A three-dimensional diagram of the AFTI F-16 is at Figure 10.23 though the aircraft external appearance yields no clue as to the major systems modifications which are contained therein.

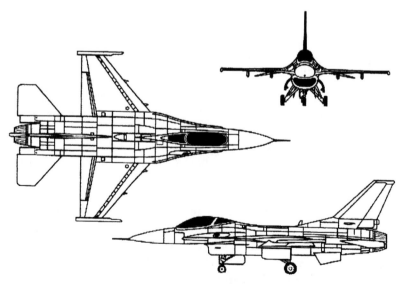

Figure 10.23 NASA Dryden AFTI F-16 (Courtesy of NASA)

An important system that is emerging from future system studies is that of prognostics. For some while it has been practice to log failures as they occur in flight to aid rapid detection and repair on the ground. However, increasing demands to reduce support costs and improve turnaround times have led to a demand for something more sophisticated – the ability to predict and plan for failures.

The modern aircraft computing architecture contains a wealth of information that characterises the normal and potentially degrading performance of a system and its components. Knowledge of such information as flow rates, pressures, loss rates and actuator positions and states, number of excursion, and elapsed operating time can be compared with input data of known wear characteristics to form the basis of an analysis of degrading performance.

The introduction of knowledge-based systems and the application of Bayesian statistics allows models to be constructed that can draw inferences from measured performance. This inferential data can be used to predict the time at which system or component performance becomes unacceptable or to estimate time to failure.

This is of importance to operators who can support the aircraft by arranging for maintenance to be performed at preferred maintenance centres before failure occurs. This allow them to plan for parts, tools and staff to be available for a rapid repair. This leads on to a concept of Maintenance Free Operating Periods (MFOPs) as a contractual requirement rather than working to a scheduled maintenance period.

References

[1] Tuttle, F.L., Kisslinger, R.L., Ritzema, D.F. (1990) 'F-15 S/MTD IFPC Fault Tolerant Design', IEE.
[2] Moir, I., Seabridge, A.G. (1986) 'Management of Utilities Systems in the Experimental Aircraft Programme', *Aerospace*, September.
[3] Lowry, J.M., Moir, I., Seabridge, A.G. (1987) 'Integration of Secondary Power Systems on the EAP'. *SEA Aerotech 87*, Long Beach, California.
[4] Cronin, M.J. (1951) 'The Development of Electrical System in the Bristol Brabazon Mk1' Institution of Electrical Engineers, London, 5 April.
[5] Cronin, M.J., 'All Electric Technologies in Future Advanced Aircraft'.
[6] Cronin, M.J., 'The Role of Avionics in the All Electric Airplane', *American Institution of Aeronautics and Astronautics*.
[7] Cronin, M.J. (1982) 'All-Electric vs Conventional Aircraft: The Production/Operational Aspects', *American Institution of Aeronautics and Astronautics*, Long Beach, 12–14 May.
[8] Cronin, M.J. (1983) 'Advanced Electric Power Systems for All-Electric Aircraft', *IEEE*.
[9] Cronin, M.J. 'The De-Ja Vuof All Electric/All Digital Aircraft', *AIAA/IEEE* 6th *Digital Avionics Systems Conference*.
[10] Cronin, M.J. 'The All Electric Airplane Revisited', *Society of Automative Engineers*.
[11] All Electric Aircraft, IEE Colloquium, London, June 1998.
[12] Electrical Machines and Systems for the More-Electric Aircraft, IEE Colloquium, London, November 1999.
[13] The More Electric Aircraft and Beyond, I Mech E Conference, May 2000.
[14] Schaefer, W.S., Inderhees, L.J., Moynes, J.F., 'Flight Control Actuation System for the B-2 Advanced Technology Bomber', SAE Paper 911112.
[15] Schley, W.R., Kotalik, R.J. (2000) 'Implementation of Flightworthy Electrical Actuators for the F-16', I Mech E Conference.

11

System Design
and Development

11.1 Introduction

As the reader will judge from the contents of this book, aircraft systems are becoming more complex and more sophisticated for a number of technology and performance reasons. In addition, avionics technology, while bringing the benefits of improved control by using digital computing and greatly increased integration by the adoption of digital data buses, is also bringing greater levels of complexity to the development process. The disciplines of avionics system development – including hardware and software integration – are now being applied to virtually every aircraft system.

The increasing level of system sophistication and the increased inter-relation of systems is also making the development process more difficult. The ability to capture all of the system requirements and interdependencies between systems has to be established at an early stage in the programme. Safety and integrity analyses have to be undertaken to ensure that the system meets the necessary safety gaols, and a variety of other trades studies and analytical activities have to be carried out.

These increasing strictures need to be met by following a set of rules and this chapter gives a brief overview of the regulations, development processes and analyses which are employed in the development of modern aircraft systems; particularly where avionics technology is also extensively employed.

The design of an aircraft system is subject to many rigours and has to satisfy a multitude of requirements derived from specifications and regulations. There are also many development processes to be embraced. The purpose of this chapter is not to document these ad nauseam but to give the reader an appreciation of the depth and breadth of the issues which need to be addressed.

Aircraft Systems: Mechanical, electrical, and avionics subsystems integration. 3rd Edition   Ian Moir and Allan Seabridge
© 2008 John Wiley & Sons, Ltd

11.1.1 Systems Design

There are references to some of the better known specifications and requirements, but this chapter also attempts to act as a tutorial in terms of giving examples of how the various design techniques and methods are applied. As the complexity and increasing interrelationship and reliance between aircraft systems has progressed it has become necessary to provide a framework of documents for the designer of complex aircraft systems.

11.1.2 Development Processes

An overview of a typical life cycle for an aircraft or equipment is given and the various activities described. Further, some of the some of the programme management disciplines are briefly visited.

11.2 System Design

Key documentation is applied under the auspices of a number of agencies. A list of the major documents which apply are included in the reference section of this chapter and it is not intended to dwell on chapter and verse of those documents in this brief overview. There are several agencies who provide material in the form of regulations, advisory information and design guidelines whereby aircraft and system designers may satisfy mandatory requirements.

11.2.1 Key Agencies and Documentation

These agencies include:

- Society of Automobile Engineers (SAE):
  ARP 4754 [1]
  ARP 4761 [2]

- Federal Aviation Authority (FAA):
  AC 25.1309-1A [3]

- Joint Airworthiness Authority (JAA):
  AMJ 25.1309 [4]

- Air Transport Association (ATA):
  ATA-100 [5]

- Radio Technical Committee Association (RTCA):
  DO-178b [6]
  DO-254 [7]

  This list should not be regarded as exhaustive but merely indicative of the range of documentation which exists.

11.2.2 Design Guidelines and Certification Techniques

References 1 and 2 offer a useful starting point in understanding the interrelationships of the design and development process:

- Def Stan 00-970 for military aircraft

Figure 11.1 shows the interplay between the major techniques and processes associated with the design and development process.

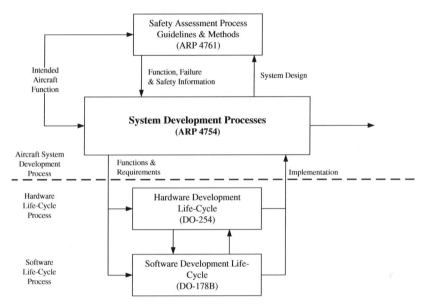

Figure 11.1   ARP 4754 system development process

This figure which is presented as part of the SAE ARP 4761 document gives an overview of the interplay between some of the major references/working documents which apply to the deign and development process. In summary:

- ARP 4761 represents a set of tools and techniques
- ARP 4754 is a set of design processes
- DO-254 offers guidance for hardware design and development
- DO-178B offers advice for the design and certification of software

The key elements of these documents may be summarised by listing the main subject headings.

11.2.3 *Key Elements of the Development Process*

System Development Processes – ARP 4754

- System development
- Certification process and coordination
- Requirements determination and assignment of development assurance level
- Safety assessment process
- Validation of requirements
- Implementation verification
- Configuration management
- Process assurance
- Modified aircraft

Methodologies and Techniques – ARP 4761

- Functional Hazard Assessment (FHA)
- Preliminary System Safety Analysis (PSSA)
- System Safety Analysis (SSA)
- Fault Tree Analysis (FTA)
- Dependency Diagrams
- Markov Analysis (MA)
- Failure Modes and Effects Analysis (FMEA)
- Failures Modes and Effects Summary (FMES)
- Zonal Safety Analysis (ZSA)
- Particular Risks Analysis (PRA)
- Common Mode Analysis (CMA)
- Contiguous safety assessment process example

DO-178B Overview Design Assurance for Airborne Software (1 December 1992)

- Introduction
- System Aspects relating to Software Development
- Software Life Cycle
- Software Planning Process
- Software Development Process
- Software Verification Process
- Software Configuration Management Process
- Software Quality Assurance Process
- Certification Liaison Process
- Overview of Aircraft and Engine Certification
- Software Life Cycle Data
- Additional Considerations

DO-254 Overview – Design Assurance Guidance for Airborne Electronic Hardware (April 2000)

- Introduction
- System Aspects of Hardware Design Assurance
- Hardware Design Life Cycle
- Planning Process
- Validation and Verification Process
- Configuration Management Process
- Process (Quality) Assurance
- Certification Liaison
- Hardware Design Life Cycle Data
- Additional Considerations

The correspondence of these documents to EUROCAE ED documents is given in Table 11.1.

Table 11.1 Correspondence of US RTCA and EUROCAE documents

Specification Topic	US RTCA (1) Specification	European EUROCAE (2) Specification
Systems Development Processes	SAE 4754	ED-79
Safety Assessment Process Guidelines & Methods	SAE 4761	
Software Design	DO-178B	ED-12
Hardware Design	DO-254	ED-80
Environmental Test	DO-160	ED-14

Notes: (1) RTCA Inc
     (2) European Organisation for Civil Aviation Equipment

Serious students or potential users of this process are advised to procure an updated set of these documents from the appropriate authorities.

11.3 Major Safety Processes

There are a number of interrelated processes that are applied most frequently during the safety assessment of an aircraft system. These are:

- Functional Hazard Analysis (FHA)
- Preliminary System Safety Analysis (PSSA)
- System Safety Analysis (SSA)
- Common Cause Analysis (CCA)

Figure 11.2 shows a simplified version of the interplay between these processes as the system design evolves and eventually the system achieves certification.

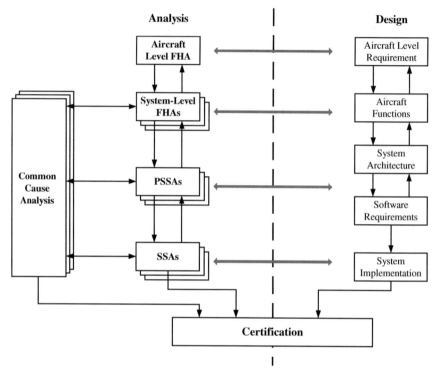

Figure 11.2 Simplified portrayal of safety processes

The diagram effectively splits into two sections: design activities on the right and analysis on the left. As the system evolves from aircraft level requirements, aircraft functions are evolved. These lead in turn to system architectures which in turn define software requirements and the eventual system implementation. At corresponding stages of the design, various analyses are conducted which examine the design in the light of the mandated and recommended practices. At every stage the analyses and the design interact in an evolutionary manner as the design converges upon a solution which is both cost-effective and which meets all the safety requirements.

11.3.1 Functional Hazard Analysis (FHA)

A FHA is carried out at both aircraft and system levels; one flows down from the other. The FHA identifies system failures and identifies the effects of these failures. Failures are tabulated and classified according the effects which that failure may cause and the safety objectives assigned according to the criteria briefly listed in Table 11.2.

The FHA identifies the data in first two columns of the table: the failure condition classification and the development assurance level. These allow the safety objectives to assigned for that particular condition and a quantitative probability requirement assigned.

For a failure which is identified as having a catastrophic effect, the highest assurance level A will be assigned. The system designer will be required to implement fail-safe features in his design and will have to demonstrate by appropriate analysis that the design is capable of meeting or exceeding the probability offailure less than 1×10^{-9} per flight hour. In other words, the particular failure should occur less than once per 1 000 000 000 flight hours or once per 1000 million flight hours. This category of failure is assigned to systems such as flight controls, structure etc. where a failure could lead to the loss of the aircraft. The vast majority of aircraft systems are categorised at much lower levels where little or no safety concerns apply.

Table 11.2 Overview of failure classification and safety objectives

Failure condition classification	Development assurance level	Safety objectives	Safety objectives quantitative requirement (probability per flight hour)
Catastrophic	A	Required	$< 1 \times 10^{-9}$
Hazardous/Severe	B	May be required	$< 1 \times 10^{-7}$
Major	C	May be required	$< 1 \times 10^{-5}$
Minor	D	Not required	None
No safety effect	E	Not required	None

A more user friendly definition quoted in words as used by the Civil Airworthiness Authority (CAA) may be:

Catastrophic: less than 1×10^{-9}; extremely improbable
Hazardous: between 1×10^{-9} and 1×10^{-7}; extremely remote
Major: between 1×10^{-7} and 1×10^{-5}; remote
Minor: between 1×10^{-5} and 1×10^{-3}; reasonably probable
greater than 1×10^{-3}; frequent

11.3.2 Preliminary System Safety Analysis (PSSA)

The PSSA examines the failure conditions established by the FHA(s) and demonstrates how the system design will meet the specified requirements. Various techniques such as Fault Tree Analysis (FTA), Markov diagrams etc. may be used to identify how the design counters the effects of various failures and may point towards design strategies which need to be incorporated in the system design to meet the safety requirements. Typical analyses may include the identification of system redundancy requirements, e.g. how many channels, what control strategies could be employed and the need for dissimilarity of control; e.g. dissimilar hardware and/or dissimilar software implementation. The PSSA is therefore part of an iterative process which scrutinises the system design and assists the system designers in ascribing and meeting risk budgets across one or a number of systems. Increasingly, given the high degree of integration and interrelationship between major aircraft systems, this is likely to be a multi-system, multi-disciplinary exercise coordinating the input of many systems specialists.

11.3.3 System Safety Analysis (SSA)

The SSA is a systematic and comprehensive evaluation of the system design using similar techniques to those employed during the PSSA activities. However, whereas the PSSA identifies the requirements, the SSA is intended to verify the that the proposed design does in fact meet the specified requirements as identified during the FHA and PSSA analyses conducted previously. As may be seen in the early Figure 11.2, the SSA occurs at the point in the design cycle where the system implementation is concluded or finalised and prior to system certification.

11.3.4 Common Cause Analysis (CCA)

The CCA begins concurrently with the system FHA and is interactive with this activity and subsequent PSSA and SSA analyses. The purpose of the CCA is – as the name suggests – to identify common cause or common mode failures in the proposed design and assist in directing the designers towards strategies which will obviate the possibility of such failures. Such common cause failures may include:

- Failure to correctly identify the requirement
- Failure to correctly specify the system
- Hardware design errors
- Component failures
- Software design and implementation errors
- Software tool deficiencies
- Maintenance errors
- Operational errors

The CCA is therefore intended to scrutinise a far wider range of issues than the system hardware or software process. Rather it is meant to embrace the whole process of developing, certifying, operating and maintaining the system throughout the life cycle.

11.4 Requirements Capture

It can be seen from the foregoing that requirements capture is a key activity in identifying and quantifying all the necessary strands of information which contribute to a complete and coherent system design. There are a number of ways in which the requirements capture may be addressed. Two main methods are commonly used:

- Top-down approach
- Bottom-up approach

11.4.1 Top-Down Approach

The top-down approach is shown in Figure 11.3.

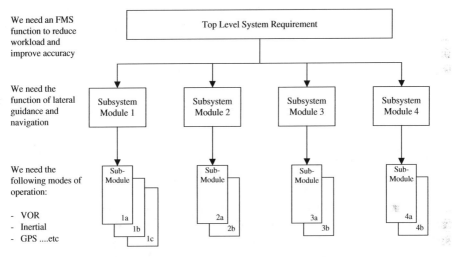

Figure 11.3 Top-down approach

This represents a classical way of tackling the requirements capture by decomposing the system requirements into smaller functional modules. These functional modules may be further decomposed into functional submodules. This approach tends to be suited to the decomposition of large software tasks where overall requirements may be flowed down into smaller functional software tasks or modules. This would apply to a task where the hardware boundaries are fairly well understood or inferred by the overall system requirement. An example might be the definition of the requirements for an avionics system such as a Flight Management System (FMS). In such a system basic requirement – the need to improve the navigation function is well understood – and the means by which the various navigation modes are implemented: INS, GPS, VOR, etc. are well defined.

11.4.2 Bottom-Up Approach

The bottom-up method is shown in Figure 11.4.

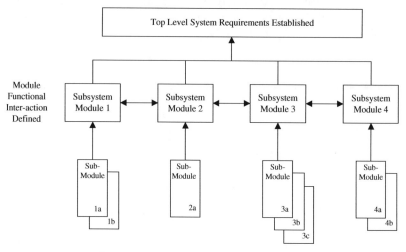

Figure 11.4　Bottom-up approach

The bottom-up approach is best applied to systems where some of the lower level functions may be well understood and documented and represented by a number of submodules. However, the process of integrating these modules into a higher subset presents difficulties as the interaction between the individual subsystems is not fully understood. In this case building up the top level requirements from the bottom may well enable the requirements to be fully captured. An example of this type might be the integration of aircraft systems into an integrated utilities management system. In this case the individual requirements of the fuel system, hydraulic system, environmental control system etc. may be well understood. However, the inter-relationships between the candidate systems and the implications of adopting integration may better understood and documented by working bottom up.

In fact most development projects may use a combination of both of these approaches to best capture the requirements.

11.4.3 Requirements Capture Example

The example given in Figure 11.5 shows a functional mapping process which identifies the elements or threads necessary to implement a fuel jettison function. Two main functional subsystems are involved: the fuel quantity measurement function and the fuel management function. Note that this technique

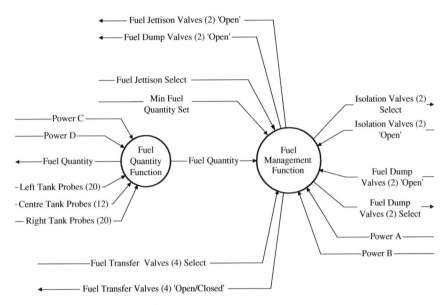

Figure 11.5 System requirements capture example

merely identifies the data threads which are necessary to perform the system function. No attempt is made at this stage to ascribe particular functions to particular hardware or software entities. Neither is any attempt made to determine whether signals are hard-wired or whether they may be transmitted as multiplexed data as part of an aircraft system data bus network.

The system requirements from the flight crew perspective are:

- The flight crew need to jettison excess fuel in an emergency situation in order that the aircraft may land under the maximum landing weight
- The flight crew wish to be able to jettison down to a preselected fuel quantity
- The crew wish to be given indications that fuel jettison is underway

The information threads associated with the flight crew requirements are shown in the upper centre portion of the diagram. It may be seen that although the system requirements are relatively simple when stated from the flight crew viewpoint, many other subsystem information strands have to be considered to achieve a cogent system design:

Fuel Quantity Function

The fuel quantity function measures the aircraft fuel quantity by sensing fuel in the aircraft fuel tanks; in this example a total of 52 probes are required to measure the fuel held in three tanks. The fuel quantity calculations measure the amount of fuel which the aircraft has onboard taking account of fuel density and temperature. It is usual in this system, as in many others, to have dual power supply inputs to the fuel quantity function to assure availability in the event of an aircraft electrical system busbar failure. Finally, when the calculations have been completed they are passed to the flight deck where the aircraft fuel quantity is available for display to the flight crew. Fuel quantity is also relayed to the fuel management function so that in the event of fuel jettison, the amount of fuel onboard may be compared with the preset jettison value. The fuel quantity function interfaces to:

• The fuel quantity system measurement probes and sensors
• The flight deck multi-function displays
• The fuel management system
• The aircraft electrical system

Fuel Management Function

The fuel management function accepts information regarding the aircraft fuel state from the fuel quantity function. The flight crew inputs a 'Fuel Jettison Select' command and the minimum fuel quantity which the crew wishes to have available at the end of fuel jettison. The fuel management function accepts flight crew commands for the fuel transfer valves [4], fuel dump (jettison) valves [2], and fuel isolation valves [2]. It also provides 'Open'/'Closed' status information on the fuel system valves to the flight crew. As before two separate power inputs are received from the aircraft electrical system. The fuel management function interfaces with:

• The fuel system valves
• The flight deck displays multi-function displays and overhead panel
• The fuel quantity function
• The aircraft electrical system

This example shows how a relatively simple function interfaces to various aircraft systems and underlines some of the difficulties which exist in correctly capturing system requirements in a modern integrated aircraft system.

11.5 Fault Tree Analysis (FTA)

The Fault Tree Analysis (FTA) is one of the tools described in SAE document ARP 4761 [1]. This analysis technique uses probability to assess whether a particular system configuration or architecture will meet the mandated requirements. For example, assume that the total loss of aircraft electrical power onboard an aircraft has catastrophic failure consequences as identified by the Functional Hazard Analysis – see Figure 11.2 and Table 11.1 above. Then the safety objective quantitative requirement established by FAR/JAR25.1309 and as amplified in ARP 4754 will be such that this event cannot occur with a probability of greater than 1×10^{-9} per flight hour (or once per 1000 million flight hours). The ability of a system design to meet these requirements is established by a FTA which uses the following probability techniques.

In the example it is assumed:

- That the aircraft has two independent electrical power generation systems, the main components of which are the generator and the Generator Control Unit (GCU) which governs voltage regulation and system protection
- The aircraft has an independent emergency system such as a Ram Air Turbine (RAT)
- That the failure rates of these components may be established and agreed due to the availability of in-service component reliability data or sound engineering rationale which will provide a figure acceptable to the certification authorities

The FTA analysis – very much simplified – for this example is shown in Figure 11.6.

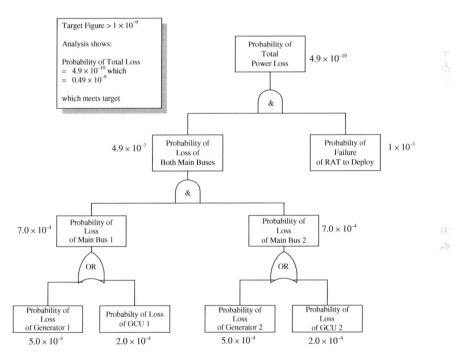

Figure 11.6 Simplified FTA for an aircraft electrical power system

Starting in the bottom left hand portion of the diagram: the Mean Time Between Failure (MTBF) of a generator is 2000 hours – this means that the failure rate of Generator 1 is 1/2000 or 5.0×10^{-4} per flight hour. Similarly if the MTBF of the generator controller GCU 1 is 5000 hours then the failure rate of GCU 1 is 1/5000 or 2.0×10^{-4} per flight hour. The combined failure rate gives the probability of loss of electrical power to Main Bus 1. This is calculated by summing the failure rates of generator and controller as either failing will cause the loss of Main Bus 1:

$= 5.0 \times 10^{-4} + 2.0 \times 10^{-4} = \textbf{7} \times \textbf{10}^{-\textbf{4}}$ per flight hour
(Generator 1)(GCU 1)(Main Bus 1)

Similarly, assuming generator channels 1 and 2 are identical the failure rate of Main Bus 2 is given by:

$= 5.0 \times 10^{-4} + 2.0 \times 10^{-4} = \textbf{7} \times \textbf{10}^{-\textbf{4}}$ per flight hour
(Generator 2) (GCU 2)(Main Bus 2)

(Note that at this state the experienced aircraft systems designer would be considering the effect of a common cause or common mode failure.)

The probability of two independent channels failing (assuming no common cause failure) is derived by multiplying the respective failure rates. The probability of both Main Buses failing is:

$= 7 \times 10^{-4} \times 7 \times 10^{-4} = \textbf{49} \times \textbf{10}^{-\textbf{8}}$ or $\textbf{4.9} \times \textbf{10}^{-\textbf{7}}$ per flight hour
(Main Bus 1) (Main Bus 2)

Therefore the two independent electrical power channels alone will not meet the requirement. Assuming the addition of the Ran Air Turbine (RAT) emergency channel as shown in the figure, the probability of total loss of electrical power

$= 4.9 \times 10^{-7} \times 1 \times 10^{-3} = \textbf{4.9} \times \textbf{10}^{-\textbf{10}}$ per flight hour which meets the requirements
(Main Bus) (RAT failure) (Main Bus 2)

This very simple example is illustrative of the FTA which is one of the techniques used during the PSSA and SSA processes. However, even this simple example outlines some of the issues and interactions which need to be considered. Real systems are very much complex with many more system variables and interlinks between a number of aircraft systems.

11.6 Dependency Diagram

The dependency diagram offers an alternative tool to the FTA for the analysis of architectural alternatives and also to establish whether a particular architecture will meet its mandated integrity goal. As for the FTA, the approach offered by the dependency diagram is best served by using a simple system example – in this case the electrical system analysis.

The dependency diagram has the superficial advantage that its structure maps readily on to a system architecture diagram as shown in Fig 11.7. Dependencies are shown in series or parallel form:

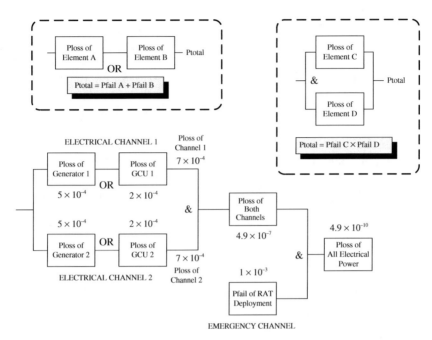

Figure 11.7 Electrical system dependency diagram example

- In a situation where both elements of a particular function need to be opera-tive theses are shown in series; a failure of either element A or element B will deny the total function. The analysis adds the contribution of each element to give the function total
- Where elements are replicated for reasons of redundancy or backup the elements are shown in parallel; a failure of element C and element D is required to deny the overall function. The analysis multiplies the element contributions to give the function total

The values that populate the analysis are the element predicted or actual failure rates. Taking the electrical system failure rates used in the previous example:

Failure rate of Main Bus 1 =	$5.0 \times 10^{-4} + 2.0 \times 10^{-4}$	$= 7.0 \times 10^{-4}$
	Generator 1Control Unit 1	Main Bus 1
Failure rate of Main Bus 2 =	$5.0 \times 10^{-4} + 2.0 \times 10^{-4}$	$= 7.0 \times 10^{-4}$
	Generator 2Control Unit 2	Main Bus 2
Failure rate of Both Electric Channels =	$7 \times 10^{-4} \times 7 \times 10^{-4}$	$= 4.9 \times 10^{-7}$ Both Buses
	Main Bus 1Main Bus 2	
Failure of RAT Emergency Source =	1.0×10^{-3}	
Total failure of Electrical System =	$4.9 \times 10^{-7} \times 1.0 \times 10^{-3}$	$= 4.9 \times 10^{-10}$
	Channel 1 + Emergency Channel 2 Source	

This exceeds (is lower than) the mandated requirement of 1×10^{-9} per flight hour

11.7 Failure Modes and Effects Analysis (FMEA)

The example given is a useful tool to examine total system integrity using a bottom-up approach. Certain parts of systems may be subject to scrutiny as they represent single point failures and as such more detailed analysis is warranted. The analysis used in this situation is the Failure Modes and Effects Analysis (FMEA).

Again, the process used in the FMEA is best illustrated by the use of a simple example. In this case the failure of an electrical generator feeding an aircraft main electrical busbar via an electrical power line contactor. The line contactor is operated under the control of the Generator Control Unit (GCU) as shown in Figure 11.8.

An FMEA on this portion of the aircraft electrical system will examine the possible failures of all the elements:

- The generator failures and effects; in other words, examine in detail all the failures which contribute to the generator failure rate of 5×10^{-4} per flight hour as used in the previous analysis and the effects of those failures
- The GCU failures and effects: examining all the failures which contributed to the overall failure rate of 2×10^{-4} per flight hour as used above and the effects of those failures
- The line contactor failures and effects. If a line contactor has an MTBF of 100 000 hours/failure rate of 1×10^{-5} per flight hour, the ways in which the contactor may fail are ascribed portions of this failure rate for the different failures and effects:

  - The contactor may fail open
  - The contactor may fail closed
  - The contactor may fail with one contact welded shut but the others open and so on until all the failures have been allocated a budget

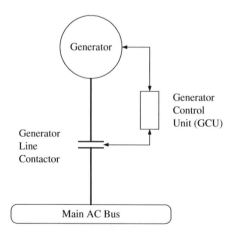

Figure 11.8 Main generator, GCU and power contactor relationship

This process is conducted in a tabular form such that:

- Failure modes are identified
- Mode failure rates are ascribed
- Failure effects are identified
- The means by which the failure is detected is identified

An FMEA should therefore respond to the questions asked of the system or element under examination in a quantitative manner.

11.8 Component Reliability

In the analyses described, a great deal of emphasis is placed upon the failure rate of a component or element within the system under review. This clearly calls into question how reliability values for different type of component are established. There are two main methods of determining component reliability:

- Analytical by component count
- Historical by means of accumulated in-service experience

11.8.1 Analytical Methods

MIL-STD-781E is a standard developed by the US military over a period of years to use an analytical bottom-up approach to predicting reliability. This method uses a component count to build up an analysis of the reliability of a unit. This approach has probably best been applied to electronics over the years as the use of electronic components within a design tends to be replicated within a design and across a family of designs. This method uses type of component, environment and quality factor as major discriminators in predicting the failure of a particular component, module and ultimately subsystem. Component failure rates are extracted from the US military standard and then applied with the appropriate factors to establish the predicted value as shown in the simplified example below:

$$\text{Failure Rate, } \lambda = \pi_Q \times (_{K1}\pi_T +\, _{K2} \times \pi) \times \pi_L$$

Where π_Q is a device quality factor
π_T is a temperature factor
π_E is an environmental factor
π_L is a maturity factor
K1 and K2 are constants

There are a number of issues associated with this method:

- It is only as good as the data base of components and the factors used
- Experience has generally shown that – if anything – predicted values are generally pessimistic thereby generating predicted failure rates worse than might be expected in real life

- The technique has merit in comparing competing design options in a quantitative manner when using a common baseline for each design
- It is difficult to continue to update the data base; particularly with the growing levels of integration with Integrated Circuits (ICs) which makes device failure rates difficult to establish
- The increasing number of Commercial Off-The-Shelf (COTS) components also confuses the comparison
- The technique is particularly valuable when it can be compared with in-service experience and appropriate correction factors applied

Reference [8] is a paper presented at an international aerospace conference several years ago and gives a very good overview of this technique when applied to power electronics.

11.8.2 In-Service Data

The use of in-service data is the best way of approaching the assessment of mechanical components used in the same environment. It does depend upon correspondence between the components which the design is contemplating with the in-service data base being used. Any significant variation in component usage, technology baseline or location in the aircraft/environment may nullify the comparison. Nevertheless, when used in conjunction with other methods this is a valid method. The manufacturers of civil, fighter aircraft and helicopters and their associated suppliers will generally be able to made 'industry standard' estimates using this technique.

11.9 Dispatch Reliability

Dispatch availability is key to an aircraft fulfilling its mission, whether a military or civil aircraft. The ability to be able to continue to dispatch an aircraft with given faults has been given impetus by the commercial pressures of the air transport environment where the use of dual-redundancy for integrity reasons has been also used to aid aircraft dispatch. On the Boeing 777 the need for high rates of dispatch availability was specified in many systems and in some systems this lead to the adoption of dual redundancy for dispatch availability reasons rather than for reasons of integrity. A simplified version of the dispatch requirements is shown in Figure 11.9.

This means of specifying the dispatch requirement of part of an aircraft system leads to an operational philosophy far beyond a 'get-you-home' mode of operation. In fact it is the first step towards a philosophy of no unscheduled maintenance. For an aircraft flying 12 hours per day – a typical utilisation for a wide-bodied civil transport – this definition dictates a high level of availability for up to a 120 hour flying period. The ability to stretch this period in the future – perhaps to 500 hour operating period – as more reliable systems become available, could lead to a true system of unscheduled maintenance. A

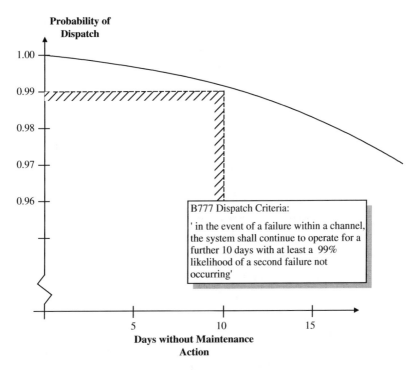

Figure 11.9 Simplified Dispatch Criteria

500 hour operating period roughly equates to 8–9 weeks of flying, at which time the aircraft will probably be entering the hangar for other specified maintenance checks.

This leads to a more subtle requirement to examine the system's ability to meet integrity requirements when several failures have already occurred and this requires different techniques to be utilised.

11.10 Markov Analysis

Another technique used to assist in system analysis is the Markov Analysis (MA). This approach is useful when investigating systems where a number of states may be valid and also are inter-related. This could be the case in a multi-channel system where certain failures may be tolerated but not in conjunction with some failure conditions. The question of whether as system is air-worthy is not a simple mathematical calculation as in previous analyses but depends upon the relative states of parts of the system. The simple methods used are insufficient in this case and another approach is required. The Markov Analysis is the technique to be applied in these circumstances.

As before a simple example will be used to illustrate the MA technique. In this case the dual channel Full Authority Digital Engine Control (FADEC) example outlined in Figure 11.10.

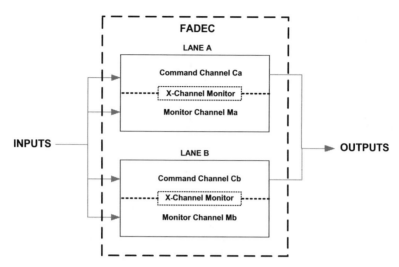

Figure 11.10 Simplified FADEC architecture

This simplified architecture is typical of many dual-channel FADECs. There are two independent lanes: Lane A and Lane B. Each lane comprises a Command and Monitor portion which are interconnected for cross-monitoring purposes and undertakes the task of metering the fuel flow to engine in accordance with the necessary control laws to satisfy the flight crew thrust command. The analysis required to decide upon the impact of certain failures in conjunction with others requires a Markov model in order to be able to understand the dependencies.

Figure 11.11 depicts a simple Markov model which equates to this architecture. By using this model the effects of interrelated failures can be examined. The model has a total of 16 states as shown by the number in the bottom right hand corner of the appropriate box. Each box relates to the serviceability state of the Lane A Command (Ca) and Monitor (Ma) channels and Lane B Command (Cb) and Monitor (Mb) channels. These range from the fully serviceable state in box 1 through a series of failure conditions to the totally failed state in box 16. Clearly most normal operating conditions are going to be in the left hand region of the model.

Concentrating on the left hand side of the model it can be seen that the fully serviceable state in box 1 can migrate to any one of six states:

- Failure of Command channel A results in state 2 being reached
- Failure of Monitor channel A results in state 3 being reached
- Failure of Command channel B results in state 4 being reached
- Failure of Monitor channel B results in state 5 being reached
- Failure of the cross-monitor between Command A and Monitor A results in both being lost simultaneously and reaching state 6
- Failure of the cross-monitor between Command B and Monitor B results in both being lost simultaneously and reaching state 11

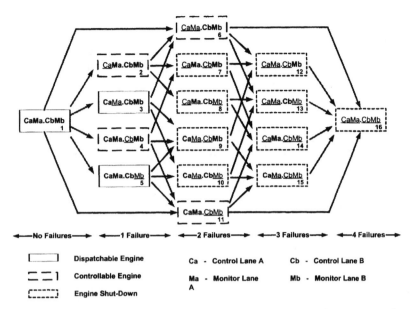

Figure 11.11 Use of Markov analysis to examine engine dispatch reliability

All of these failure states result in an engine which may still be controlled by the FADEC

However, further failures beyond this point may result in an engine which may not be controllable either because both control channels are inoperative or because the 'good' control and monitor lanes are in opposing channels. The model shown above is constructed according to the following rules: an engine may be dispatched as a 'get-you-home' measure provided that only one monitor channel has failed. This means that states 3 and 5 are dispatchable, but not states 2, 4, 6 or 11 as subsequent failures could result in engine shut down.

By knowing the failure rates of the command channels, monitor channels and cross-monitors, quantitative values may be inserted into the model and probabilities assigned to the various states. By summing the probabilities so calculated enables numerical values to be derived.

11.11 Development Processes

11.11.1 The Product Life Cycle

Figure 11.12. shows a typical aircraft product life cycle from concept through to disposal at the end of the product's useful life.

Individual products or equipment may vary from this model, however, it is a sufficiently good portrayal to illustrate the role of systems engineering and the equipment life cycle. The major phases of this model are:

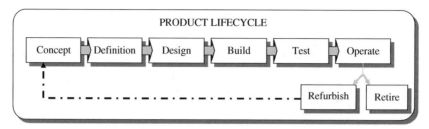

Figure 11.12 Typical aircraft product life cycle

- Concept Phase
- Definition Phase
- Design Phase
- Build Phase
- Test Phase
- Operate Phase
- Refurbish or Retire

This model closely approximates to the Downey cycle used by the UK Ministry of Defence, for the competitive procurement of defence systems. The model is equally applicable for systems used in commercial aircraft as it is for military applications. It is used describe the role of systems engineering in each phase of the product life cycle.

11.11.2 Concept Phase

The concept phase is about understanding the customer's emerging needs and arriving at a conceptual model of a solution to address those needs. The customer continuously assesses his current assets and determines their effectiveness to meet future requirements. The need for a new military system can arise from a change in the local or world political scene that requires a change in defence policy. The need for a new commercial system may be driven by changing national and global travel patterns resulting from business or leisure traveller demands.

The customer's requirement will be made available to industry so that solutions can be developed specifically for that purpose, or that can be adapted from the current research and development (R&D) base. This is an ideal opportunity for industry to discuss and understand the requirements to the mutual benefit of the customer and his industrial suppliers, to understand the implications of providing a fully compliant solution or one which is aggressive and sympathetic to marketplace requirements. See Figure 11.13.

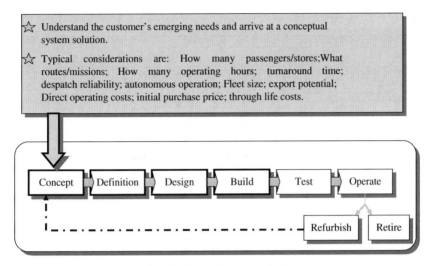

Figure 11.13 Concept phase

Typical considerations at this phase are:

- Establishing and understanding the primary role and functions of the required system
- Establishing and understanding desired performance and market drivers such as:
  - range
  - endurance
  - routes or missions
  - technology baseline
  - operational roles
  - number of passengers
  - mass, number and type of weapons
  - availability and dispatch reliability
  - fleet size to perform the role or satisfy the routes
  - purchase budget available
  - operating or through-life costs
  - commonality or model range
  - market size and export potential
  - customer preference

This phase is focussed on establishing confidence that the requirement can be met within acceptable commercial or technological risk. The establishment of a baseline of mature technologies may be first solicited by means of a Request for Information (RFI). This process allows possible vendors to establish their technical and other capabilities and represents an opportunity for the platform integrator to assess and quantify the relative strengths of competing vendors and also to capture mature technology of which he was previously unaware for the benefit of the programme.

11.11.3 Definition Phase

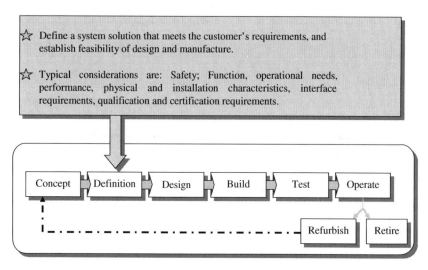

☆ Define a system solution that meets the customer's requirements, and establish feasibility of design and manufacture.

☆ Typical considerations are: Safety; Function, operational needs, performance, physical and installation characteristics, interface requirements, qualification and certification requirements.

Concept → Definition → Design → Build → Test → Operate → Refurbish → Retire

Figure 11.14 Definition phase

See Figure 11.14. The customer will usually consolidate all the information gathered during the concept phase to firm up his requirement. A common feature used more frequently by platform integrators is to establish engineering joint concept teams to establish the major system requirements. These teams are sometimes called Integrated Product Teams (IPTs). They may develop a cardinal points specification; perhaps even undertake a preliminary system or baseline design against which all vendors might bid. This results in the issue of a specification or a Request for Proposal (RFP). This allows industry to develop their concepts into a firm definition, to evaluate the technical, technological and commercial risks, and to examine the feasibility of completing the design and moving to a series production solution. Typical considerations at this stage are:

- Developing the concept into a firm definition of a solution
- Developing system architectures and system configurations
- Re-evaluating the supplier base to establish what equipment, components and materials are available or may be needed to support the emerging design
- Ensuring that materials are selected with knowledge of appropriate legislation determining their use to control Health & Safety and environmental issues
- Defining physical and installation characteristics and interface requirements
- Developing operational and initial safety models of the individual systems
- Quantifying key systems performance such as:
  - mass
  - volume
  - growth capability
  - range/endurance

The output from this phase is usually in the form of feasibility study reports, performance estimates, sets of mathematical models of individual system's behaviour and an operational performance model. This may be complemented by breadboard or experimental models or laboratory or technology demonstrators. Preliminary design is also likely to examine installation issues with mock-ups in three-dimensional computer model form (CATIA) which replaces in the main the former wooden and metal models.

11.11.4 Design Phase

If the outcome of the Definition phase is successful and a decision is made to proceed further, then industry embarks on the design phase within the programme constraints as described later in the chapter. Design takes the Definition phase architectures and schemes and refines them to a standard that can be manufactured. See Figure 11.15.

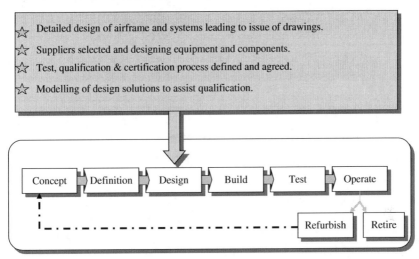

Figure 11.15   Design phase

Detailed design of the airframe ensures that the structure is aerodynamically sound, is of appropriate strength, and is able to carry the crew, passengers, fuel and systems that are required to turn it into a useful product. As part of the detailed design, cognisance needs to be made of mandated rules and regulations which apply to the design of an aircraft or airborne equipment. The processes and techniques used to conduct the necessary safety assessments and analyses and described a little later in the chapter.

Three-dimensional solid modelling tools are used to produce the design drawings, in a format that can be used to drive machine tools to manufacture parts for assembly.

Systems are developed beyond the block diagram architectural drawings into detailed wiring diagrams. Suppliers of bought-in equipment are selected and they become an inherent part of the process of starting to design equipment that can be used in the aircraft or systems. Indeed in order to achieve a fully certifiable design, many of the complex and integrated systems found on aircraft today, an integrated design team comprising platform integrators and supplier(s) is essential. In fact many of these processes are iterative extending into and even beyond the build and test phases.

11.11.5 Build Phase

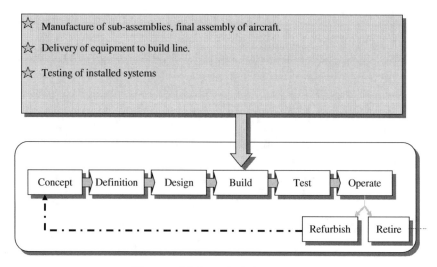

Figure 11.16   Build phase

The aircraft is manufactured to the drawings and data issued by design as shown in Figure 11.16. During the early stages of the programme, a delivery schedule would have been established. Some long-lead time items – those which take a long time to build – may need to be ordered well ahead of aircraft build commencing. In the case of some of the more complex, software-driven equipment, design will be overlapping well into the test phase. This usually accommodate by a phased equipment delivery embracing the following:

- Electrical models – equipment electrically equivalent to the final product but not physically representative
- Red label hardware – equipment which is physically representative but not cleared for flight
- Black label hardware – equipment which is physically representative and is cleared for flight either by virtue of the flight-worthy testing carried out and/or the software load incorporated

These standards are usually accompanied by a staged software release which enables a software load which progressively becomes more representative of the final functionality.

11.11.6 Test Phase (Qualification Phase)

The aircraft and its components are subject to a rigorous test programme to verify its fitness for purpose as shown in Figure 11.17. This phase – usually referred to as qualification, includes testing and integration of equipment, components, sub-assemblies, and eventually; the complete aircraft. The qualification process may include analysis, modelling, analogy with existing or similar systems as well as functional testing. Functional testing of equipment and systems on the ground and flight trials verifies that the performance and the operation of the equipment is as specified. Conclusion of the test programme and the associated design, analysis and documentation process leads to certification of the aircraft or equipment.

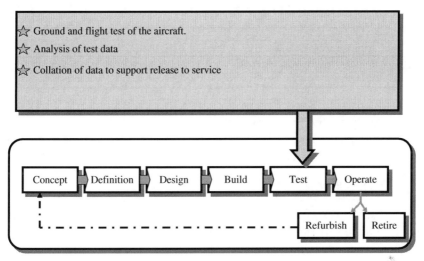

Figure 11.17   Test phase

   In the event of a new aircraft, responsibility for the certification of the aircraft lies with the aircraft manufacturer. However, where an equipment is to be improved or modified in the civil arena, equipment suppliers or other agencies can certify the equipment by means of the Supplementary Type Certificate (STC) in a process defined by the certification authorities. This permits discrete equipment – for example, a more accurate fuel quantity gauging to a particular aircraft model to be changed without affecting other equipment.

11.11.7 Operate Phase

During this phase the customer is operating the aircraft on a routine basis. Its performance will be monitored by means of a formal defect reporting process, so that any defects or faults that arise are analysed by the manufacturer.

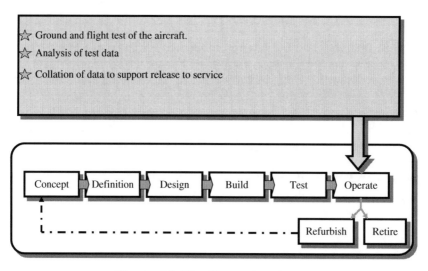

Figure 11.18 Operate phase

It is possible to attribute causes to faults such as random component failures, operator mis-handling, or design errors. The aircraft manufacturer and his suppliers are expected to participate in the attribution and rectification of problems arising during aircraft operations, as determined by the contract.

11.11.8 Disposal or Refurbish

At the end of the useful or predicted life of the aircraft, decisions have to be made about its future as depicted in Figure 11.19. The end of life may be determined by unacceptably high operating costs, unacceptable environmental

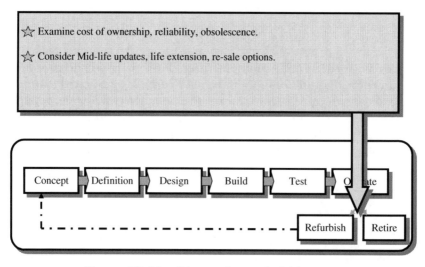

Figure 11.19 Disposal or refurbishment

considerations – noise, pollution etc. – or by predicted failure of mechanical or structural components determined by the supplier's test rigs. If it is not possible to continue to operate the aircraft, then it may be disposed – sold for scrap or alternative use, such as aircraft enthusiast or gate guardian.

The process of disposal of aircraft and equipment needs care to be taken in the safe removal of hazardous materials and the most appropriate method of destruction, storage and reuse of materials. This will include materials that were once safe and permissible to use when the aircraft and equipment was originally designed, but may not be when the aircraft is withdrawn from service up to 50 years later.

If the aircraft still has some residual and commercially viable life, then it may be refurbished. This latter activity is often known as a mid-life update, or even a conversion to a different role, e.g. VC10 passenger aircraft converted to in-flight refuelling use as has happened with the Royal Air Force. Similarly, in the civil arena, many former passenger aircraft are being converted to the cargo role.

11.11.9 Development Programme

So far, the processes, methods and techniques used during aircraft system design have been described. However these need to applied and controlled within an overall programme management framework. Figure 11.20 shows the major milestones associated with the aircraft systems development process. It is assumed – as is the case for the majority of aircraft systems developed

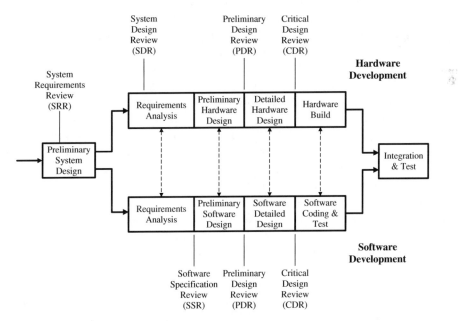

Figure 11.20 Typical development programme

today – that the system has electronics associated with the control function and that the electronics has a software development content.

The main characteristic of the development is the bifurcation of hardware and software development processes into two separate paths though it can be seen that there is considerable interaction between the two. The key steps in the avionics development programme which are primarily designed to contain and mitigate against risk are:

- *System Requirements Review (SRR).* The SRR is the first top-level, multi-disciplinary review of the perceived system requirements. It is effectively a sanity check upon what the system is required to achieve; a top-level overview of requirements and review against the original objectives. Successful attainment of this milestone leads to a preliminary system design leading in turn to the parallel development of the hardware and software requirements analysis, albeit with significant coordination between the two
- *System Design Review (SDR).* The hardware SDR immediately follows the preliminary design phase and will encompass a top-level review of the system hardware characteristics such that preliminary design may proceed with confidence. Key hardware characteristics will be reviewed at this stage to ensure that there are no major mismatches between the system require-ments and what the hardware is capable of supporting
- *Software Specification Review (SSR).* The SSR is essentially a similar process to the hardware SDR but applying to the software when a better appreciation of the software requirements has become apparent and possibly embracing any limitations such as throughput, timing or memory which the adopted hardware solution may impose. Both the SDR and SSR allow the preliminary design to be developed up to the Preliminary Design Review (PDR)
- *Preliminary Design Review (PDR).* The preliminary design review process is the first detailed review of the initial design (both hardware and soft-ware) versus the derived requirements. This is usually the last review before committing major design resource to the detailed design process. This stage in the design process is the last before major commitment to providing the necessary programme resources and investment
- *Critical Design Review (CDR).* By the time of the CDR major effort will have been committed to the programme design effort. The CDR offers the possi-bility of identifying final design flaws, or more likely, trading the risks of one implementation path versus another. The CDR represents the last opportunity to review and alter the direction of the design before very large commitments and final design decisions are taken. Major changes in system design – both hardware and software – after the CDR will be very costly in terms of cost and schedule loss, to the total detriment of the programme

The final stages following CDR will realise the hardware build and software coding and test processes which bring together the hardware and software into the eventual product realisation. Even following system validation and equipment certification it is unusual for there to be a period free of modification either at this stage or later in service when airlines may demand equipment changes for performance, reliability or maintainability reasons.

This formal process – originally invoked for the development of avionic equipment – has been extended to aircraft systems at the systems and subsystems level, particularly where much of the system functionality resides in computers and software (avionics).

11.11.10 'V' Diagram

The rigours of software development are particularly strict and are dictated by DO-178B Software Considerations in Airborne systems and Equipment Certification [6]. For obvious reasons, the level of criticality of software used in avionics systems determines the rigour applied to the development process. Reference [6] also defines three levels of software:

- Level 1: Used in critical systems application and subject to the greatest levels of control in terms of methodology: quality, design, test, certification, tools and documentation
- Level 2: Used for essential applications with standards comparable to Level 1 but less stringent in terms of test and documentation
- Level 3: Used in non-essential applications and with less stringent standards generally equivalent to a good standard of commercial software

The software development process is generally of the form shown in Figure 11.21 which shows the development activities evolving down the right of the diagram and the verification activities down the left. This shows how the activities eventually converge in the software validation test at the foot of the diagram that is the confluence of hardware and software design and development activities. Down the centre of the diagram the various development software stages are shown. It can be seen that there is considerable interaction between all the processes that represent the validation of the requirements and of the hardware and software design at each level. Any problems or issues discovered during the software validation tests are fed back up the chain, if necessary back into the top level. Therefore any minor deviations are reflected back into all the requirements stages to maintain a consistent documentation set and a consistent hardware and software design.

Whereas the earlier stages of software development and test might be hosted in a synthetic software environment it is increasingly important as testing proceeds to undertake testing in a representative hardware environment. This testing represents the culmination of functional testing for the LRU or equipment short of flight test.

While the 'V' diagram as described here is generally relating to a subsystem, it is also used in an overall systems sense where a family of such diagrams for components may be considered to represent the foundation of an overall

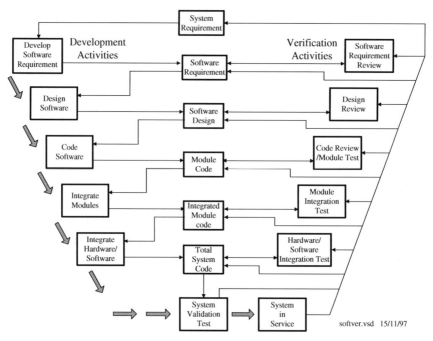

Figure 11.21 'V' diagram

system. Each component is therefore proved as fit for purpose as a component within the overall system.

11.12 Extended Operations (ETOPS)

Extended Operations (ETOPS) of multi-engine aircraft was introduced in response to calls for the relaxation of operations of two-engine aircraft allowing them to be operated further from diversion airports than had previously been allowed. The driver for this desire was the considerable increase in reliability of later generation turbo-fan engines with the earlier propeller and jet engines. Starting in 1985 the FAA introduced advisory circulars (AC 120-42, 6 June 1985; AC 120-40A, 30 December 1988) that provided guidance for the operation of air carrier two-engine aircraft certified under FAR Part 121. These documents introduced the term ETOPS and introduced guidance for these extended operations including the consideration of aircraft and engine design, maintenance programmes and flight operations. Under this guidance two-engine aircraft are allowed to fly up to 180 minutes from an airport suitable to receive the aircraft provided necessary criteria are met. It has been estimated that the number of ETOPS operations have increased from 1000 per month in 1985 to in excess of 1000 per day in 2004. In the meantime engine reliability as measured by the In-Flight Shut-Down (IFSD) has reduced to less than half that experienced in the mid-1980s.

The guidance document utilises a relative risk model to demonstrate the expansion of the maximum diversion time for up to 180 minutes. The major premise is based upon the aircraft-engine combination maintaining a target IFSD at or below 0.02 per 1000 engine hours which the model shows allows safe ETOPS flight for a 180 minute diversion. An applicant seeking approval who is able to demonstrate the 0.02 per 1000 hour IFSD rate will be granted approval. Applicants may also get approval for 120 minute ETOPS in the candidate aircraft-engine combination is ⌐0.05 per 1,000 engine hours. In this situation the FAA conducts an analysis of in flight shut-downs and determines the corrective actions required of the operator to bring the rate down to the 0.02 level. A 12 month successful operating period will lead to granting of 180 minute approval as described below. The IFSD is measured as a rolling or moving average over a 12 month period. In more recent regulatory discussions the application of an IFSD of 0.01 per 1000 engine hours has been under review.

The existing ETOPS requirements have subsequently been evolved and refined. The operational approval for the granting of ETOPS to an operator depends upon the demonstrated in-service experience of the operator. The operator also has to demonstrate its ability to maintain and operate the aircraft so as to achieve the necessary reliability and to train its personnel to achieve competence in ETOPS. The current regulations as promulgated by the FAA may be summarised as follows:

- Unless otherwise authorised, the operation of two-engine aircraft is limited to 60 minutes flying from an adequate airport. The range to the airport is defined by the single engine cruise speed in still air. This is the baseline operating condition
- 75 minute ETOPS – no minimum level of experience required
- 120 minute ETOPS – 12 consecutive months of operational experience with the aircraft-engine combination listed in its application
- 180 minute ETOPS – 12 consecutive months of operational experience at 120 minute ETOPS with the aircraft-engine combination listed in its application
- 207 minute ETOPS – Hold current approval for 180 minutes ETOPS
  In certain cases 207 minutes' approval has in the past been granted for use on specific routes for nominated operators in the North Pacific. 207 minutes represents the mandated 180 minutes plus 15% or 180 + 27 minutes.

It can be seen that the factors affecting ETOPS clearance depends in part upon the design and technology utilised in the aircraft and engine systems, i.e. the aircraft-engine combination. Modern engines are demonstrating IFSD rates of 0.01 per 1,000 engine hours. However operator maintenance practices affecting system reliability and operating practices also have a significant effect. Operators with larger fleets of the same aircraft-engine model have an advantage as they can more quickly generate the in-service engine hours necessary to secure more stringent ETOPS approvals. For example a large airline operator with a fleet of 50 aircraft flying 4000 hours per year would establish an annual useage of ⌐400 000 engine hours; 4 IFSDs or less in a 12 month operating period would demonstrate an IFSD of < 0.01 per 1000 engine hours. A smaller airline with 10 aircraft flying similar rates would only accumulate 80 000 engine hours and a single IFSD in a 12 month operating period would reduce the IFSD to 0.0125 per 1000 engine hours. Furthermore this rate would persist for a year until the offending incident dropped out of the moving average data set.

More recently, following considerable discussion, the FAA has issued regulations permitting 240 minutes ETOPS for specific geographical areas such as polar routes, with the notable exception of certain South polar regions. These flights in the most severe operating conditions place demands not only upon the aircraft-engine combination but upon other systems such as fuel; ECS and pressurisation, cargo fire hold suppression, oxygen and others. Another requirement is for aircraft to be fitted with SATCOM when operating for more than 180 minutes to ensure that the flight crew can remain in contact with air traffic control throughout the ETOPS segment.

A very recent FAA publication on the final rule following extensive ETOPS review may be found at reference [9].

References

[1] ARP4761 Guidelines and Methods for Conducting the Safety Assessment Process on Civil Airborne Systems.
[2] ARP4754 Certification Considerations for Highly-Integrated or Complex Aircraft Systems, 1996.
[3] AC 25.1309-1A System Design and Analysis, Advisory Circular, 1998.
[4] AMJ 25.1309 System Design and Analysis, Advisory Material Joint, 1994.
[5] ATA-100 ATA Specification for Manufacturer's Technical Data.
[6] DO-178B Software Considerations in Airborne systems and Equipment Certification.
[7] DO-254Design Assurance Guidance for Airborne Electronic Hardware.
[8] Dual use of Variable Speed Constant Frequency (VSCF) cyclo-converter technology, V Bonneau, FITEC 98.
[9] Federal Register Vol. 72, No. 9, Tuesday, 16 January 2007, Rules and Regulations- RIN2120 – A103, Extended Operations (ETOPS) of Multi-Engine Airplanes.

12

Avionics Technology

12.1 Introduction

The first major impetus for use of electronics in aviation occurred during World
War II. Communications were maturing and the development of airborne radar
using the magnetron and associated technology occurred at a furious pace
throughout the conflict [1].

Transistors followed in the late 1950s and 1960s and supplanted thermionic
valves for many applications. The improved cost-effectiveness of transistors
led to the development of digital aircraft systems throughout the 1960s and
1970s, initially in the military combat aircraft where it was used for Nav/Attack
systems. See Figure 12.1.

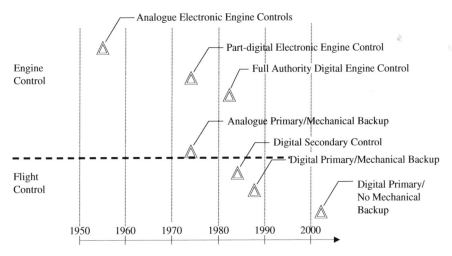

Figure 12.1 Major electronics developments in aviation since 1930

Aircraft Systems: Mechanical, electrical, and avionics subsystems integration. 3rd Edition   Ian Moir and Allan Seabridge
© 2008 John Wiley & Sons, Ltd

For many years the application of electronics to airborne systems was limited to analogue devices and systems with signal levels and voltages generally being related in some linear or predictive way. This type of system was generally prone to heat soak, drift and other nonlinearities. The principles of digital computing had been understood for a number of years before the techniques were applied to aircraft. The development of thermionic valves enabled digital computing to be accomplished but at the expense of vast amounts of hardware. During the World War II a code-breaking machine called Colossus employed thermionic valves on a large scale. The machine was physically enormous and quite impracticable for use in any airborne application.

The first aircraft to be developed in the US using digital techniques was the North American A-5 Vigilante, a US Navy carrier-borne bomber which became operational in the 1960s. The first aircraft to be developed in the UK intended to use digital techniques on any meaningful scale was the ill-fated TSR 2 which was cancelled by the UK Government in 1965. The technology employed by the TRS 2 was largely based upon solid-state transistors, then in comparative infancy. In the UK, it was not until the development of the Anglo-French Jaguar and the Hawker Siddeley Nimrod in the 1960s that weapon systems began to seriously embody digital computing for use in any airborne application, albeit on a meagre scale compared to the 1980s.

Since the late 1970s/early 1980s digital technology has become increasingly used in the control of aircraft systems as well as just for mission related systems. A key driver in this application has been the availability of cost-effective digital data buses such as ARINC 429, MIL-STD-1553B and ARINC 629. This technology, coupled with the availability of cheap microprocessors and more advanced software development tools, has lead to the widespread application of avionics technology throughout the aircraft. This has advanced to the point that virtually no aircraft system – including the toilet system – has been left untouched.

The evolution and increasing use of avionics technology for civil applications of engine controls and flight controls since the 1950s is shown in Figure 12.2.

Engine analogue controls were introduced by Ultra in the 1950s which comprised electrical throttle signalling used on aircraft such as the Bristol Britannia. Full-Authority Digital Engine Control became commonly used in the 1980s.

Digital primary flight control with a mechanical backup has been used on the Airbus A320 family, A330/A340 using side-stick controllers and on the B777 using a conventional control yoke. Aircraft such as the Airbus A380 and Boeing 787 appear to be adopting flight control without any mechanical backup but with electrically signalled backup.

The application of digital techniques to other aircraft systems – utilities systems – began later as will be described in this chapter. Today, avionics technology is firmly embedded in the control of virtually all aircraft systems. Therefore an understanding of the nature of avionics technology is crucial in understanding how the control of aircraft systems is achieved.

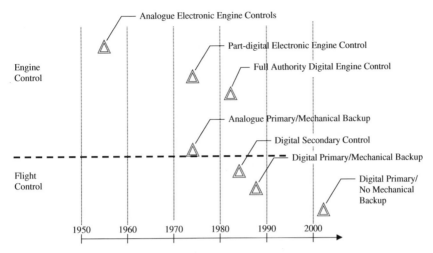

Figure 12.2 Evolution of electronics in flight and engine control

12.2 The Nature of Microelectronic Devices

The development of a wholly digital system control system has to accommodate interfaces with the 'real world' which is analogue in nature. The figure
shows how the range of microelectronic devices is used in different applications within a digital system.

Hybrid chips and Input/Output (I/O) Application Specific Integrated
Circuits (ASICs) are key technologies associated with interfacing to the
analogue world. A/D and D/A devices undertake the conversion from
analogue to digital and digital to analogue signals respectively. Processor and
memory devices, together with digital ASICS perform the digital processing
tasks associated with the application. See Figure 12.3.

Microelectronic devices are produced from a series of masks that shield
various parts of the semiconductor during various processing stages. The
resolution of most technology is of the order of 1–3 microns (1 micron is
10^{-6} metres or one-millionth of a metre, or one-thousandth of a millimetre)
so the physical attributes are very minute. Thus a device or die about
0.4 inches square could have hundreds of thousands of transistors/gates
to produce the functionality required of the chip. Devices are produced
many at a time on a large circular semiconductor wafer, some devices at
the periphery of the wafer will be incomplete and some of the remaining
devices may be flawed and defective. However, the remainder of the good
die may be trimmed to size, tested and mounted within the device package.
The size of the die, complexity and maturity of the overall semiconductor
process and the quality of the material will determine the number of good
die yielded by the wafer and this yield will eventually reflect in the cost
and availability of the particular device. Therefore the overall maturity
of a particular device will be decided by these factors which determines

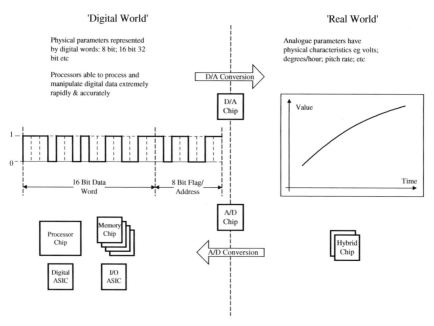

Figure 12.3 The nature of microelectronic devices

the cost-effectiveness of the application. Therefore standard devices used for the interfacing of ARINC 429 and MIL-STD-1553B have had very wide industry application and acceptance; others have not been so successful. See Figure 12.4.

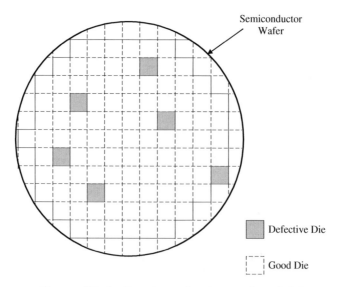

Figure 12.4 Semiconductor wafer yield

Microelectronics devices are environmentally screened according to the severity of the intended application; usually three levels of screening are applied, in increasing levels of test severity:

- Commercial Grade
- Industrial Grade
- Aerospace Military Grade – also used in many cases for civil aerospace applications

There is little doubt that this screening technique has helped to improve the maturity of the manufacturing process and quality of the devices in the past. However as an increasingly small proportion of devices overall are used for aerospace applications, full military screening is difficult to assure for all devices. There is a body of opinion that believes that screening is not beneficial, and adds only to the cost of the device. It is likely that avionics vendors will have to take more responsibility for the quality of devices used in their product in future. There is an increasing and accelerating trend for aerospace micro-electronics to be driven by the computer and telecommunications industries.

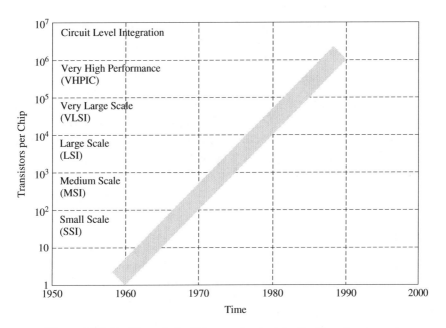

Figure 12.5 Trends in integrated circuit development

The extent of the explosion in IC developments can be judged by reference to Figure 12.5. This shows a tenfold increase per decade in the number of transistors per chip. Another factor to consider is the increase in the speed of device switching. The speed of operation is referred to as gate delay; gate delay for a thermionic valve is of the order of 1000 nanoseconds (1 nanosecond is 10^{-9} or one-thousandth of one-millionth of a second); transistors are about ten times quicker at 100 nanoseconds. Silicon chips are faster again at ~ 1 nanosecond). This gives an indication of how powerful these devices are and why they have had such an impact upon our daily life.

Another area of major impact for the IC relates to power consumption.ICs consume minuscule amounts. Consumption is related to the technology type and speed of operation. The quicker the speed of operation then the greater the power required and vice-versa. The main areas where avionics component technology have developed are:

- Processors
- Memory
- Data buses

12.2.1 Processors

Digital processor devices became available in the early 1970s as 4-bit devices. By the late 1970s 8-bit processors had been superseded by 16-bit devices; these led in turn to 32-bit devices such as the Motorola 68000 which have been widely used on the European Fighter Aircraft and Boeing 777. The pace of evolution of processor devices does present a significant concern due to the risk of the chips becoming obsolescent, leading to the prospect of an expensive redesign.

Following adverse experiences with its initial ownership of microprocessor based systems, the US Air Force pressed strong standardisation initiatives based upon the MIL-STD-1750A microprocessor with a standardised Instruction Set Architecture (ISA) though this found few applications in aircraft systems computing. For these types of application, starting with the adoption of the Motorola 68020 on the prototype version of Typhoon, the industry is making extensive use of commercial developed microprocessor or microcontroller products.

12.2.2 Memory Devices

Memory devices have experienced a similar explosion in capability. Memory devices comprise two main categories: Read Only Memory (ROM) represents the memory used to host the application software for a particular function; as the term suggests this type of memory may only be read but not written to. A particular version of ROM used frequently was Electrically Programmable Read Only Memory (EPROM); however, this suffered the disadvantage that memory could only be erased by irradiating the device with Ultra-Violet (UV) light. For the last few years EPROM has been superceded by the more user-friendly Electrically Erasable Programmable Read Only Memory (E^2PROM). This type of memory may be reprogrammed electrically with the memory module still resident within the LRU, using this capability it is now possible to reprogram many units in-situ on the aircraft via the aircraft digital data buses.

Random Access Memory (RAM) is read-write memory that is used as program working memory storing variable data. Early versions required a power backup in case the aircraft power supply was lost. More recent devices are less demanding in this regard.

12.2.3 Digital Data Buses

The advent of standard digital data buses began in 1974 with the specification by the US Air Force of MIL-STD-1553. The ARINC 429 data bus became the first standard data bus to be specified and widely used for civil aircraft being widely used on the Boeing 757 and 767 and Airbus A300/A310 in the late 1970s and early 1980s. ARINC 429 (A429) is widely used on a range of civil aircraft today as will become apparent during this chapter. In the early 1980s Boeing developed a more capable digital data bus termed Digital Autonomous Terminal Access Communication (DATAC) which later became an ARINC standard as A629; the Boeing 777 is the first and at present the only aircraft to use this more capable data bus. At the same time, these advances in digital data bus technology were matched by advancements in processor, memory and other microelectronic devices such as analogue-to-digital and digital-to-analogue devices, logic devices etc. which made the application of digital technology to aircraft systems possible.

The largest single impact of microelectronics on avionic systems has been the introduction of standardised digital data buses to greatly improve the intercommunication between aircraft systems. Previously, large amounts of aircraft wiring were required to connect each signal with the other equipment. As systems became more complex and more integrated so this problem was aggravated. Digital data transmission techniques use links which send streams of digital data between equipment. These data links comprise only two or four twisted wires and therefore the interconnecting wiring is greatly reduced.

Common types of digital data transmission are:

- *Single Source – Single Sink*. This is the earliest application and comprises a dedicated link from one equipment to another. This was developed in the 1970s for use on Tornado and Sea Harrier avionics systems. This technique is not used for the integration of aircraft systems
- *Single Source – Multiple Sink*. This describes a technique where one transmitting equipment can send data to a number of recipient equipment (sinks). ARINC 429 is an example of this data bus which is widely used by civil transports and business jets
- *Multiple Source – Multiple Sink*. In this system multiple transmitting sources may transmit data to multiple receivers. This is known as a full-duplex system and is widely employed by military users (MIL-STD-1553B) and by the B777 (ARINC 629)

The use of digital data buses to integrated aircraft systems has increased enormously over the last few years. A huge impetus has resulted from the introduction of Commercial-off-the-Shelf (COTS) technology – adopting those buses that have been designed for the computer and telecommunications industries. This technology has been adopted for reasons of cost, speed, component obsolesence and availability though care must be exercised to ensure that deterministic variants are selected for aerospace applications. Figure 12.6 lists most if not all of the data buses used on aircraft today in ascending order of data transmission.

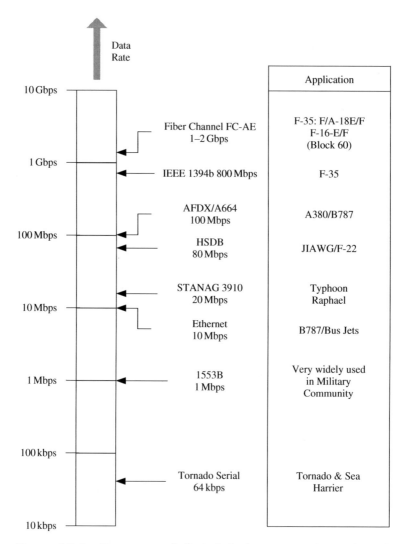

Figure 12.6   Common digital data buses used on aircraft

In the area of aircraft systems integration the most recent to be adopted are IEEE 1394b on the Joint Strike Fighter (JSF)/F-35 and AFDX/ARINC 664 on the Airbus A380 and Boeing 787. These system architectures will be outlined later in this chapter. The following buses will be briefly described as they represent those most commonly used of being introduced for large-scale aircraft systems integration:

- ARINC 429
- MIL-STD-1553B
- ARINC 629
- AFDX/ARINC 664
- IEEE 1394b

More detailed information on data rates, protocol and transmission media options may be found in the appropriate specifications.

12.2.4 A 429 Data Bus

The characteristics of ARINC 429 were agreed among the airlines in 1977/78 and the data bus first used on the B757/B767 and Airbus A300 and A310 aircraft. ARINC, short for Aeronautical Radio Inc. is a corporation in the US whose stakeholders comprise US and foreign airlines, and aircraft manufacturers. As such it is a powerful organisation central to the specification of equipment standards for known and perceived technical requirements.

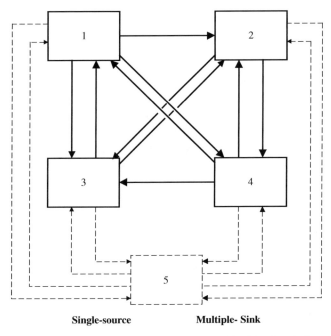

Single-source   **Multiple- Sink**

Figure 12.7 A429 topology and the effect of adding units

The ARINC 429 (A429) bus operates in a single-source-multiple sink mode which is a source that may transmit to a number of different terminals or sinks, each of which may receive the data message. However if any of the sink equipment need to reply then they will each require their own transmitter to do so, and cannot reply down the same wire pair. This half-duplex mode of operation has certain disadvantages. If it is desired to add additional equipment as shown in Figure 12.7, a new set of buses may be required – up to a maximum of eight new buses in this example if each new link has to possess bi-directional qualities.

A429 is by far the most common data bus in use on civil transport aircraft, regional jets and executive business jets today. Since introduction on the Boeing 757/767 and Airbus aircraft in the early 1980s hardly an aircraft has been produced which does not utilise this data bus.

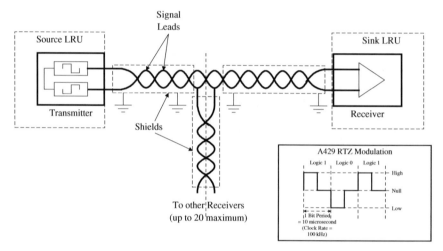

Figure 12.8 A429 data bus and encoding format

The physical implementation of the A429 data bus is a screened, twisted wire pair with the screen earthed at both ends and at all intermediate breaks. The transmitting element shown on the left in Figure 12.8 is embedded in the source equipment and may interface with up to 20 receiving terminals in the sink equipment. Information may be transmitted at a low rate of 12–14 Kbytes per second or a higher rate of 100 kbits per second; the higher rate is by far the most commonly used. The modulation technique is bipolar Return to Zero (RTZ) as shown in the box in Figure 12.7. The RTZ modulation technique has three signal levels: high, null and low. A logic state 1 is represented by a high state returning to zero; a logic state 0 is represented by a low state returning to null. Information is transmitted down the bus as 32 bit words as shown in Figure 12.9.

The standard embraces many fixed labels and formats so that a particular type of equipment always transmits data in a particular way. This standardisation has the advantage that all manufacturers of particular equipment know what data to expect. Where necessary, additions to the standard may also be implemented. Further reading for A429 may be found at references [2], [3] and [4].

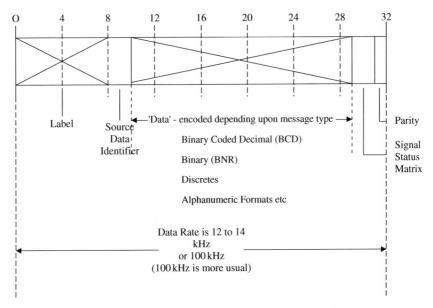

Figure 12.9   A429 data word format

12.2.5 MIL-STD-1553B

MIL-STD-1553B has evolved since the original publication of MIL-STD-1553 in 1973. The standard has developed through 1553A standard issued in 1975 to the present 1553B standard issued in September 1978. The basic layout of a MIL-STD-1553B data bus is shown in Figure 12.10. The data bus comprises a

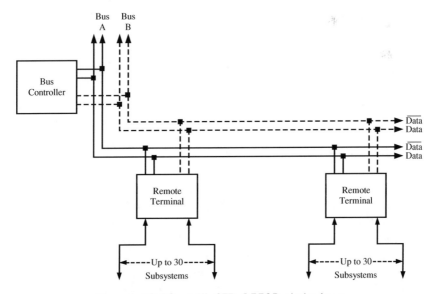

Figure 12.10   MIL-STD-1553B data bus

twin twisted wire pair along which DATA and DATA complement are passed. The standard generally allows for dual-redundant operation.

Control of the bus is executed by a Bus Controller (BC) which is connected to a number of Remote Terminals (RTs) (up to a maximum of 31) via the data bus. RTs may be processors in their own right or may interface a number of subsystems (up to a maximum of 30) with the data bus. Data is transmitted at 1 MHz using a self-clocked Manchester bi-phase digital format. The transmission of data in true and complement form down a twisted screened pair, together with a message error correction capability, offers a digital data link which is highly resistant to message corruption. Words may be formatted as data words, command words or status words as shown in Figure 12.11. Data words encompass a 16-bit digital word while the command and status words are associated with the data bus transmission protocol. Command and status words are compartmented to include various address, sub-address and control functions as shown in the figure.

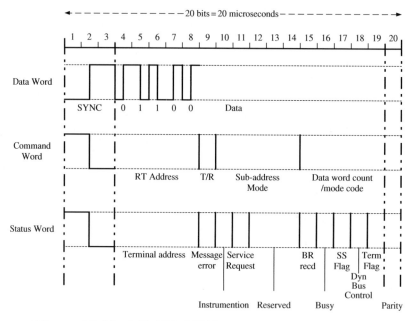

Figure 12.11   MIL-STD-1553B data bus word formats

MIL-STD-1553B is a command-response system in which all transmissions are conducted under the control of the bus controller; although only one bus controller is shown in these examples a practical system will employ two bus controllers to provide control redundancy.

Two typical transactions are shown in Figure 12.12. In a simple transfer of data from RT A to the BC, the BC sends a transmit to RT A, which replies after a short interval with a status word, followed immediately by one or

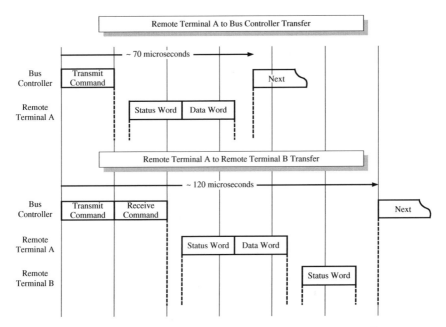

Figure 12.12   MIL-STD-1553B typical data transactions

more data words up to a maximum of 32 words. In the example shown in the upper part of Figure 12.12, transfer of one data word from the RT A to the BC will take approximately elapsed time of about 70 μseconds. For the transfer of data from between RTs as shown from RT A to RT B, the BC sends a receive word to RT B followed by a transmit word to RT A. RT A will send a status word plus a data word (up to a maximum of 32 words) to RT B which responds by sending a status word to the BC, thereby concluding the transaction. In the simple RT to RT transaction shown in Figure 12.12 the total elapsed time is around 120 μseconds for a single data word which appears to be rather expensive in overheads. However if the maximum of words had been transferred the overheads would be the same, though now representing a much lower percentage of the overall message time. For further reading on MIL-STD-1553B see reference [5].

12.2.6 ARINC 629 Data Bus

ARINC 629 (A629) – like MIL-STD-1553B – is a true data bus in that the bus operates as a Multiple-Source, Multiple Sink system (see Figure 12.13). That is, each terminal can transmit data to, and receive data from, every other terminal on the data bus. This allows much more freedom in the exchange of data between units in the avionics system than the Single-Source, Multiple Sink A429 topology. Furthermore the data rates are much higher than for A429 where the highest data rate is 100 kbits/sec. The A629 data bus operates at 2 Mbits/sec or 20 times that of A429. The true data bus topology is much more flexible in that additional units can be fairly readily accepted physically on the data bus. A further attractive feature is the ability to accommodate up to a total of 131 terminals on a data bus, though in a realistic implementation data bus traffic would probably preclude the use of this large number of terminals. The protocol utilised by A629 is a time-based, collision-avoidance concept in which each terminal is allocated a particular time slot access to transmit data on to the bus. Each terminal autonomously decides when the appropriate time slot is available and transmits the necessary data. This protocol was the civil aircraft industry's response to the military MIL-STD-1553B data bus that utilises a dedicated controller to decide what traffic passes down the data bus.

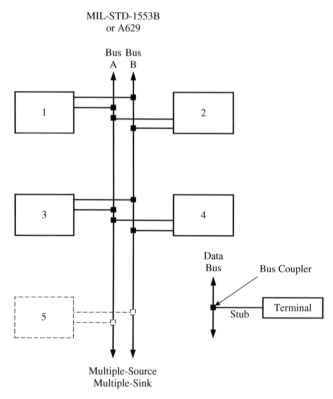

Figure 12.13   MIL-STD-1553 and A629 data bus topology

Because of the higher data rates and higher technology baseline, the A629 bus coupler arrangement is slightly more involved than for A429. Figure 12.14. shows how the host LRU connects to the A629 data bus via the Serial Interface Module (SIM), embedded in the LRU, and via a stanchion connector to the coupler itself. Due to the transmit/receive nature of the A629 protocol there are separate channels for transmit and receive. Current coupling is used due to concern that a single bus failure could bring down all the terminals associated with the data bus. Somewhat oddly, the bus couplers are all grouped in a fairly low number of locations to ease the installation issues.

Also shown within the box in Figure 12.14 is a simplified portrayal of the Manchester Bi-phase encoding which the A629 data bus (and MIL-STD-1553B) uses. In this protocol a logic 0 is signified when there is a negative to positive change of signal; this change of state occurs midway during the particular bit duration. Similarly, logic 1 is denoted when there is a positive to negative change of signal during the bit period. This timing is aided by the fact that the first three bits in a particular data word act as a means of synchronisation for the whole of the word. The data is said to be 'self-clocked' on a word-by-word basis and therefore these rapid changes of signal state may be accurately and consistently recognised with minimal risk of mis-reads.

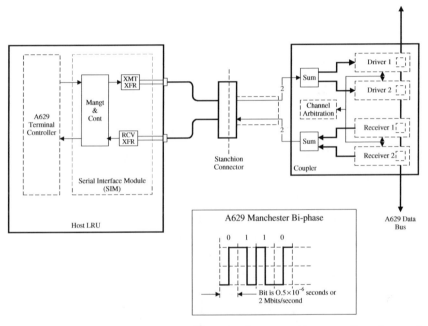

Figure 12.14   A629 bus coupler interface and encoding format

Figure 12.15 shows the typical A629 20 bit data word format. The first three bits are related to word time synchronisation as already described. The next 16 bits are data related and the final bit is a parity bit. The data words may have a variety of formats depending upon the word function; there is provision for general formats, systems status, function status, parameter validity, binary and discrete data words. Therefore although the data format is simpler than A429, the system capabilities are more advanced as the bit rate is some 20 times faster than the fastest (100 kbit/sec) option for A429.

The only aircraft utilising A629 data buses so far is the Boeing 777.The widespread application of technology such as A629 is important as more widespread application drives component prices down and makes the technology more cost-effective. Certainly that has been the case with A429 and MIL-STD-1553B.

For more detail on A629 see references [6], [7] and [8].

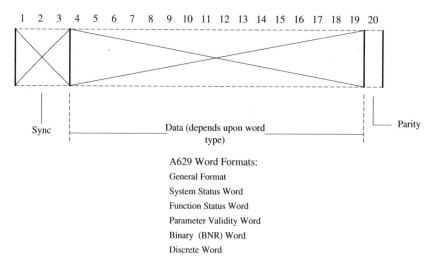

A629 Word Formats:
General Format
System Status Word
Function Status Word
Parameter Validity Word
Binary (BNR) Word
Discrete Word

Figure 12.15   A629 digital word format

12.2.7 COTS Data Buses

COTS Data Buses – AFDX

AFDX is a derivative of the commercial fast switched Ethernet standard that operates at a data transmission rate of 100 Mbits/sec that is finding extensive application in aircraft such as the Airbus A380 and Boeing 787.

AFDX may be succinctly described as:

Avionics	Network adapted to avionics constraints
Full-Duplex	Subscribers transmit and receive data at 100 Mbits/sec
Switched	Exchanges between subscribers is achieved by switching data
Ethernet	Based upon Ethernet 100 Base T(100 Mbit/sec over twisted wire pairs)

This describes the Airbus implementation; more generally this data bus standard is referred to as ARINC 664. The implementation is summarised in the Figure 12.16 shown below.

The figure shows a number of LRUs interconnected by means of two AFDX switches. In many respects the AFDX data transmission operates more like a telephone exchange compared to contemporary aerospace buses with data being switched from 'subscriber' to 'subscriber' except that the subscribers are aircraft LRUs. Data is passed half-duplex over twisted wire pairs at 100 Mbits/sec to avoid collisions or contention.

The advantage of AFDX over the A429 data buses commonly used on civil aircraft may be seen by reference to Figure 12.17.

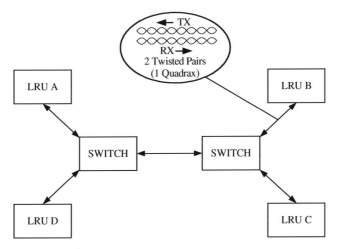

Figure 12.16 Overview of AFDX

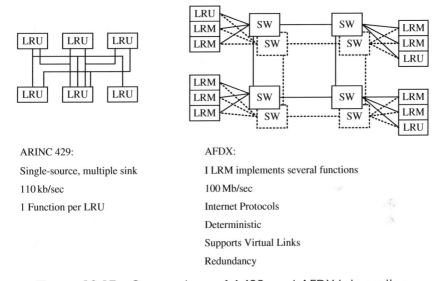

ARINC 429:

Single-source, multiple sink

110 kb/sec

1 Function per LRU

AFDX:

I LRM implements several functions

100 Mb/sec

Internet Protocols

Deterministic

Supports Virtual Links

Redundancy

Figure 12.17 Comparison of A429 and AFDX integration

A429 is a single-source multiple sink topology using a 110 kbits/sec data transmission between dedicated function LRUs.

AFDX/A664 has the following advantages:

- An Integrated Modular Avionics (IMA) architecture allows the integration of multi-function LRMs in avionics cabinets or racks
- Data rate is 100 Mbits/sec full duplex; no collision avoidance required as links are half-duplex and independent
- Internet protocols are used throughout: no dedicated aerospace implementations

- Deterministic data transfers
- Supports the concept of virtual links where multi-layered transmissions may be used
- Provides dual redundancy for fault tolerance

AFDX also permits the integration of IMA based systems with other systems that may remain as standalone LRUs for sound system engineering reasons. One may be the use of proven legacy systems for reasons of cost and risk reduction. Another may be the need to segregate high integrity systems from the IMA solution: typical systems in this category include flight control, engine control and electrical systems control.

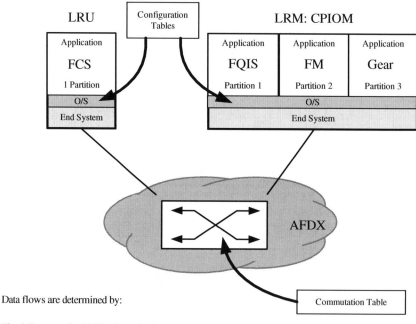

Figure 12.18 Integration of high integrity and IMA functions

Figure 12.18 shows an example how the integration of standalone or legacy subsystem and IMA resources may be achieved using AFDX. The left hand LRU is associated with the Flight Control System (FCS) which is a high integrity system and which needs to be segregated from other units. The right hand cabinet uses a partitioned IMA approach with fuel quantity indication, fuel management and gear functions. The FCS may communicate with any of these

IMA hosted functions and vice versa though the AFDX switches. The control for these data exchanges resides in tables as follows:

- Configuration tables that reside within the Operating System (OS) of the LRU or IMA Cabinet. The configuration tables dictate the formatting of the data received and donated by the LRU
- Commutation tables hosted within the AFDX switch that determine the data to be communicated by means of virtual links. The commutation tables specify the composition of the virtual links that provide the intersystem data transfers

COTS Data Buses – IEEE 1394

IEEE 1394 – often referred to as 'Firewire' – is a commercial data bus originally designed to serve the domestic electronics market, linking personal electronic items such as laptops, routers and peripherals such as scanners, printers, CADCAMs and TVs as shown. In its original form the standard was a scaleable bus capable of operating at speeds from 50 to 200 Mbits/sec. The standard has also found considerable application in In-Flight Entertainment (IFE) systems onboard civil aircraft.

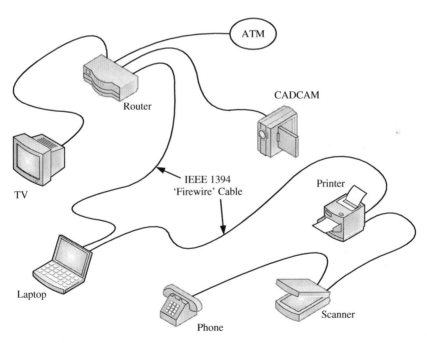

Figure 12.19 Original IEEE 1394 'Firewire' concept

The latest IEEE 1394b standard allows transmission up to 800 Mbits/sec. IEEE 1394 defines a media, topology and protocol for both a backplane physical layer implementation or a point-to-point serial cable interface. In aircraft system interconnections the latter is used. The interface is also called the High Performance Serial Bus (HPSB).

12.3 Data Bus Integration of Aircraft Systems

The increasing cost-effectiveness which system integration using digital data buses and microelectronic processing technologies offer has lead to a rapid migration of the technology into the control of aircraft systems. Several examples are described below which highlight how all-embracing this process has become:

- MIL-STD-1553B -EAP Utilities Management System
- A429 – Airbus A330/340 Aircraft Systems
- A629 – B777 aircraft Systems
- Ruggedised Ethernet 10BaseT – Business Jets
- AFDX – Airbus A380
- IEEE 1394b – F-35 Lightning II

12.3.1 Experimental Aircraft Programme (EAP)

The first aircraft to utilise MIL-STD-1553B for the integration of aircraft as opposed to avionics systems was the UK Experimental Aircraft Programme (EAP) which was a technology demonstrator fore-runner to the Eurofighter Typhoon. This aircraft first flew in August 1986 and was demonstrated at the Farnborough Air Show the same year, also being flown at the Paris Air Show the following year. This system is believed to be the first integrated system of its type; given purely to the integration of aircraft utility systems. The system encompassed the following functions:

- Engine control and indication interfacing
- Engine starting
- Fuel management and fuel gauging
- Hydraulic system control and indication, undercarriage indication and monitoring, wheel brakes
- Environmental control systems, cabin temperature control – and later an On-Board Oxygen Generating System (OBOGS)
- Secondary power system
- LOX contents, electrical generation and battery monitoring, probe heating, emergency power unit

The system comprised four LRUs – Systems Management Processors (SMPs) – which also housed the power switching devices associated with operating motorised valves, solenoid valves etc. These four units comprised a set of common modules or building blocks replaced a total of 20–25 dedicated controllers and six power switching relay units which a conventional system would use. The system comprised several novel features; offering a level of integration which has not been equalled to date. See Figure 12.15.

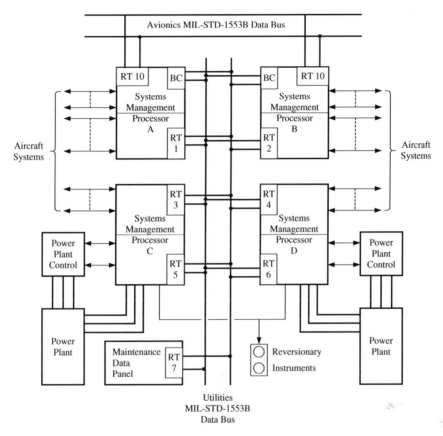

Figure 12.20 EAP utilities system architecture

The technology and techniques applied to aircraft utilities systems demonstrated on EAP have been used successfully on Eurofighter Typhoon and Nimrod MR4. The lessons learned on each aircraft have been passed on through the generations of utilities management systems, and are now being used on a number of aircraft projects under the heading of Vehicle Management Systems.

12.3.2 Airbus A330/340

The two-engined A330 and four-engined A340 make extensive use of A429 data buses to integrated aircraft systems control units which each other and with the avionics and displays. Table 12.1 lists some of the major subsystems and control units.

Table 12.1   A330/A340 typical aircraft system controllers

Control Unit	A330	A340	Remarks
Bleed air control	2	4	One per engine
Fuel control	2	2	
Landing gear control	2	2	
Flight control:			
– Flight control primary computer	2	2	
– Flight control secondary computer	2	2	
– Flight control data concentrator	2	2	
– Slat/flap control computer	2	2	
Probe heat	3	3	
Zone controller	1	1	
Window heat control	2	2	
Cabin pressure control	2	2	
Pack controller	2	2	
Avionics ventilation computer	1	1	
Generator control unit	2	4	One per engine
Full-Authority Digital Engine Control (FADEC)	2	4	One per engine
Flight warning computer	2	2	
Central maintenance computer	2	2	
Hydraulic control	1	1	

12.3.3 Boeing 777

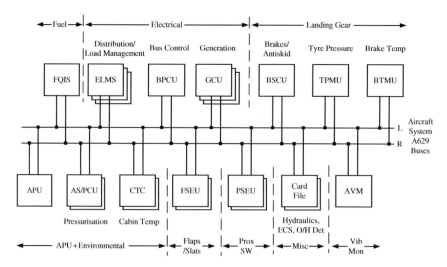

Figure 12.21   B777 Aircraft systems integration using A629 data buses

The B777 makes extensive use of the A629 digital data bus to integrate the avionics, flight controls and aircraft systems. Figure 12.21 depicts a simplified version of a number of B777 aircraft systems which are integrated using the aircraft system A629 buses. Most equipment is connected to the left and right aircraft system buses but some are also connected to a centre bus. Exceptionally, the engine Electronic Engine Controllers (EECs) are connected to left, right, centre 1 and centre 2 buses to give true dual-dual interface to the engines. The systems so connected embrace the following:

- Fuel
- Electrical:

  - electrical load management
  - bus control
  - generation control

- Landing gear

  - brakes and anti-skid
  - tyre pressure monitoring
  - brake temperature monitoring

- APU and environmental control:

  - APU controller
  - air pressure and pressurisation control
  - cabin temperature control

- Flaps and slats
- Proximity switches
- Card files: Boeing produced modules used for the management of hydraulics, overheat detection, environmental and other functions
- Vibration monitoring

12.3.4 Regional Aircraft/Business Jets

The foregoing examples relate to fighter and civil transport aircraft. The development of regional aircraft and business jet integrated avionics systems is rapidly expanding to include the aircraft utilities functions. The Honeywell EPIC system as being developed for the Hawker Horizon, Dornier 728 family and Embraer ERJ-170/190 is an example of how higher levels of system integration have been achieved. See Figure 12.22.

This is a much more closed architecture than the ones already described which utilise open, internationally agreed standards. This architecture uses a proprietary Avionics Standard Communication Bus (ASCB) – Variant D data bus developed exclusively by Honeywell, originally for General Aviation (GA) applications. Previous users of ASCB have been Cessna Citation, Dassault Falcon 900, DeHavilland Dash 8 and Gulfstream GIV.

The key characteristics of ASCB D are:

- Dual data bus architecture
- 10 Mhz bit rate – effectively a hundred times faster than the fastest A429 rate (100 kbits/sec)
- Up to 48 terminals may be supported
- The architecture has been certified for flight critical applications

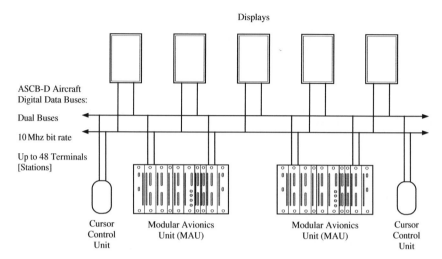

Figure 12.22 Honeywell EPIC system – typical

This example shows two Modular Avionics Units (MAUs); however, it is more typical to use four such units to host all the avionics and utilities functions. It can be seen from this example that the ambitions of Honeywell in wishing to maximise the return on their EPIC investment is driving the levels of system integration in the regional aircraft and business jets to higher levels than the major OEMs such as Boeing and Airbus. A publication which addresses ASCB and compares it with other data buses is at reference [9]. This reference contains much useful data regarding the certification of data bus systems. For an example of a fuel system integrated into the EPIC system – see the paper at reference [10] which describes the integration of the Hawker Horizon fuel system into the Honeywell EPIC system.

12.3.5 A380 Avionics Architecture

The A380 is the first civil aircraft to make large scale use of COTS technology for the integration of avionics and aircraft systems. The A380 utilises the 100 Mbits/sec AFDX as the central communications spine; though aerospace buses such as A429 and COTS buses such as CANbus are also used. The A380 architecture is divided into a number of functional domains supporting the displays suite; the key elements are:

- Displays suite – 8 × colour glass panel displays
- Cockpit domain integrated cabinets
- Cabin domain integrated cabinets
- Energy domain integrated cabinets
- Utilities domain integrated cabinets

These domains are interconnected by a dual redundant AFDX switch network that provides a high capacity data transmission system across the aircraft. See Figure 12.23.

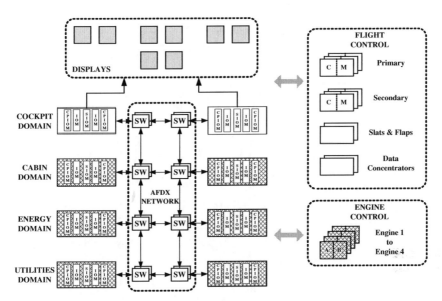

Figure 12.23 A380 top-level avionics architecture

The aircraft systems functions are hosted in the following domains:

- *Cabin domain:*

    - overheat detection system
    - supplementary cooling system
    - engine bleed air system
    - pneumatic air distribution system
    - ventilation control system
    - avionic ventilation control system
    - cabin pressure control system
    - temperature control system

These systems are the responsibility of Airbus Gmbh who also have responsibility for the integration of the cabin domain integrated cabinets.

- *Energy domain:*

    - hydraulics control system
    - hydraulics monitoring
    - primary electrical power
    - secondary electrical power

These systems are the responsibility of Airbus France who also have responsibility for the integration of the energy domain integrated cabinets.

- *Utilities domain*:

  - fuel quantity indication system
  - fuel management
  - landing gear monitoring system
  - landing gear extension and retraction system
  - brake control system
  - steering control system

These systems are the responsibility of Airbus UK who also have responsibility for the integration of the utilities domain integrated cabinets.

The split between the AFDX network and IMA computing modules comprising the core of the avionics system is shown is Figure 12.24.

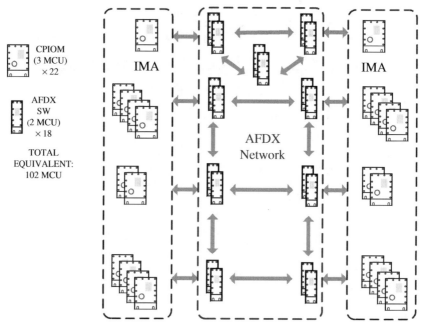

Figure 12.24   A380 AFDX network core

The AFDX network comprises 18 AFDX switch units of 2 MCU size.

The core computing function is provided by a total of 22 Common Processor Input Output Modules (CPIOMs) of 3 MCU size. The CPIOM hardware set is comprised of a total of seven different types that are distributed among the functional domains as follows:

- CPIOM A – Cabin [4]
- CPIOM B – Cabin [4]

- CPIOM C – Displays and Energy [2]
- CPIOM D – Displays and Energy [2]
- CPIOM E – Energy, dealing with electrical system [2]
- CPIOM F – Utilities, dealing with fuel [4]
- CPIOM F – Utilities, dealing with landing gear [4]

Although the CPIOMs are subtly different they share a common set of design and support tools and have a similar form factor. The primary difference between them is in the Input/Output (I/O) configuration that differs between different aircraft subsystems.

Specific high integrity systems are implemented outside the common avionics core but interface with it. These systems are partitioned for reasons of integrity and typical examples are:

- Generator control units
- Flight control system
- Main engine FADECs

The advantages of this approach are:

- Common set of core modules used across several functional domains
- Standardised processing elements
- Common use of software tools, standards and languages
- Dispenses with a multitude of dedicated and specialised LRUs though some LRUs remain as standalone
- Ability to accommodate specialised interfaces
- Benefits of scale across entire aircraft
- Improved logistics for OEM and airlines
- Scaleable architecture with scope for implementation across entire model range for future programmes. The A380 AFDX/IMA approach has already been adopted for A400M and may conceivably also be used for the A350XWB

12.3.6 Boeing 787 Avionics Architecture

The Boeing 787 aircraft which is a follow-on to the Boeing 757/767 family has also adopted 100 Mbit/sec A664 as the data transmission media for the avionics 'spine' of the aircraft. While similar to the Airbus A380 in terms of selecting a COTS derivative data bus, Boeing has chosen a different architecture to provide the integration of avionic and aircraft functions.

The Boeing 787 architecture is shown in Figure 12.25 above. The main avionics system and computation tasks are undertaken by two Common Computing Resource (CCR) cabinets. These cabinets interface with the flight deck and the rest of the avionics and aircraft systems and embrace the functions undertaken by the Airplane Information Management System (AIMS) on the Boeing 777. Each contains four general processing modules, network

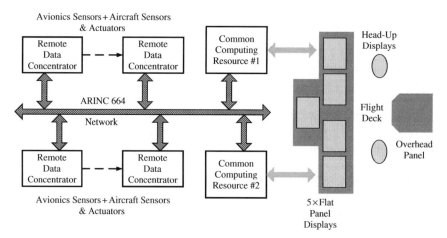

Figure 12.25 Boeing 787 top-level avionics architecture

switches and two fibre-optic translation modules. Application specific modules provided by third party suppliers may also be accommodated – an example of these is the display processor modules supplied by Rockwell Collins.

The Boeing approach differs in utilising ~20 Remote Data Concentrator (RDC) units situated throughout the aircraft to perform a data gathering function for analogue, discrete signals and serial digital data from remote avionics and aircraft systems sensors and effectors. In addition to the RDCs, there are ~20 Remote Power Distribution Units (RPDUs) to distribute electrical power locally to the aircraft electrical loads. The approach is therefore to distribute the sensing and control loops and electrical power distribution. In some cases dedicated single function LRUs are still used for functions such as the electrical system Generator Control Units (GCUs).

The dual redundant Central Data Network (CDN) uses deterministic A664 capable of supporting both copper and fibre-optic interfaces with connection speeds of 10 Mbit/sec and 100 Mbit/sec respectively.

12.3.7 COTS Data Buses – IEEE 1394

IEEE 1394 as outlined previously defines a media, topology and protocol for both a backplane physical layer or point-to-point serial cable interface; in aircraft system interconnections the latter is used. The interface is also called the High Performance Serial Bus (HPSB).

The cable (differential) version operates at 100 Mbits/sec, 200 Mbits/sec, or 400 Mbits/sec; this increases to 800 Mbits/sec for 1394b. The baseline 1394 uses half-duplex or unidirectional transmission whereas 1394b is capable of full duplex or bi-directional transmission. The latest IEEE 1394b standard also allows transmission up to 800 Mbits/sec.

The bus supports up to 63 devices at a maximum cable distance between devices of 4.5 metres. When the number of devices on the bus is limited to 16, a maximum cable distance of 72 metres is possible. When 1394a is transmitted over CAT5 cable 100 Mbits/sec is possible over 100 metres.

IEEE 1394b is used on the F-35 Lightning II to interconnect several Remote Input/Output Units (RIUs) that act as data concentrators in the Vehicle Management System (VMS). The RIUs allow data to be transferred to the VMS computers and other major controllers within the VMS system.

12.4 Fibre Optic Buses

The examples described thus far relate to electrically signalled data buses. Fibre-optic interconnections offer an alternative to the electrically signalled bus that is much faster and more robust in terms of Electro-Magnetic Interference (EMI). Fibre-optic techniques are widely used in the telecommunications industry and those used in cable networks serving domestic applications may typically operate at around 50–100 MHz.

A major problem with fibre-optic communication is that it is uni-directional. That is the signal may only pass in one direction and if bi-directional communication is required then two fibres are needed. There is also no 'T-junction' in fibre-optics and communication networks have to be formed by 'Y-junctions' or ring topologies. An example of the ring topology is shown in Figure 12.18 in which the bi-directional interconnection between four terminals requires a total of eight uni-directional fibres. This network does have the property that inter unit communication is maintained should any terminal or fibre fail.

This particular topology is similar to that adopted by the Raytheon Control-By-Light™ (CBLTM) system that has been demonstrated in flight controlling the engine and thrust reversers of a Raytheon Business Jet. In this application

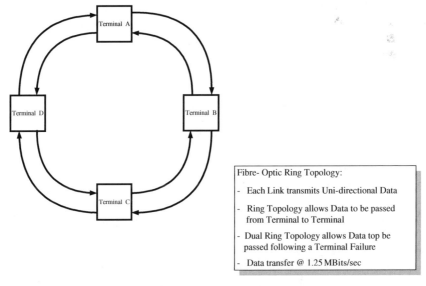

Figure 12.26   Fibre-optic ring topology

the data rate is a modest 1.25 Mbits/sec which is no real improvement over conventional buses such as MIL-STD-1553B and indeed is slower than A629. A fibre-optic bus does have the capability of operating at much higher data rates. It appears that the data rate in this case may have been limited by the protocol (control philosophy) which is an adaptation of a US PC/Industrial Local Area Network (LAN) protocol widely used in the US.

Fibre-optic standards have been agreed and utilised on a small scale within the avionics community, usually for On-Board Maintenance System (OMS) applications.

12.5 Avionics Packaging Standards

Line Replaceable Units (LRUs) were developed as a way of removing functional elements from an avionics system with minimum disruption. LRUs have logical functional boundaries associated with the task they perform in the aircraft. LRU formats were standardised to the following standards:

12.5.1 Air Transport Radio (ATR)

The origins of ATR standardisation may be traced back to the 1930s when United Airlines and ARINC established a standard racking system called Air Transport Radio (ATR) unit case. ARINC 11 identified three sizes: ½ ATR, 1 ATR, 1½ ATR with the same height and length. In a similar timescale, standard connector and pin sizes were specified for the wiring connections at the rear of the unit. The US military and the military authorities in UK adopted these standards although to differing degrees and they are still in use in military parlance today. Over the period of usage ATR 'short' ~ 12.5 inch length and ATR 'long' ~ 19.5 inch length have also been derived. ARINC 404A developed the standard to the point where connector and cooling duct positioning were specified to give true interchangeability between units from different suppliers. The relatively dense packaging of modern electronics means the ATR 'long' boxes are seldom used.

12.5.2 Modular Concept Unit (MCU)

The civil airline community developed the standardisation argument further which was to develop the Modular concept Unit (MCU). An 8 MCU box is virtually equivalent to 1 ATR and boxes are sized in MCU units. A typical small aircraft systems control unit today might be 2 MCU while a larger avionics unit such as an Air Data and Inertial Reference System (ADIRS) combining the Inertial Reference System (IRS) with the air data computer function may be 8 or 10 MCU; 1 MCU is roughly equivalent to ~ 1¼ inch but the true method of sizing an MCU unit is given in Figure 12.27 below.

An 8 MCU box will therefore be 7.64 in high x 12.76 in deep x 10.37 in wide. The adoption of this concept was in conjunction with ARINC 600 which specifies connectors, cooling air inlets etc. in the same way that ARINC 404A did earlier.

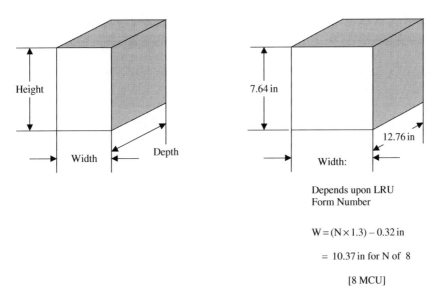

Height

Width   Depth

7.64 in

12.76 in

Width:

Depends upon LRU
Form Number

$W = (N \times 1.3) - 0.32\,\text{in}$

$= 10.37\,\text{in for N of }8$

[8 MCU]

Figure 12.27   MCU sizing

12.6 Typical LRU Architecture

The architecture of a typical avionics Line Replaceable Unit (LRU) is shown in Figure 12.28. This shows the usual interfaces and component elements. The unit is powered by a Power Supply Unit (PSU) which converts either 115V AC or 28V DC aircraft electrical power to low voltage DC levels − +5V and + or

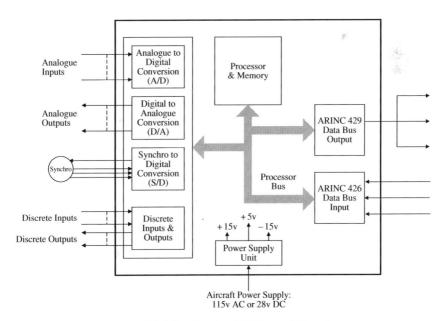

Figure 12.28   Typical LRU architecture

−15V are typical – for the predominant microelectronic devices. In some cases where commercially driven technology is used +3.3V may also be required. The processor/memory module communicates with the various I/O modules via the processor bus. The 'real world' to the left of the LRU interfaces with the processor bus via a variety of I/O devices which convert true analogue values to/from a digital format. The right portion of the LRU interfaces with other LRUs by means of digital data buses; in this example A429 is shown and it is certainly the most common data bus in use in civil avionics systems today.

One of the shortcomings exhibited by microelectronics is their susceptibility to external voltage surges and static electricity. Extreme care must be taken when handling the devices outside the LRU as the release of static electricity can irrevocably damage the devices. The environment that the modern avionics LRU has to withstand and be tested to withstand is onerous as will be seen later.

The environmental and EMI challenges faced by the LRU in the aircraft can be quite severe, typically including the following:

- *Electro-Magnetic Interference*

  – EMI produced by sources external to the aircraft; surveillance radars, high power

- *Radio and radar stations and communications*

  – Internal EMI: interference between equipment or by passenger carried laptops, gaming machines or mobile phones
  – Lightning effects

  MIL-STD-461 & 462 are useful military references
- *Physical effects due to one or more of the following (see Chapter 13):*

  – vibration: sinusoidal or random in three orthogonal axes
  – temperature
  – altitude
  – temperature and altitude
  – temperature, altitude and humidity
  – salt fog
  – dust
  – sand
  – fungi

Figure 12.29 shows the construction of a typical LRU; most avionics suppliers adopt this or similar techniques to meet the EMI requirements being mandated today. The EMI sensitive electronics is located in an enclosure on the left which effectively forms a 'Faraday cage'. This enclosed EMI 'clean' area is shielded from EMI effects such that the sensitive microelectronics can operate in a protected environment. All signals entering this area are filtered to remove voltage spikes and surges. To the right of the EMI boundary are the EMI filters and other 'dirty' components such as the Power Supply Unit (PSU). These components are more robust than the sensitive electronics and can successfully

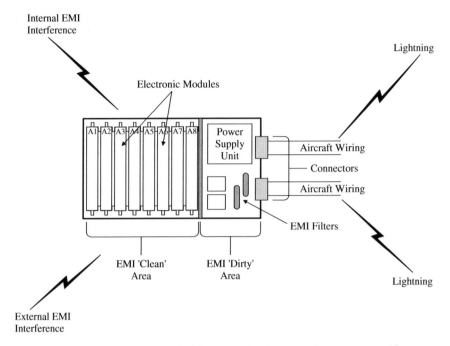

Figure 12.29 LRU EMI hazards

operate in this environment. Finally, in many cases the external wiring will be shielded and grounded to screen the wiring from external surges or interference induced by lightning – and more recently and perhaps more certainly – from radiated and conducted emissions from passenger's laptop computers and hand held computer games.

A typical test plan for modern avionics units will include many or all of the above tests as part of the LRU/system, as opposed to the aircraft certification process. Additionally, production units may be required to undergo an Environmental Stress Screening (ESS) during production testing which typically includes 50 hours of testing involving temperature cycling and/or vibration testing to detect 'infant mortality' prior to units entering full-time service.

12.7 Integrated Modular Avionics

Integrated Modular Avionics (IMA) is a new packaging technique which could move electronic packaging beyond the ARINC 600 era. ARINC 600 as described earlier relates to the specification of in recent transport aircraft LRUs and this is the packaging technique used by many aircraft flying today. However the move towards a more integrated solution is being sought as the avionics technology increasingly becomes smaller and the benefits to be attained by greater integration become very attractive. Therefore the advent of Integrated Modular Avionics introduces an integrated cabinet approach where the conventional ARINC 600 LRUs are replaced by fewer units.

The IMA concept is shown in Figure 12.30. The diagram depicts how the functionality of seven ARINC 600 LRUs (LRUs A through to J) may instead be installed in an integrated rack or cabinet as seven Line Replaceable Modules (LRMs) (LRMs A through to J). In fact the integration process is likely to be more aggressive than this, specifying common modules and interleaving multiple processing tasks within common processor modules.

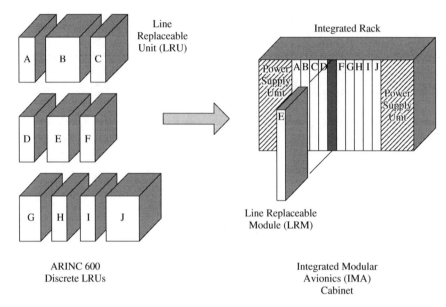

Figure 12.30   LRU and integrated modular cabinet comparison

The US military were the first to implement modular avionics, starting with the Pave Pillar program and then applying the principles to the F-22 Raptor integrated avionics suite. In this implementation dynamic reconfiguration is employed which enables the remaining computer resources to take over computational tasks should a computing module fail.

As the diagram suggests there are a number of obvious potential advantages to be realised by this integration:

- Volume and weight savings
- Sharing of resources, such as power supplies, across a number of functional module
- More unified approach to equipment design
- LRMs are more reliable than LRUs

These advantages must be weighed against the disadvantages:

- Possibly more expensive overall to procure
- Possibly more risky

- May pose proprietary problems by having differing vendors working more closely together
- Segregation considerations (more eggs in one basket)
- Will an 'open' or 'closed' architecture prevail?
- What standards will apply – given the fact that a lot of effort has been invested in ARINC 600?
- Possibly more difficult to certify
- Who takes responsibility for systems integration?

Clearly there are some difficult issues to be resolved. However, the application of modular architectures is now widely established within the avionics community.

References

[1] Sir Bernard Lovell (1991) *Echoes of War: The Story of H2 S Radar*. Bristol: Adam Hilger.
[2] B. Middleton *et al.* (1989) *Avionics Systems*. Harlow: Longman Scientific & Technical.
[3] Cary R Spitzer (1993) *Digital Avionics Systems: Principles & Practice*, McGraw-Hill.
[4] ARINC Specification 429: Mk 33 Digital Information transfer System, Aeronautical Radio, Inc., 1977.
[5] MIL-STD-1553B Digital Time Division Command/Response Multiplex data Bus, Notice 2, 8 September 1986.
[6] ARINC Characteristic 629, Multi-Transmitter Data Bus, Aeronautical Radio, Inc., November 1989.
[7] Boeing 777 ARINC 629 Data Bus – Principles, Development and Application, RAeS Conference – Advanced Avionics on the Airbus A330/A340 and the Boeing 777 Aircraft, November 1993.
[8] Aplin, Newton & Warburton (1995) 'A Brief Overview of Databus Technology', *RAeS Conference The Design and Maintenance of Complex Systems on Modern Aircraft*, April.
[9] Principles of Avionics Data Buses, Avionics Communications Inc., 1995
[10] Tully, T. (1998) 'Fuel Systems as an Aircraft Utility', *International Conference – Civil Aerospace Technologies*, FITEC '98, London, September.

13

Environmental Conditions

13.1 Introduction

Environmental conditions exert a major influence over the complete product and its component parts. The product must operate in a specific environment, in the case of an aircraft that environment will change throughout the mission. Commercial aircraft will travel regularly from one climatic condition at the start of their journey, to a totally different environment at the end of the journey. This they must do day in and day out without an impact on their availability. Military aircraft do much the same thing, but their operating environment is influenced by their speed and manoeuvrability as well as the very specific impact of the battle space.

Thus the design of an aircraft must take into account a set of environmental conditions that may be specific to a type of aircraft, while the equipment used to perform the system functions is increasingly being procured from off the shelf stocks which may have been designed for an entirely different set of conditions. The system designer, therefore, must pay scrupulous attention to the specifications of systems and equipment, and must further ensure that they are tested to enable qualification evidence to be collected to certify the type.

The operating environment of any item of equipment that is installed in an aircraft is determined by a number of factors:

- The location in which it is installed and in some cases – its orientation
- The level of protection offered to the item by the aircraft
- The external environment that influences the aircraft and its contents

Aircraft Systems: Mechanical, electrical, and avionics subsystems integration. 3rd Edition   Ian Moir and Allan Seabridge
© 2008 John Wiley & Sons, Ltd

Many of these factors can be quantified from the basic requirements of the product, and from prevailing international, national and company standards, in particular from:

- The customer's requirement
- National and international standards used by the industry
- Engineering knowledge of the aircraft design

The customer will define the areas of the world in which the aircraft is expected to operate, and this will largely determine the climatic conditions to which the aircraft will be exposed. However, to design specifically for that operating environment may restrict sales to other areas of the world, and hence it may be cost effective to design for worldwide operations and use at the outset. The conditions that aircraft and systems must withstand are well understood and there are standards of testing that have evolved to verify designs under a wide variety of environmental conditions – many of them extremely severe.

The conditions of use of the aircraft will determine the local environment that will affect structure, systems and inhabitants, introducing such aspects as vibration, shock, temperature etc. It is important to recognise also that equipment needs to be transported to aircraft operating bases and stored locally. This means that it is manhandled, transported from place to place and may be stored in less than suitable locations; as such it may be subject to additional environmental hazards during transportation and storage. It may indeed be dropped out of the cargo bay of a military transport aircraft onto the battlefield. All this adds to the need to rigorously test equipment in a non-operating mode so that it is able to work when installed.

Combinations of these environmental aspects are used by systems engineers to understand and specify their design and test requirements. A handbook or database of such conditions will be of use to systems engineering teams to ensure a consistency of approach within any one project. Typical considerations include:

- Consider what areas of the world in which the aircraft will be used
- Consider the impact of designing for worldwide operations to increase the market
- Determine what impact the conditions of use will have on internal equipment and inhabitants and translate this into engineering parameters
- Understand the various environmental conditions that exist for different zones or compartments in the aircraft
- Understand how equipment will be transported and stored as spares
- Define all engineering requirements in a handbook or data base
- Gain an understanding of the operating spectrum in terms of ambient temperature, aircraft speed and altitude This drives other environmental parameters and it avoids the tendency to over-design, i.e. it is too easy to assume that the system or equipment is always at the most adverse conditions

Having determined what parameters are important to specify for the design of equipment, it is usual to make use of a standard set of tests to obtain evidence that the equipment will meet its design in that environment. Advice and guidance can be found in references [1–4].

13.2 Environmental Factors

Figure 13.1 shows some factors that affect the behaviour of the aircraft and the equipment and systems contained in it. Many of the factors are generated externally in the environment surrounding the equipment. There are some factors that are generated internally that have an impact in their own right or may exacerbate the external factor. An example of this is the climatic condition of high temperature, whether generated by solar energy or by aircraft speed, compounded by heat generated within the aircraft itself. Both of these factors combine to exert a maximum temperature in equipment bays.

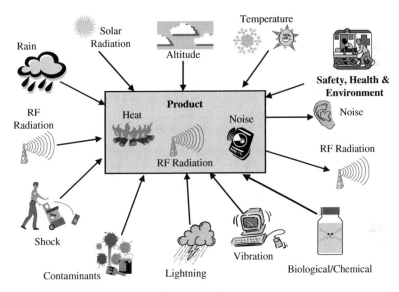

Figure 13.1 Some environmental factors affecting the product

This section will present a summary of design considerations – things that the systems engineer must take into account when designing or specifying a systems and its components – followed by a summary of test requirements and sources of test cases.

13.2.1 Altitude

Design Considerations

Many aircraft operate between sea level and 40 000 ft, Concorde used to routinely operate up to 50 000 ft, and some military aircraft routinely operate well above this altitude, if only briefly. Missiles and spacecraft will enter the stratosphere or operate in a vacuum. The interior of conventional aircraft is maintained at a pressure that is tolerable to crew and passengers, so that there is a pressure differential across the aircraft skin. Any disturbance of this pressure differential may cause a rapid or explosive decompression that is a potential cause of damage. Typical considerations include:

- The aircraft can routinely operate at altitudes up to 50 000 ft and sometimes beyond
- Equipment and aircrew must be capable of operating at pressures representative of altitudes from sea level to 50 000 ft
- Differential pressures can affect the performance of sealed components
- Although the cockpit, cabin and equipment bays are normally pressurised, rapid or sudden decompression (rapid rates of change of pressure) can lead to component failures. This can occur as a result of damage, failed seals, canopy loss or battle damage

Test Considerations

- It is important to understand the flight envelope of the aircraft, to define the maximum altitude and the number of excursions to this altitude that the aircraft is likely to undergo. This will be used to determine the appropriate test conditions
- Military fast jets will undergo rapid rates of change of altitude, which lead to rapid rates of change of pressure
- Equipment mounted in pressurised zones such as the cockpit or passenger cabin will need to be tested for explosive or rapid decompression. Determine how the zone will respond to damage. For example a fighter canopy failure will cause a very rapid rate of reduction in pressure, whereas a large volume passenger aircraft cabin will depressurise more slowly

MIL-STD-810: Method 500 Low Pressure Altitude – to determine if materiel can withstand and/or operate in a low pressure environment and/or withstand rapid pressure changes.

13.2.2 Temperature

Design Considerations

All aircraft are expected to operate in a wide range of temperatures from arctic to desert conditions. The effect of ambient temperature on internal temperatures can affect equipment, especially when powered. The effect of external ambient temperatures is a key design consideration. Typical considerations include:

- The aircraft will be expected to operate in extremes of temperature ranging from $-55\,°C$ to $+90\,°C$. The range depends on the part of the world for which the product is expected to be deployed. In some cases the environment may be even more severe after a hot or cold soak. In some parts of the world $-70\,°C$ is not uncommon
- An aircraft is expected to operate in worldwide conditions and to experience temperature extremes, and in some cases experience gross deviations during normal operating regimes between different climatic zones e.g. Northern Canada, Iceland, Norway, Saudi Arabia, Arizona
- It may be economical to design and develop the system for worldwide operation to increase market potential, avoiding redesign or retest
- The aircraft may be parked for extended periods of time in hot or cold conditions (hot or cold soak) or subject to direct sunlight. Key equipment may be expected to operate immediately as soon as it is powered up in such conditions, but not necessarily entire systems
- Equipment mounted in undercarriage bays and bomb bays may be subject to rapid changes of temperature. These bays will naturally be subject to a flow of air within the aircraft which will be relatively warm because of the dissipation from internally carried equipment and exhaust cooling air flows from avionics bays. Opening the bays in flight will cause a rapid change of temperature
- It is important to understand the range of temperatures that equipment will be subjected to when not operating (not powered up) and operating. It is also important to understand the temperature variation in the bay in which equipment is to be installed. For consistency it is sensible to identify all installation locations in the aircraft and to allocate each of them a temperature range

As well as the external climatic conditions, the interior of the aircraft has its own microclimate as a result of the energy conversion processes taking place. Heat is a waste commodity generated by inefficiencies of power sources, by equipment using power, by solar radiation, by crew and passengers, and by friction of air over the aircraft surface especially during high speed flight. Thus all human and physical occupants of the aircraft are subject to the effects of heat. These effects range from those affecting the comfort of human occupants to those that cause irreparable damage to components of equipment. Typical considerations include:

- If the system or systems component is likely to be affected by heat, then it should not be installed near to a major heat source or it should be provided with cooling
- The aircraft Environmental Control System (ECS) can cool equipment using air or a liquid coolant
- Some systems produce heat as a waste product in performing their function and must be isolated or insulated from other systems. Examples: engines, high power transmitters, computing equipment

- Some systems produce heat which is useful/essential, e.g. hydraulics for flight control systems
- It is vital that the physical limitations of the active components in the equipment is known, and that self-heating is taken into account. For example the maximum junction temperature of silicon semiconductors is 120 °C at which point damage occurs. It is important to allow for at least 15 °C for self-heating

A useful mechanism for ensuring that the most appropriate test conditions are specified is to divide the aircraft into zones, and to specify the temperature conditions for each zone. The aircraft will most likely be zoned by the installation design team, and it will be sensible to adopt these zones.

Test Considerations

- World climatic charts are available that will specify maximum and minimum temperatures and the duration of exposure. Bear in mind the market for the product, and remember that market expansion may require additional testing
- The impact of internally generated heat must be taken into account in the test cases to allow for additional heating which will set the upper limit of the temperature range
- Understand the nature of the equipment and its components to determine the impact of self-heating
- Are there any circumstances where the equipment must operate after cooling system failure, and how long before a catastrophic failure of the equipment occurs? This is especially important where the system is flight safety critical

MIL-STD-810: Method 501 High Temperature – to obtain data and to help evaluate effects of high temperature conditions on materiel safety, integrity and performance.

MIL-STD-810: Method 502 Low Temperature – to measure how low temperature conditions during storage, operation and manipulation affect materiel safety, integrity and performance.

MIL-STD-810: Method 503 Temperature Shock – to determine if materiel can withstand sudden changes (> 10 °C per minute) in temperature of the surrounding atmosphere without experiencing physical damage or deterioration in performance.

Alternative Standards: BS3G100, Part 2, Sub 3.11.

13.2.3 Contamination by Fluids

Design Considerations

The aircraft exterior and interior surfaces, and the installed equipment can be contaminated by substances that in a normal environment can cause corrosion damage or malfunction. Contamination may occur by direct means such as spillage, leakage or spray; or indirectly by being handled with contaminated hands or tools. Equipment and furnishings must be specified and designed to minimise the effects of contamination. A contaminant can be specific to an aircraft type either due to role (e.g. biological or chemical agents) or systems on the aircraft itself (e.g. radar cooling fluid). Typical contaminants to be taken into account include:

- Fuel (different types)
- Oils and greases
- De-icing fluid
- Windscreen wash fluid
- Hydraulic fluid
- Beverages – coffee, tea, soft drinks
- Cleaning fluids

Test Considerations

MIL-STD-810: Method 504 Contamination by fluids – to determine if materiel is unacceptably affected by temporary exposure to contaminating fluids (liquids) such as may be encountered during its life cycle either occasionally (extraordinary/unusual circumstances occurring once or twice a year), intermittently (regular basis under normal operation; possibly seasonally over the life of the materiel) or over extended periods (long periods such that the materiel is thoroughly exposed).

Alternative standards: BS3G100, Part 2, Section 3 or RTCA/DO160C, Sect 11, Cat F.

13.2.4 Solar Radiation

Design Considerations

Sunlight will impinge on the surface of the aircraft and will enter through windows and canopies, thereby exposing some parts of the interior. Prolonged exposure at high altitudes to unfiltered ultraviolet (UV) and infrared (IR) is likely to damage some materials. UV exposure is also experienced when parked for long periods on the tarmac. Typical considerations include:

- The UV and IR content of solar radiation can cause damage to plastic materials such as discolouration, cracking and brittleness. This can affect interior furnishings such as display bezels and switch/knob handles
- Items most affected are those situated on the aircraft outer skin, e.g. antennas, where high altitude, long duration exposure is experienced
- Cockpit items are also vulnerable if likely to be in direct sunlight in flight or while the aircraft is parked – cockpit temperatures have been known to reach over 100 °C in some parts of the world

- All such items must be designed to withstand such effects and must be tested
- Consideration should be given to the use of exterior cockpit covers when parked
- Glare and reflection will affect crew visual performance, and may adversely affect display visibility

Test Considerations

MIL-STD-810: Method 505 Solar Radiation (Sunshine) – to determine the heating effects of direct solar radiation on materiel and/or to help identify the actinic (photo-degradation) effects of direct solar radiation.

Alternative Standards: Def Stan 07-55, Part 2, Test 3B, Procedure C.

13.2.5 Rain, Humidity, Moisture

Design Considerations

Although the aircraft is ostensibly a sealed vessel, there are many opportunities for moisture to enter the aircraft and its equipment.

- The aircraft exterior will be subject to rain when it is parked in the open and seals must be provided to prevent ingress when the aircraft is closed up
- In flight rain is experienced as driving rain and may have an erosive effect on externally mounted items
- Military pilots often taxy with the canopy open and water will pool on the cockpit floor unless drains are provided which do not compromise pressurisation
- Rapid descent from high altitude may cause mist or excessive humidity to form
- Operation in tropical climates with high relative humidity may lead to moisture or free water in the aircraft
- Equipment installed in the undercarriage bays will be subject to contaminated water from the rotating wheel as the gear is stowed, and may also be subject to spray during take-off and landing
- All equipment must be provided with drain holes of suitable size to allow drainage

Test Considerations

MIL-STD-810: Method 506 Rain – to determine:

- The effectiveness of protective covers, cases, seals in preventing penetration of water
- The capability of the materiel to satisfy its performance requirements during and after exposure to water

- Any physical deterioration of the materiel caused by rain
- The effectiveness of any water removal system
- The effectiveness of protection offered to a packaged materiel

Alternative Standards: BS3G100, Part 2, Sub 3.11; RTCA/DO160C, Sect 10.
   MIL-STD-810: Method 507 Humidity – to determine the resistance of materiel to the effects of a warm humid atmosphere.
   Alternative Standards: BS3G100, Part 2, Sub 3.7; RTCA/DO160C, Sect 6.

13.2.6 Fungus

Design Considerations

Fungus is a particular issue in tropical climates or for equipment in long-term storage in unconditioned buildings.

Test Considerations

MIL-STD-810: Method 508 Fungus – to assess the extent to which materiel will support fungal growth or how any fungal growth may affect performance or the use of the materiel. The primary objectives of the fungus test are to determine:

- If the materials comprising the materiel, or the assembled combination of same, will support fungal growth
- How rapidly fungus will grow on the materiel
- How fungus affects the materiel, its mission and its safety for use, following the growth of fungus on the materiel
- If the materiel can be stored effectively in a field environment
- If there are simple reversal processes, e.g. wiping off fungal growth

Alternative Standards: RTCA/DO160C, Sect 13; BS2011 Part 2 [7].

13.2.7 Salt Fog/Salt Mist

Design Considerations

Aircraft that operate in maritime conditions such as maritime patrol aircraft or carrier borne aircraft may be subject to salt laden atmospheres for extended periods of time. This can lead to corrosion of metal parts, deterioration in electrical bonds, corrosion of equipment internals if ingested and the potential for short circuits because of the increased conductivity of salt water in severe cases of ingestion.

Test Considerations

MIL-STD-810: Method 509 Salt Fog – to determine the effectiveness of protective coatings and finishes on materials. It may also be applied to determine the effects of salt deposits on the physical and electrical aspects of materiel.

Alternative Standards: RTCA/DO160C, Sect 14; BS2011 Part 2.1 [7].

13.2.8 Sand and Dust

Design Considerations

There are some climates where sand and dust are prevalent, and even in hangar environments there may be dust. Sand is usually encountered in desert climates where damage and erosion may occur, particularly if encountered as a sand storm. Sand is usually encountered close to the ground because of the mass of the particles and will predominantly affect components such as landing gear, wheels, brakes.

Test Considerations

MIL-STD-810: Method 510 Sand and Dust – the small particle dust (< 149 micro m) procedures are performed to help evaluate the ability of materiel to resist the effects of dust that may obstruct opening, penetrate into cracks, crevices, bearings and joints, and to evaluate the effectiveness of filters. The blowing sand (150–850 μm) procedures are performed to help evaluate if materiel can be stored and operated under blowing sand conditions without degrading performance, effectiveness, reliability and maintainability due to abrasion (erosion) or clogging effects of large, sharp edged particles.

Alternative Standards: Def Stan 07-55, Part 2, Sect 4/1.

13.2.9 Explosive Atmosphere

Design Considerations

There may be situations where equipment is mounted in bays that may fill with an explosive mixture of vapour and air. This is most likely to be as the result of a prior failure, for example a ruptured hydraulic component leaking hydraulic fluid into a bay. If such an event occurs then an arcing switch contact could ignite the vapour. If equipment is likely to be mounted in such a location then consideration must be given to sealing the equipment or any components likely to cause arcs or sparks.

For ground equipment or test equipment used in, on or near an aircraft in an enclosed hangar it may be necessary to consult ATEX European Directive 94/9/EC [9].

Test Considerations

MIL-STD-810: Method 511 Explosive Atmosphere

- Demonstrate the ability of the materiel to operate in fuel-air explosive atmospheres without causing ignition *or*
- Demonstrate that an explosive or burning reaction occurring within encased equipment will be contained and will not propagate outside the test item

Alternative Standards: BS3G100, Part 2, Sub 3.5; RTCA/DO160C, Section 9.

13.2.10 Acceleration

Design Considerations

To test the unit under test to ensure that it will survive accelerations due to operation during in flight or during transit. Identical units located at different locations around the airframe need to take account of the installation orientation of the equipment since the acceleration is not the same in all aircraft axes or all locations.

Test Considerations

MIL-STD-810: Method 513 Acceleration – to assure that materiel can structurally withstand the g forces that are expected to be induced by acceleration in the service environment and function without degradation during and following exposure to these forces.

13.2.11 Immersion

Design Considerations

In some instances it may be essential that equipment continues to perform when immersed, for example sonobuoy locator beacons and crew survival equipment. In such cases it is necessary to specify an immersion test.

Test Considerations

Method 512 Leakage (Immersion) – to determine if the materiel can:

- withstand immersion or partial immersion (e.g. fording) in water and operate as required during or following immersion
- resist unacceptable amounts of penetration of water into an enclosure

13.2.12 Vibration

Design Considerations

All equipment is subject to vibration coupled into the mountings from the airframe. This vibration can, in turn, be coupled into circuit cards and components leading to fractures of wiring, connector pins and circuit boards. The effects are more severe if resonant modes occur. Typical considerations are:

- Vibration encountered in normal operation
- 3-axis vibration that can be randomly or continuously applied:
  - sinusoidal vibration at fixed frequencies and directions
  - specific vibration regimes as determined by the aircraft zone in which equipment is installed
- Gunfire vibration in fighter aircraft and attack helicopters
- Anti-vibration mountings in certain installations
- Flexible equipment racks

Test Considerations

MIL-STD-810: Method 514 Vibration – vibration tests are performed to:

- Develop materiel to function in and withstand the vibration exposure of a life cycle including synergistic effects of other environmental factors, materiel duty cycle and maintenance; combine the guidance of this method with the guidance of Pt1 and other methods to account for environmental synergism
- Verify that materiel will function in and withstand the vibration exposures of a life cycle

Alternative Standards: BS3G100, Part 2, Section 3, Sub 3.1; RTCA/DO160C Para 8.6.2.

MIL-STD-810: Method 519 Gunfire Vibration – to replicate the materiel response to a gunfire environment that is incurred by materiel during the specified operational conditions.

For a representative energy of vibration the distance from the gun muzzle needs to be measured and stated in the test requirement.

13.2.13 Acoustic Noise

Design Considerations

Noise is ever present in an aircraft environment. It is produced by the engines or auxiliary power units, by motor-driven units such as fans and motors and by air flow over the fuselage. It can cause discomfort to passenger and crew, while high noise levels external to the aircraft can cause physical damage. Typical considerations include:

- High sound pressures or acoustic noise levels can damage equipment. Installation in areas subject to high noise levels should be avoided. Typical areas are engine bays, external areas subject to engine exhaust or bays likely to be opened in high-speed flight, e.g. bomb bays
- Equipment can produce noise that is likely to be a nuisance to aircrew, contributing to fatigue and loss of concentration. Examples are fans and pumps/motors installed in the cockpit. Measures must be taken to install equipment so that excessive noise can be avoided and crew efficiency maintained
- It is also important to consult contemporary Health & Safety legislation to ensure that air crew and ground crew hearing is not endangered

MIL-STD-810: Method 515 Acoustic Noise – to demonstrate the adequacy of materiel to resist the specified acoustic environment without unacceptable degradation of its functional and/or structural performance.
Alternative Standards: BS3G100, Part 2, Sub 3.14.

13.2.14 Shock

Design Considerations

Violent or sharp shock can cause equipment and components to become detached from its mountings. It may then become a loose article hazard capable of casing secondary damage to other items of equipment or to occupants. Shock may also cause internal components of equipment to become detached leading to malfunction. Typical causes of shock are:

- Violent aircraft manoeuvres
- Heavy landings
- Crash conditions
- Accidental drop during manual handling
- Deliberate air-drop by military transport

Test Considerations

MIL-STD-810: Method 516 Shock- shock tests are performed to:

- Provide a degree of confidence that materiel can physically and functionally withstand the relatively infrequent, nonrepetitive shocks encountered in handling, transportation and service environments; this may include an assessment of the overall materiel system integrity for safety purposes in any one or all of the handling, transportation and service environments
- Determine the materiel's fragility level, in order that packaging may be designed to protect the materiel's physical and functional integrity
- Test the strength of devices that attach materiel to platforms that can crash

Alternative Standards: RTCA/DO160C, Section 7.

13.2.15 Pyroshock

Design Considerations

These tests are relevant to equipment containing pyrotechnic devices such as armaments, chaff and flares, emergency release units, explosive canopy release mechanisms and ejection seats.

Test Considerations

MIL-STD-810: Method 517 – Pyroshock tests involving pyrotechnic (explosive or propellant activated) devices are performed to:

- Provide a degree of confidence that materiel can structurally and function-ally withstand the infrequent shock effects caused by the detonation of a pyrotechnic device on a structural configuration to which the material is mounted
- Experimentally estimate the materiel's fragility level in relation to pyroshock in order that shock mitigation procedures may be employed to protect the materiel's structural and functional integrity

13.2.16 Acidic Atmosphere

Design Considerations

To be taken into account if the aircraft is likely to enter unusual acidic condi-tions or if a particular zone of the aircraft may be subject to acidic atmospheres, Examples are battery bays in case of battery damage or overcharging, or trans-port aircraft where unsual cargo may be located.

Test Considerations

MIL-STD-810: Method 518 Acidic atmosphere – to determine the resistance of materials and protective coatings to acidic atmospheres.

13.2.17 Temperature, Humidity, Vibration, Altitude

Design Considerations

This is a combined test to ensure that equipment can withstand the conditions of use. The test conditions will contain a realistic profile of an aircraft mission or sortie.

Test Considerations

MIL-STD-810: Method 520 Temperature, Humidity, Vibration, Altitude – the purpose of this test is to help determine the combined effects of temperature, humidity, vibration and altitude on airborne electronic and electro-mechanical materiel safety, integrity and performance during ground and flight operations.

13.2.18 Icing/Freezing Rain

Design Considerations

This test applies to equipment subject to icing conditions where ice formation may have an adverse effect such as jamming. This includes externally mounted actuation mechanisms and hinges where failure to operate may cause a safety hazard.

Test Considerations

Method 521– Icing/freezing rain to evaluate the effect of icing on the operational capability of materiel. This method also provides tests for evaluating the effectiveness of de-icing equipment and techniques, including means to be used in the field.

Alternative Tests: BS3G100, Part 2, Sub 3.9.; RTCA/DO160C, Sect 2.4.

13.2.19 Vibro-Acoustic, Temperature

Design Considerations

A condition that may affect externally carried items on a military aircraft such as weapons, external fuel tanks or sensor pods.

Test Considerations

MIL-STD-810: Method 523 Vibro-acoustic, temperature – to determine the synergistic effects of combined vibration, acoustic noise and temperature on externally carried aircraft stores during captive carry flight.

13.2.20 RF Radiation

Design Considerations

Radio Frequencies (RF) are radiated from equipment and from the aircraft, either deliberately or accidentally. As far as aircraft systems are concerned, RF emission generally occur in the electromagnetic spectrum from 10 MHz to tens of GHz. Accidental radiation occurs when equipment or wiring is badly installed, or inadequately or incorrectly screened. Deliberate radiation occurs during radio transmissions, navigation equipment transmissions and operation of radars and other communication equipment. RF radiation can cause interruption or corruption of a system function by affecting system component operation or by corrupting data. Requirements are placed on equipments in terms of both emission of and susceptibility to conducted and radiated emissions. The equipment needs to be tested against the totality of the requirement for its platform. Other tests address the issue of conducted susceptibility.

Typical considerations include:

- Equipment should be protected from the effects of RF radiation by the application of an Electromagnetic Health (EMH) strategy. This involves the use of signal wire segregation, screening, bonding, separation of wiring and equipment, and RF sealing of equipment. This will obviate the effects of some of the key electromagnetic effects:
    - Electromagnetic Interference (EMI) resulting from the effects of local equipment on board the aircraft
    - Lightning strike on the structure or in the vicinity of the aircraft
    - High Intensity Radio Frequency (HIRF) from local high power transmitters such as airfield primary surveillance radar or domestic radio transmitters
- Radiated transmissions can disclose the presence of an aircraft to enemy forces, which can be used as intelligence or as a means of identifying a target for attack
- In the military field, analysis of signals by an Electronic Support Measures (ESM) team can provide valuable intelligence about deployment of military assets
- It is generally acknowledged that Signals Intelligence (SIGINT) is one of the most prolific sources of intelligence during peacetime, periods of tension or conflict. Its contribution to diplomatic or military success can have an effect far outweighing the relatively small investment required to gather and analyse information
- Incorrectly screened secure communications can cause 'leakage' of classified information from the aircraft that can be detected by enemy forces
- Equipment must be designed to prevent radiated emissions that will affect other adjacent equipment. This includes interconnecting wiring
- Equipment must be protected against receiving or emitting interference by conduction on power cables
- The EMH strategy is also intended to reduce the risk of equipment producing RF emissions from local onboard equipment and suppliers must be fully aware of the need to demonstrate compliance
- Each project will have an EMH plan defining the strategy to be adopted for that project
- There is a risk of mutual interference between transmitters and receivers. Care must be taken in the design of RF systems to prevent this

13.2.21 Lightning

Design Considerations

Most aircraft are expected to operate in all weather conditions, and it may not be possible to schedule flights or routes to avoid lightning conditions. Measures must be taken to limit the impact of lightning strike and associated structural damage and induced electrical effects. Typical considerations include:

- Lightning can be expected at all and any time of the year
- Lightning strike can damage structure locally, and induce very high transient voltages in aircraft cables
- Lightning induced effects can destroy entire systems
- All equipment must be bonded and on carbon surfaces, special foil inlay is used to provide a conductive path
- Equipment and complete aircraft are lightning strike tested

13.2.22 Nuclear, Biological and Chemical

Design Considerations

Military aircraft in particular may enter a theatre of combat in which deliberate contamination by chemical agents is a real possibility. The aircraft and its equipment must survive such contamination and the decontamination process. Typical considerations include:

- *Biological agent* – a living microorganism or toxin delivered by bomb, missile or spray device. Contamination of the aircraft and its equipment can harm air and ground crew
- *Chemical agent* – a compound which, when suitably disseminated, produces incapacitating, damaging or lethal effects delivered by bomb, missile or spray device. Contamination of the aircraft and its equipment can harm air and ground crew
- *Nuclear effects* – blast, radiation and electro-magnetic pulse which can damage aircraft, equipment, communications and personnel

13.3 Testing and Validation Process

Standards for environmental testing define methods for conducting tests of components or items of equipment, either in isolation or combined in order to verify their ability to operate correctly in extremes of conditions. The standards have evolved over a long period of time and have been updated to reflect lessons learned during many years of test experience. They provide a consistent and internationally recognised set of test methods used by many test houses and accepted as evidence by many customers.

It is important for tests to be selected that will reflect the conditions that equipment will see in service – these conditions should have been used to set the conditions of use in the original equipment specification. The test parameters such as ranges, rates of change, absolute values etc. can and should be determined to reflect the conditions in which equipment is installed and to reflect the actual exposure to internal and external environmental conditions, as well as the expected duty cycle of exposure.

Selecting the most appropriate standard depends on a number of factors:

- It is not uncommon for a customer to declare the standards to be used in designing a product
- A customer will usually declare the use of national standards
- Military aircraft customers will usually declare the use of military standards, while commercial aircraft customers will usually declare the use of commercial standards
- Inclusion of existing COTS equipment will mean that that the standards used in its qualification must be understood and accepted

This means that an aircraft may be designed and qualified to one set of particular standards – for example some military aircraft are designed exclusively to US Military Standards, and some commercial aircraft are designed exclusively to RTCA standards. Yet other aircraft are a mixture of military, commercial, national and international standards. As long as the standards used are well understood, the qualification evidence is accepted as being equivalent whatever the source, and that the customer certification authorities accept the situation, then the product has been demonstrated to be fit for purpose.

It is becoming important to be pragmatic rather than prescriptive in the selection and use of standards. This means that system designers must be familiar with a range of standards and must rigorously scrutinise test evidence.

Commonly used standards include:

- BS 3G 100: General Requirements for Equipment for use in Aircraft [3]
- MIL-STD-810B/C/D/E Test Method Standard for Environmental Engineering Considerations and Laboratory Tests [5]
- RTCA/DO-160 A/B/C/D Environmental Conditions and Test Procedures for Airborne Equipment [6]
- AIR 7304 Conditions D'Essais D'Environmental [8]

Environmental testing is usually the responsibility of the supplier of equipment. The supplier responds to the specification placed by the aircraft company and will design suitable equipment. As part of the test process the supplier will arrange for experienced test houses to perform environmental testing. This testing generates a number items of evidence that become part of the qualification of the equipment. Figure 13.2 shows how environmental testing fits into the testing process.

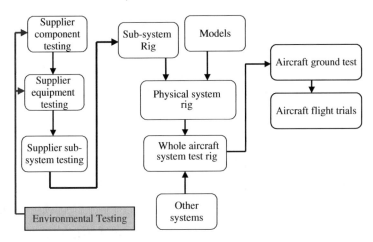

Figure 13.2 Environmental testing in the validation process

Table 13.1 Sources of Information

Condition	MIL-STD-810	RTCA DO/160	Chapter 2 BS3G100 Part 2	Other
Low pressure altitude	Method 500	Sect. 4.	Sect 3.2.	
High temperature	Method 501			
Low temperature	Method 502			
Temperature shock	Method 503		Sect. 3.15.	
Contamination by fluids	Method 504	Sect. 11.	Sect 3.12.	
Solar radiation	Method 505			Def Stan 00-55 Part 2 Test B3.
Rain/Water proofness	Method 506	Sect. 10.	Sect 3.11.	
Humidity	Method 507	Sect. 6.	Sect 3.7.	
Fungus	Method 508	Sect. 13.	Sect 3.3.	BS2011 Pt 2 Test J
Salt fog	Method 509	Sect. 14.	Sect 3.8.	BS2011 Pt 2.1 Test K
Sand and dust	Method 510			Def Stan 07-55 Pt 2 Set. 14.
Explosive atmosphere	Method 511	Sect. 9.	Sect 3.5.	
Leakage	Method 512			
Acceleration	Method 513	Sect. 7.	Sect 3.6.	
Vibration	Method 514		Sect 3.1.	

Table 13.1 (Continued)

Condition	MIL-STD-810	RTCA DO/160	Chapter 2 BS3G100 Part 2	Other
Acoustic noise	Method 515		Sect. 3.14.	
Shock	Method 516	Sect. 7.		
Pyroshock	Method 517			
Acidic atmosphere	Method 518			
Gunfire vibration	Method 519			
Temperature, humidity, vibration, altitude	Method 520			
Icing/freezing rain	Method 521		Sect 3.9.	
Ballistic shock	Method 522			
Vibro-acoustic, temperature	Method 523			

Evidence of testing will be included in a Declaration Of Design and Performance (DDP) for each item of equipment. This will make reference to the following:

- Specification test requirements
- Test procedures
- Test results validated by engineers
- Any special conditions of testing
- Any deviations or waivers

It may also prove necessary to supplement the individual equipment environmental testing by conducting whole aircraft trials, especially for extensive operations in cold weather, hot weather or tropical conditions. Such trials are difficult to conduct since weather is notoriously unpredictable and the repeatability of trials may be compromised. In addition to trials performed at suitable locations to give the best chance of achieving the required conditions, whole aircraft environmental hangar facilities such as those at Boscombe Down in the UK or the Mckinley facility in the USA can be used for suitable aircraft.

Table 13.1 provides a list of sources of information on testing and test conditions.

References

[1] DEF STAN 00-970 Volume 1 Amendment 12: Design and Airworthiness Requirements for Service Aircraft.
[2] DEF STAN 00-35 Environmental Handbook for Defence Material.
[3] BS 3G 100: General Requirements for Equipment for use in Aircraft – Part 2:All Equipment.
[4] DEF STAN 07-55 Environmental Testing of Service Material.

[5] MIL-STD-810B/C/D/E Test Method Standard for Environmental Engineering Considerations and Laboratory Tests.
[6] RTCA/DO-160 A/B/C/D Environmental Conditions and Test Procedures for Airborne Equipment.
[7] BS2011 Basic Environmental Testing.
[8] AIR 7304 Conditions D'Essais D'Environmental.
[9] ATEX European Directive 94/9/EC – Equipment and Protection Systems Intended for Use in Potentially Explosive Atmospheres.

Further Reading

Egbert, Herbert W. (2005) *The History and Rationale of MIL-STD-810.* Institute of Environmental Sciences and Technology.

Index